Clinical Assessment of the Autonomic Nervous System

Satoshi Iwase • Junichiro Hayano • Satoshi Orimo
Editors

Clinical Assessment of the Autonomic Nervous System

Springer

Editors
Satoshi Iwase
Department of Phyisiology
Aichi Medical University
Nagakute, Japan

Junichiro Hayano
Nagoya City University
Graduate School of Medical Science
Nagoya, Japan

Satoshi Orimo
Department of Neurology
Kanto Central Hospital
Tokyo, Japan

ISBN 978-4-431-56010-4 ISBN 978-4-431-56012-8 (eBook)
DOI 10.1007/978-4-431-56012-8

Library of Congress Control Number: 2016948709

Printed on acid-free paper

This Springer imprint is published by Springer Nature
The registered company is Springer Japan KK

Preface

This book examines the present status of the three types of methods used to evaluate autonomic nerve activity in humans and describes newly developed methods for assessing the autonomic nervous system that were mainly developed in Japan.

Microneurography was originally developed as an analytical technique for identifying muscle spindle afferents; however, it began to be used to analyze sympathetic nerve activity in the mid-1980s. At present, this technique is widely used all over the world, especially in analyzing the pathophysiology of human cardiovascular function. It also has become a significant tool in investigating the thermoregulatory function in humans. This section of the book describes the current frontiers of research into microneurographic analysis of human cardiovascular and thermoregulatory function.

Analyzing heart rate variability is a useful approach for investigating human vagal function. This technique was first developed in a study of SD/mean values of R-R interval, but it was significantly improved by the use of frequency domain analysis in Akselrod's study in 1987. The introduction of the maximum entropy method by Japanese researchers further improved power spectral studies of heart rate variability. This technique can be used to both assess the present status of the patient and predict future cardiovascular disease. This section of the book describes important information about heart rate variability.

^{131}I-Metaiodobenzylguanidine scintigraphy has now become a decisive tool for cardiac noradrenergic function, especially reductions in integration seen in Parkinson's disease, Lewy body disease, and multisystem atrophy. The method was first used to evaluate cardiac adrenergic function by Dr. Shigetaka Hakusui and subsequently was developed by Dr. Satoshi Orimo. It is now widely used by Japanese neurologists. The section about this topic was edited by Dr. Orimo.

This book will help readers to understand the current status of our knowledge about the autonomic nervous system. The authors are in debt to our mentors, especially Professor Tadaaki Mano, who leads Japanese researchers in the field of the autonomic nervous system.

Nagakute, Japan Satoshi Iwase

Contents

Part I
Sympathetic Microneurography

Chapter 1
Introduction to Sympathetic Microneurography

Tadaaki Mano

Abstract Sympathetic microneurography is an electrophysiological method to record directly from human peripheral nerves' sympathetic neural traffic leading to the muscle and skin called muscle sympathetic nerve activity (MSNA) and skin sympathetic nerve activity (SSNA). In this chapter, the author explains (1) what is microneurography, (2) the recording technique of microneurography, (3) how to identify MSNA and SSNA, and (4) applications of sympathetic microneurography. Sympathetic microneurography is very useful to analyze sympathetic neural functions in humans by observing directly neural traffic in postganglionic multiple and single sympathetic efferent fibers innervating the muscle and skin under physiological and pathological conditions. MSNA which regulates peripheral vascular resistance in skeletal muscles is particularly important for controlling blood pressure homeostasis, while SSNA which regulates mainly sweat glands and skin blood vessels plays important roles in thermoregulation. Recordings of MSNA and SSNA have been widely applied to analyze sympathetic mechanisms in various disease conditions, as well as stressful situations when the human body is exposed to various environmental conditions.

Keywords Microneurography • Muscle sympathetic nerve activity (MSNA) • Skin sympathetic nerve activity (SSNA) • Blood pressure control • Thermoregulation • Environmental stress • Clinical application

1.1 What Is Microneurography?

Microneurography is an electrophysiological method to record directly neural traffic from intact human peripheral nerves. Using this method we can observe directly neural impulses in sensory afferent and sympathetic efferent nerve fibers.

T. Mano (✉)
Gifu University of Medical Science, 795-1 Ichihiraga-Nagamine, Seki, Gifu 501-3892, Japan
e-mail: tadaaki.mano@nifty.com

S. Iwase et al. (eds.), *Clinical Assessment of the Autonomic Nervous System*,
DOI 10.1007/978-4-431-56012-8_1

1.1.1 Short History of Recordings of Neural Traffic from Intact Human Peripheral Nerves

Hensel and Bowman [1] in Marburg, Germany, reported in 1960 the first successful recording of neural impulses from human peripheral nerve in situ. They applied a glass microelectrode, which has been widely used in animal experiments to record single unit discharges from various tissues or cells, and succeeded to record skin sensory afferent discharges from multiple and single fiber preparations of exposed superficial branch of the radial nerve using micromanipulator in the arm rigidly fixed with metal holders in seven healthy human volunteers. This was rather an invasive method to approach peripheral nerves by making skin incision using ultrashort anesthetic. Seven years later in 1967, two different Swedish groups published independently less invasive techniques to record human peripheral nerve discharges, using metal microelectrodes being penetrated percutaneously without anesthesia into the peripheral nerves. One was a report by Knutsson and Widén [2] in Stockholm, Sweden, who used glass-coated platinum-iridium microelectrodes with a diameter of 400 μm and a tip of 0.5–1 μm, having impedance between 40 and 100 kΩ. They penetrated these electrodes using micromanipulator into the ulnar nerves just above the wrist immobilized by embedding half of the upper arm, the forearm, and a part of the hand in sandbags and recorded single unit discharges from skin afferent nerves in six healthy human subjects between the ages of 20 and 30 years. Another one was a paper by Hagbarth and Vallbo [3] in Uppsala, Sweden, who employed a method using epoxy resin-insulated tungsten semi-microelectrodes with an uninsulated tip of 10–40 μm and a resistance of 0.01–1 MΩ to be penetrated into intact peripheral nerves with the hand of the examiner without using a micromanipulator. They recorded mechanoreceptor afferent discharges from the ulnar nerves at the elbow, as well as the peroneal and tibial nerves at the popliteal fossa of two subjects who were the authors. The original method of Hagbarth and Vallbo who used tungsten microelectrodes, being much less invasive, has been applied all over the world and has been called microneurography.

1.1.2 What Kinds of Neural Traffic Can We Record Using Microneurography?

Using microneurography we can record neural traffic not only in large myelinated nerve fibers but also in thin unmyelinated fibers. Microneurography enables us to record afferent discharges from the muscle and skin, as well as sympathetic efferent discharges leading to the muscle and skin with the identification of the sensory receptors and the sympathetic effector organs. Thus we can observe identified afferent discharges from muscle spindles, tendon organs, muscle nociceptors, and skin mechano- and nociceptors, as well as activity of postganglionic sympathetic efferent nerves innervating the muscle and skin. Recordings of afferent discharges from the muscle spindle are useful to analyze neural mechanisms of motor control. Recordings of afferent discharges from skin mechanoreceptors provide objective signals related to skin mechanoreception such as vibrotactile sensation. Recordings of afferent discharges from nociceptors in the muscle and skin provide objective signals related to pain and other cutaneous sensations such as itch. Recordings of sympathetic efferent activity are essential to understand the neural mechanisms of autonomic functions. Thus microneurographic studies have been carried out to elucidate various problems related to neural functions in human subjects under normal and pathological conditions. An early review detailing this technique was reported by Vallbo et al. [4]. More recent microneurographic studies were reviewed by Mano et al. [5].

1.1.3 How to Record Neural Traffic Using Microneurography

For microneurographic recordings, we use generally tungsten microelectrodes with an epoxy resin-insulated shaft having a diameter of about 100–200 μm and a tip diameter of about 1 μm with an impedance around 1–5 MΩ at 1 kHz. We insert these electrodes percutaneously without local anesthesia by a hand of the examiner into nerve fascicles of the following nerves. Principally we can use any peripheral nerve approachable by the percutaneously inserted tungsten microelectrode. We use, generally, median, ulnar, or radial nerves in the upper extremities and peroneal, tibial, femoral, or sural nerves in the lower extremities for microneurographic recordings. In rare cases, we can use the facial, trigeminal, or intercostal nerve.

When the electrode tip penetrates into the muscle nerve fascicle of a nerve, we can record afferent and efferent discharges from and to the muscle. In this case, we can elicit sensory afferent discharges from peripheral muscle receptors innervated by the impaled nerve by mechanical stimulation such as tapping, squeezing, or stretching the muscle, but not by gentle touching of the skin. When the electrode tip is placed in the skin nerve fascicle, we can record afferent and efferent discharges from and to the skin. In this case, we can elicit afferent discharges from peripheral skin receptors by gentle touching or tapping of the skin area innervated by the impaled nerve.

We record nerve discharges as voltage differences between an intraneurally inserted recording electrode and a reference electrode (surface or needle electrode) placed in the vicinity of the recording electrode. The use of thin-caliber concentric needle electrodes was also reported, although not popularly applied. We use a bio-amplifier with high impedance input and band-pass filtering between 500 and 5000 Hz to record nerve discharges to be monitored on a cathode ray or digital oscilloscope with sound monitoring, which is very important to discriminate neural signals from electromyographic discharges or contaminated electromagnetic noises. We can store recorded neural discharges in an analog or digital data recorder for later analysis.

1.1.4 Multi- and Single Fiber Recordings

When the electrode tips are inserted into muscle or skin nerve fascicles, we can record at first multiple fiber discharges containing afferent and efferent fiber discharges. With careful manipulation of the electrode tip by the hand of the examiner, we can record multiple and single afferent fiber discharges from intended sensory receptors, as well as multiple and single sympathetic efferent discharges leading to the muscle or skin. We can identify the type of sensory fibers by mechanical stimulation of muscle or skin receptors such as touch, pressure, stretch, vibration, voluntary contraction, pain, hot/cold stimulation, etc. The recordings of unmyelinated C-afferent fibers from cutaneous nociceptors became accelerated by introducing a computer-assisted identification method called marking technique [6].

Identification of sympathetic efferent discharges is described in Chaps. 1, 2, and 3.

1.1.5 Intraneural Microstimulation

By applying the microneurographic technique, not only recordings of neural traffic in peripheral nerves but also intraneural electrical stimulation of afferent and efferent nerve fibers (microstimulation) became possible. Using microstimulation, we can investigate the functions of skin mechanoreceptors [7], skin nociceptors [8], muscle nociceptors [9], and sympathetic efferent nerves [10].

1.2 Sympathetic Microneurography as a Tool to Investigate Autonomic Neural Functions in Humans

1.2.1 Sympathetic Microneurography

Autonomic functions in humans have so far been analyzed by autonomic tests mainly based on observation of functions of target organs and measurement of the plasma level of noradrenaline and/or other related substances. Currently, power spectral analysis of heart rate [Part 2 in this book] and blood pressure and recordings of sympathetic skin and flow responses are widely used as autonomic tests in clinical neurophysiology. Cardiac MIBG scintigraphy has been also used to analyze sympathetic innervation of the heart [Part 3 in this book]. Instead of these indirect methods, microneurography provides direct measurement of peripheral sympathetic neural traffic to the muscle and skin, called muscle sympathetic nerve activity (MSNA) and skin sympathetic nerve activity (SSNA), respectively. Microneurographic recording of MSNA and SSNA is called sympathetic microneurography. The first recording of sympathetic microneurography was reported by Hagbarth and Vallbo [11] in 1968. They used epoxy resin-coated tungsten microelectrode with a diameter of 200 μm and an uninsulated tip with a diameter 5–15 μm and an impedance of 10–50 kΩ tested at 1000 Hz and recorded spontaneous multi-fiber grouped discharges from human muscle nerve fascicles of the tibial and peroneal nerves at the level of popliteal fossa in two subjects (authors) lying on a table in a comfortable prone position. The recorded multi-fiber grouped discharges were proved as efferent discharges and were modulated by pulse and respiration, which are characteristic feature of MSNA. This was the first report of sympathetic microneurography. Gunnar B. Wallin, who worked at Uppsala and moved later to Göteborg, has much contributed to develop sympathetic microneurography to be applied for functional analysis of human sympathetic nervous system [12, 13]. Sympathetic microneurography has become widely used as a tool to investigate directly sympathetic neural functions under various conditions in humans.

1.2.2 MSNA and SSNA

We can record MSNA by penetrating the electrode tip into the muscle nerve fascicles, while SSNA when penetrating the electrode tip into the skin nerve fascicles. We can record MSNA and SSNA by a minute adjustment of the electrode tip in the muscle and skin nerve fascicle, respectively. MSNA plays essential roles to control systemic blood pressure by regulating peripheral vascular resistance, while SSNA is important to control thermoregulation by adjusting skin blood flow and sweating. In North/South America, Europe, and Australia, the most frequently used nerve for MSNA recording is the peroneal nerve, into which the recording electrode is inserted near the fibular head. In Japan, the tibial nerve at the popliteal fossa is often used for sympathetic microneurography. The double recording technique of MSNA simultaneously from the peroneal and tibial nerves revealed that MSNAs from these two nerves which innervate a pair of antagonistic muscles had almost synchronous and similar burst discharge patterns. On the contrary, SSNA recorded from these two nerves showed different discharge patterns since the peroneal nerve innervates hairy skin dominated by thermal sweating, while the tibial nerve innervates glabrous skin dominated by mental sweating [14]. We can record MSNA and SSNA as multi-fiber discharges to be observed as full-wave rectified and integrated traces (mean voltage neurogram). More recently, single fiber recordings of MSNA have been reported [15, 16, Chap. 6 in this book]. Single fiber MSNA can provide more abundant information about sympathetic nerve functions in humans.

1.2.3 Identification of MSNA and SSNA

We can identify MSNA and SSNA based on the following discharge characteristics:

(a) MSNA:

1. Pulse-synchronous spontaneous and rhythmic efferent burst discharges recorded from the muscle nerve fascicle
2. Modulated by respiration
3. Increased by a fall and decreased by a rise in systemic blood pressure (Fig. 1.1)
4. Enhanced by maneuvers increasing intrathoracic pressure such as Valsalva's maneuver

(b) SSNA:

1. Spontaneous arrhythmic efferent burst discharges recorded from the skin nerve fascicle
2. Followed by peripheral vasoconstriction or perspiration
3. Elicited with almost constant latency by mental stress and sensory stimuli (sound, pain, electrical stimulation of the peripheral nerve trunk, etc.)

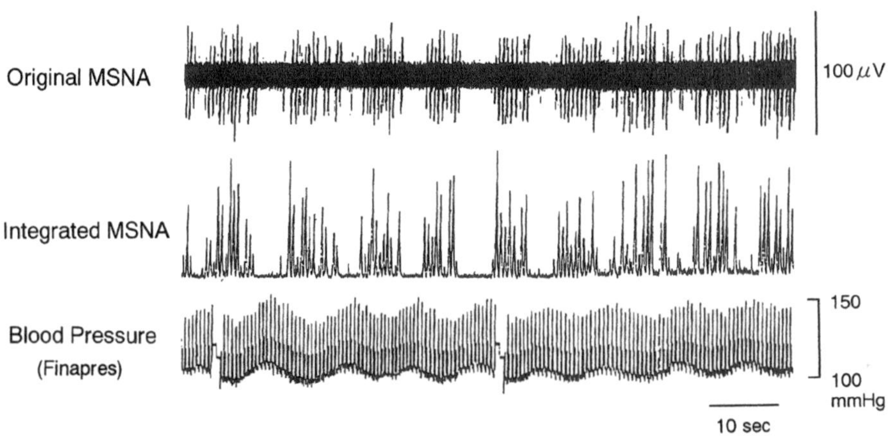

Fig. 1.1 Muscle sympathetic nerve activity (*MSNA*) and blood pressure. From the *top* to the *bottom*, original multi-fiber burst of MSNA, microneurographically recorded from the tibial nerve in a healthy subject, full-wave rectified and integrated MSNA, and systemic blood pressure recorded continuously by Finapres. MSNA discharges spontaneously responding to spontaneous fluctuation of blood pressure, discharging when blood pressure falls, while being suppressed when blood pressure rises, thus controlling blood pressure homeostasis (Figure from Published Paper [21])

Identification of the nerve fibers discharging with the above-mentioned characteristics as belonging to the efferent C-fiber group is based on the following findings: (1) The discharges are not modified by local anesthetic infiltration distal to the recording site, blocking simultaneously recorded afferent somatosensory impulses, while they are abolished rapidly by local anesthetic infiltration proximal to the recording site, with an increase in peripheral skin blood flow in the case of vasoconstrictor activity without changing somatosensory afferent activities [17]. (2) Double recording from two different sites in the same nerve clearly shows that the discharge recorded in the proximal site always precedes the discharge recorded in the distal site. The conduction velocity measured as interelectrode distance divided by the interval of two discharges shows the value of the C-fiber range to be around 1–2 m/s [18]. (3) Vasoconstrictor and sudomotor components of SSNA are identified by simultaneous recordings of skin blood flow and sweating. Vasoconstrictor bursts precede skin vasoconstriction, while sudomotor bursts precede sweating. Combined bursts of vasoconstrictor and sudomotor components precede both vasoconstriction and sweating (Fig. 1.2).

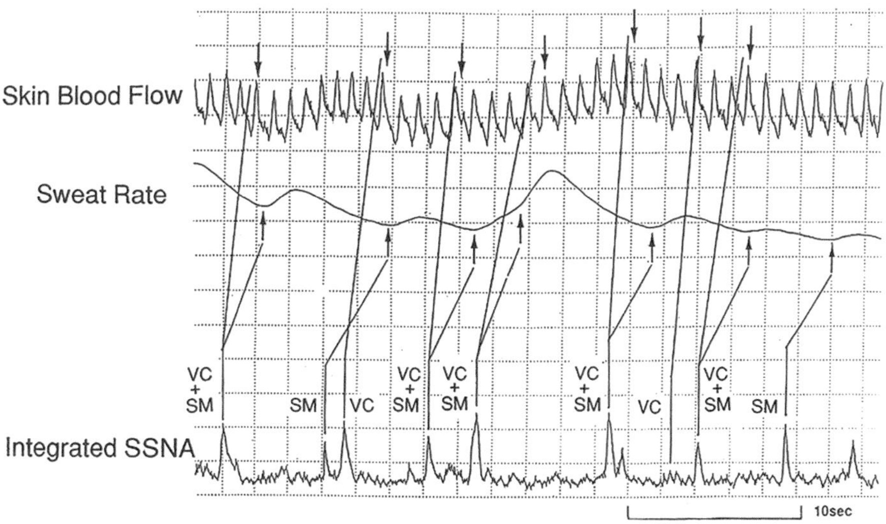

Fig. 1.2 Skin sympathetic nerve activity (*SSNA*), skin blood flow, and sweating. From the *top* to the *bottom*, skin blood flow measured by laser Doppler flowmetry, sweat rate measured by ventilated capsule method, and full-wave rectified and integrated multi-fiber SSNA bursts, microneurographically recorded from the peroneal nerve in a healthy subject. SSNA discharges spontaneously preceding vasoconstriction (*downward arrow*, *VC* vasoconstrictor burst), sweat rate (*upward arrow*, *SM* sudomotor burst), or both vasoconstriction + sweat rate (*VC* + *SM* vasoconstrictor + sudomotor burst), thus controlling skin blood flow and sweating (Figure from Published Paper [21])

1.2.4 Applications of Sympathetic Microneurography

Sympathetic microneurography has become a potent tool in clinical autonomic testing [19, Chaps. 2, 3, 4, 5, and 6 in this book]. We can record relatively easily spontaneously discharging sympathetic neural traffic in abundant fibers grouped in muscle and skin nerve fascicles as multi-fiber burst activities. Interindividual comparison of MSNA can be done as counting, for example, burst numbers per minute. The burst numbers of MSNA are reproducible in the same individuals when the recordings are made at different intervals, but increase with advancing the age [20]. Recordings of MSNA and SSNA have been applied to elucidate autonomic neural functions under various physiological and pathological conditions. MSNA is important to analyze neural mechanisms involved in the cardiovascular system, while SSNA is essential to understand neural functions related to thermoregulation [21].

We can use sympathetic microneurography to analyze neural functions not only at rest but also when exposed to various environmental stresses [22]. This method has been used to clarify physiological autonomic mechanisms as well as pathological mechanisms, not only in awake state but also during sleep [23, 24]. We used sympathetic microneurography not only in ground-based laboratories but also in various fields such as in high altitude, in cold or hot environment, during water immersion, in airplane, when loading acceleration, in space, and so on. Autonomic neural functions in space have been evaluated using various indirect methods until 1998, when microneurography was first applied during spaceflight to clarify how microgravity influences MSNA in humans. For this research, three American and one Japanese astronauts mastered perfectly the technique of sympathetic microneurography. Two of them aboard Space Shuttle Columbia could measure successfully MSNA from the peroneal nerve of their fellow astronauts. It was revealed that MSNA was rather enhanced during the 12th and 13th days of spaceflight and just after returning to Earth [25–28].

References

1. Hensel H, Boman KA. Afferent impulses in cutaneous sensory nerves in human subjects. J Neurophysiol. 1960;23:564–678.
2. Knutsson E, Widén L. Impulses from single nerve fibers recorded in man using microelectrodes. Nature. 1967;213:606–7.
3. Hagbarth K-E, Vallbo ÅB. Mechanoreceptor activity recorded percutaneously with semi-microelectrodes in human peripheral nerves. Acta Physiol Scand. 1967;69:121–2.
4. Vallbo ÅB, Hagbarth K-E, Torebjök HE, Wallin G. Somatosensory, proprioceptive, and sympathetic activity in human peripheral nerves. Physiol Rev. 1979;59:919–57.
5. Mano T, Iwase S, Toma S. Microneurography as a tool in clinical neurophysiology to investigate peripheral neural traffic in humans. Clin Neurophysiol. 2006;117:2357–84.
6. Torebjök E, Hallin RG. Identification of afferent C units in intact human skin nerves. Brain Res. 1974;67:378–403.

7. Vallbo ÅB. Sensations evoked from the glabrous skin of the human hand by electrical stimulation of unitary mechanoreceptive afferents. Brain Res. 1981;29:359–63.
8. Ochoa J, Torebjörk E. Sensation evoked by intraneural microstimulation of C nociceptor fibres in human skin nerves. J Physiol. 1989;415:583–99.
9. Simone DA, Marchettini P, Caputi G, Ochoa JL. Identification of muscle afferents subserving sensation of deep pain in humans. J Neurophysiol. 1994;72:883–9.
10. Kunimoto M, Kirno K, Elam M, Wallin BG. Neuroeffector characteristics of sweat glands in the human hand activated by regular neural stimuli. J Physiol. 1991;442:391–411.
11. Hagbarth K-E, Vallbo ÅB. Pulse and respiratory grouping of sympathetic impulses in human muscle nerves. Acta Physiol Scand. 1968;74:96–108.
12. Wallin BG. New aspects of sympathetic functions in man. In: Stålberg E, Young RR, editors. Butterworth international medical reviews, neurology 1, clinical neurophysiology. London: Butterworth; 1981. p. 145–67.
13. Wallin BG. Intraneural recording and autonomic function in man. In: Bannister R, editor. Autonomic failure. A textbook of clinical disorders of the autonomic nervous system. Oxford: Oxford University Press; 1983. p. 36–51.
14. Mano T, Iwase S, Sugiyama Y. Double recording technique of sympathetic nerve traffic in microneurography. In: Kimura J, Shibasaki H, editors. Recent advances in clinical neurophysiology. Amsterdam: Elsevier Science BV; 1996. p. 416–9.
15. Macfield VG, Wallin BG, Vallbo ÅB. The discharge behaviour of single vasoconstrictor motoneurones in human muscle nerves. J Physiol. 1994;481:799–899.
16. Tsukahara R, Mano T. The recruitment pattern of single vasoconstrictor neurons in humans. J Auton Nerv Syst. 1997;66:26–34.
17. Nagasaka T, Hirata K, Mano T, Iwase S, Rossetti Y. Heat-induced finger vasoconstriction controlled by skin sympathetic nerve activity. J Appl Physiol. 1990;68:71–5.
18. Kondo M, Iwase S, Mano T, Kuzuhara S. Direct measurement of human sympathetic nerve conduction velocity. Muscle Nerve. 2004;29:128–33.
19. Mano T. Clinical applications of microneurography. In: Robertson D, Biaggioni I, Burnstock G, Low PA, Paton JFR, editors. Primer on the autonomic nervous system. 3rd ed. Amsterdam: Elsevier and Academic Press; 2012. p. 393–7.
20. Iwase S, Mano T, Watanabe Y, Saito M, Kobayashi F. Age-related changes of sympathetic outflow to muscles in humans. J Gerontol. 1991;46:M1–5.
21. Mano T. Muscular and cutaneous sympathetic nerve activity. In: Appenzeller O, editor. The autonomic nervous system(1), Handbook of clinical neurology, vol. 72. Amsterdam: Elsevier; 1999. p. 649–65.
22. Mano T. Microneurographic research on sympathetic nerve responses to environmental stimuli in humans. Jpn J Physiol. 1998;48:99–114.
23. Hornyak M, Cejnar M, Elam M, Hornyak M, Cejnar M, Elam M, Wallin BG. Sympathetic muscle nerve activity during sleep in man. Brain. 1991;114:1281–95.
24. Okada H, Iwase S, Mano T, Sugiyama Y, Watanabe T. Changes in muscle sympathetic nerve activity during sleep in humans. Neurology. 1991;41:1961–6.
25. Cox JF, Tahvanainen KU, Kuusela TA, Levine BD, Cooke WH, Mano T, et al. Influence of microgravity on astronauts' sympathetic and vagal responses to Valsalva's manoeuvre. J Physiol. 2002;544:253–64.
26. Ertl AC, Dietrich A, Biaggioni I, Leine BD, Robertson RM, Cox JF, et al. Human sympathetic nerve activity and plasma noradrenaline kinetics in space. J Physiol. 2002;538:321–9.
27. Levine BD, Pawelczyk JA, Ertl AC, Cox JF, Zuckerman JH, Dietrich A, et al. Human muscle sympathetic neural and haemodynamic responses to tilt following spaceflight. J Physiol. 2002;538:331–40.
28. Fu Q, Levine BD, Pawelczyk JA, Ertl AC, Diedrich A, Cox JF, et al. Cardiovascular and sympathetic neural responses to handgrip and cold pressor stimuli in humans before, during and after spaceflight. J Physiol. 2002;544:653–64.

Chapter 2
Muscle Sympathetic Nerve Activity in Neurological Disorders

Kazumasa Shindo

Abstract Recently, progress in microneurographic methods and instruments has made it possible to record muscle sympathetic nerve activity (MSNA) in patients with neurological disorders in order to elucidate the pathophysiology or pathogenesis. In patients with Parkinson's disease, the frequency of MSNA bursts at rest showed a negative correlation with age in the PD patients, while age and MSNA were positively correlated in the controls. There was a negative correlation between disease duration or the severity of disability and the age-adjusted MSNA burst frequency in PD patients. In patients with amyotrophic lateral sclerosis, resting MSNA was significantly higher than in healthy subjects or patients with other neuromuscular disorders, although systolic blood pressure (BP), diastolic BP, and heart rate (HR) were similar in three groups at rest. Standardized resting MSNA showed a negative correlation with disease duration. Patients with cerebellar degeneration showed lower resting MSNA than the healthy controls, while there were no significant differences of age, resting HR, or resting BP between the patients and healthy controls. In patients with cervical spondylosis, the resting MSNA was positively correlated with age in controls and resting MSNA was significantly lower than healthy controls.

Keywords Muscle sympathetic nerve activity • Parkinson's disease • Amyotrophic lateral sclerosis • Cerebellar degeneration syndrome • Cervical spondylosis • Polyneuropathy

2.1 Introduction

In various neurological disorders associated with impairment of autonomic function, such as multiple system atrophy, cerebellar degeneration, Parkinson's disease, dementia with Lewy bodies, spinal cord disorders, pure autonomic failure, autonomic neuropathies including diabetic neuropathy, and familial amyloid

K. Shindo (✉)
Department of Neurology, Interdisciplinary Graduate School of Medicine and Engineering, University of Yamanashi, 1110 Shimogato, Chuo, Yamanashi 409-3898, Japan
e-mail: kshindo@yamanashi.ac.jp

S. Iwase et al. (eds.), *Clinical Assessment of the Autonomic Nervous System*, DOI 10.1007/978-4-431-56012-8_2

Table 2.1 Neurological disorders with autonomic failure

Dementia
Alzheimer disease, dementia with Lewy bodies, frontotemporal lobar degeneration, neuronal intranuclear inclusion disease, etc.
Cerebrovascular disorders
Cerebral infarction, intracerebral hemorrhage, subarachnoid hemorrhage, etc.
Extrapyramidal disorders
Parkinson's disease, Huntington's disease, dystonia, etc.
Cerebellar degeneration syndrome
Multiple system atrophy, cortical cerebellar atrophy, familial spinocerebellar ataxia, etc.
Motor neuron disease
Amyotrophic lateral sclerosis, primary lateral sclerosis, etc.
Demyelinating disorders
Multiple sclerosis, neuromyelitis optica, acute disseminated encephalomyelitis, central pontine myelinolysis, etc.
Inflammation of the central nervous system
Bacterial and viral meningoencephalitis, prion disease, hypertrophic pachymeningitis, limbic encephalitis, infectious myelitis, etc.
Spinal cord disorders
Cervical spondylosis, syringomyelia, spinal canal stenosis, spastic paraplegia, spinal cord AVM and AVF, spinal cord tumor, spinal cord injury, etc.
Peripheral neuropathy
Pure autonomic failure, idiopathic autonomic neuropathy, diabetic neuropathy, familial amyloid polyneuropathy, Guillain-Barré syndrome, chronic inflammatory demyelinating polyneuropathy, etc.
Myopathy and neuromuscular junction disorders
Myotonic dystrophy, mitochondrial encephalomyopathy, myasthenia gravis, Lambert-Eaton myasthenic syndrome, etc.
Functional disorders and others
Migraine, cluster headache, epilepsy, Horner's syndrome, neurally mediated syncope, Wilson disease, pellagra, neuroferritinopathy, vitamin deficiency, etc.

polyneuropathy, the detection of autonomic symptoms and autonomic dysfunction by neurophysiological examination often contributes to diagnosis and management (Table 2.1). However, after cardiovascular autonomic symptoms such as orthostatic or postprandial hypotension become clinically evident, it is often difficult to assess muscle sympathetic nerve activity (MSNA) by microneurography in these disorders. Recently, progress in microneurographic methods and instruments has made it possible to record MSNA in patients with neurological disorders, especially those with mild autonomic impairment or with early disease, in order to elucidate the pathophysiology or pathogenesis [1–3].

This chapter reviews the characteristic features of MSNA in Parkinson's disease, amyotrophic lateral sclerosis, and other neurological disorders based on our previous studies and reports by other investigators. The other autonomic dysfunction in relation to MSNA is also discussed.

2.2 Parkinson's Disease (PD)

2.2.1 PD Patients and Healthy Controls

Numerous non-motor changes or symptoms due to autonomic dysfunction are frequent in patients with PD and have been reported by many investigators [4–6]. Among autonomic abnormalities, orthostatic hypotension or prandial hypotension is often observed in patients with PD, probably resulting from the progressive impairment of preganglionic and postganglionic sympathetic neurons [7, 8]. However, clear recordings of MSNA are often hard to obtain because of advanced autonomic impairment or numerous artifacts due to involuntary tremor. Thus, only a few studies have investigated whether MSNA is actually decreased or not in patients with PD [2, 9]. The characteristic features of MSNA in patients with PD showing mild autonomic symptoms and in age-matched healthy control were reported in 2003 by our colleagues [2].

A summary of their findings is as follows:

- In a representative recording, the burst frequency of MSNA at rest and the increased response of MSNA during head-up tilting are slightly attenuated compared with healthy subjects (Fig. 2.1).
- There are no differences of the age, resting heart rate (HR), or resting blood pressure (BP) between the PD patients and the controls.

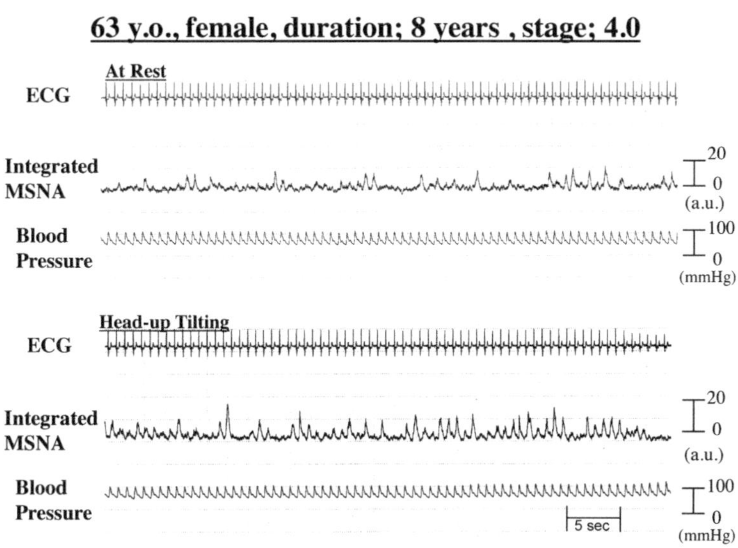

Fig. 2.1 A representative recording (*top trace*, ECG; *middle trace*, integrated neurogram of muscle sympathetic nerve activity (MSNA); *bottom trace*, blood pressure) of a patient with Parkinson's disease (PD, B)

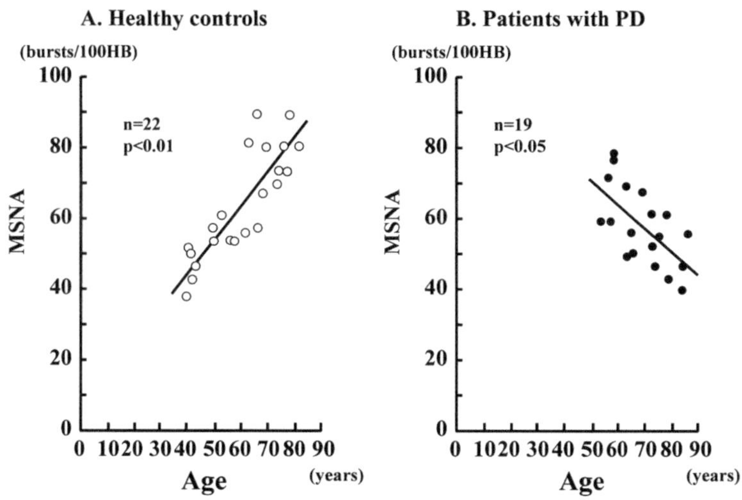

Fig. 2.2 Correlations between the burst incidence of muscle sympathetic nerve activity (*MSNA*) at rest and age in healthy controls (**a**) and in patients with Parkinson's disease (PD, **b**). *HB* heart beats

- The frequency of MSNA bursts (bursts per 100 heart beats) at rest shows a negative correlation with age in the PD patients, while age and MSNA are positively correlated in the controls (Fig. 2.2).
- There is a negative correlation between disease duration or the severity of disability and the age-adjusted MSNA burst frequency in PD patients (Fig. 2.3).
- The increase of HR, BP, and MSNA in response to head-up tilt is slightly, but not significantly, smaller in PD patients than in controls.
- There are no significant relations between the increases in MSNA, BP, or noradrenaline in response to head-up tilt or standing and the age or disease duration.

Recent microneurographic studies have confirmed that resting MSNA gradually increases with age in healthy subjects because of reduced sensitivity of the baroreceptors or reduction of parasympathetic tone with advancing age [10, 11]. In contrast, resting MSNA gradually decreases with age and also with the duration of disease in PD patients. Thus, older PD patients have lower MSNA than normal subjects, reflecting their generally longer duration of disease. The reason why the frequency of MSNA bursts in younger PD patients is similar to control values remains unclear, but it is speculated that the progression of autonomic impairment may be slower in younger patients than older patients. In the abovementioned study, none of the patients had severe manifestations of autonomic failure such as frequent syncope due to orthostatic hypotension. Symptomatic autonomic deficits would not occur as long as the MSNA response to tilting is preserved, even if resting MSNA is low. It is speculated that since peripheral nerves are impaired at the early stage and central nerves are impaired at the later stage in autonomic nerves of PD, MSNA could be elicited even because of preserved central pathway in baroreflex arc.

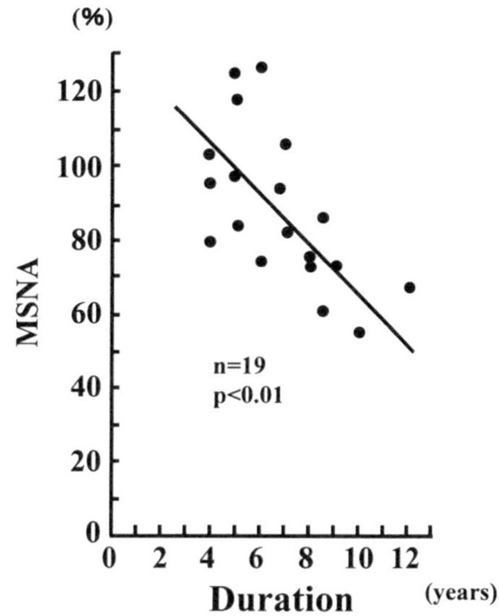

Fig. 2.3 A correlation between disease duration of Parkinson's disease and burst frequency of muscle sympathetic nerve activity (*MSNA*) at rest standardized from a control regression line in patients with Parkinson's disease

Many investigators have studied the association between PD and orthostatic intolerance or orthostatic hypotension. It was reported that 70 % of patients with PD exhibited a mild to severe orthostatic decline of blood pressure [5] and that all PD patients have the symptom of postural dizziness, with obvious postural hypotension being observed in 27 % of them [4]. Moreover, a significant decrease of the systolic blood pressure was observed in PD patients even before initiation of antiparkinsonian treatment [12]. Although a few investigators found no difference in the response of BP to head-up tilt between PD patients and healthy controls [13, 14], most recent studies have demonstrated that mild to severe orthostatic hypotension is frequent in PD, being related to both the disease duration and severity. The abovementioned MSNA findings are consistent with previous observations of sympathetic vasomotor impairment in PD, because this study confirmed that orthostatic intolerance or hypotension may reflect an age and duration-dependent reduction in sympathetic outflow to muscles and not a diminution in the vasoconstrictive response of peripheral arteries.

2.2.2 MSNA and MIBG Cardiac Scintigraphy in PD

Measurement of myocardial ^{123}I-metaiodobenzylguanidine (MIBG) uptake is useful in the assessment of cardiac sympathetic nerve function in patients with neurodegenerative disorders and diabetes mellitus [15, 16]. Recent investigations have shown that patients with PD frequently have an obvious decrease of MIBG uptake, and a relationship between MIBG findings and clinical features or disease specificity has been reported [17, 18]. However, it has not been clear whether

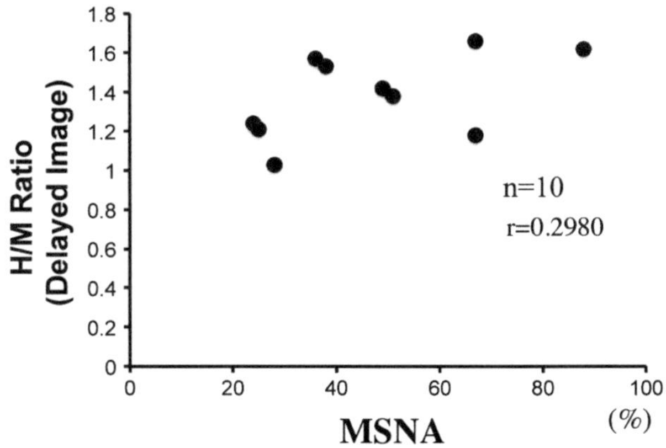

Fig. 2.4 A correlation between heart-to-mediastinum ratio (*H/M ratio*) in delayed image and burst frequency of muscle sympathetic nerve activity (*MSNA*) standardized from a control regression line in patients with Parkinson's disease

MIBG findings reflect organic or functional changes of cardiac sympathetic nerves. To resolve this issue, the relationship between abnormal cardiac MIBG uptake and quantitative MSNA data was analyzed in PD patients by our colleagues in 2005 [19].

The results they obtained were as follows:

- There is no significant correlation between MSNA and the H/M ratio on delayed images (Fig. 2.4).
- There is no significant relationship between the H/M ratio and disease duration or the level of disability, although significant negative correlations are observed between MSNA and disease duration or disability.
- There is no significant relationship between the H/M ratio or the washout ratio (WR) and disease duration or the level of disability, although disease duration is positively correlated with disability.

There may be no significant relationship between MSNA and the H/M ratio or WR because some patients with reduced MSNA have a normal H/M ratio or WR. Since the sympathetic nervous system displays characteristic regional differentiation [20], it is possible that these correlations are not observed because MSNA regulating the sympathetic tone of vessels in the leg muscles is compared with MIBG uptake that reflects cardiac sympathetic function. Another possible explanation for the missing correlation is that MIBG uptake is already very low at an early stage of PD and therefore does not decrease much further over time. Therefore, recording the MSNA of peripheral nerves is considered to be more sensitive than cardiac MIBG scintigraphy for estimating progressive changes of autonomic function in PD patients. It is speculated that cardiac MIBG scintigraphy may reflect not only organic changes of postganglionic sympathetic neurons but also other factors contributing to the metabolism of noradrenaline or MIBG at sympathetic nerve endings.

2.2.3 Other Studies of MSNA in PD

In 1994, Takeuchi et al. reported that the response of MSNA to head-up tilt was weaker after administration of levodopa than before administration, while there were no significant differences of resting MSNA and the response of MSNA to head-up tilt between PD patients and healthy controls [9]. They concluded that these findings suggested the existence of potential impairment of the medullary vasomotor center in PD. In 2014, Sverrisdottir et al. reported on recording of MSNA during ON and OFF of deep brain stimulation to the midbrain and subthalamic nuclei. During stimulation, they noted a decreased burst frequency and decreased blood pressure, along with an unchanged burst amplitude and increased vasomotor baroreflex sensitivity [21]. However, cardiac MIBG scintigraphy findings and the relationship between MSNA and age or disease duration were not described in these reports.

2.3 Amyotrophic Lateral Sclerosis (ALS)

2.3.1 ALS Patients and Healthy Controls

ALS is a fatal neurodegenerative disorder that mainly causes motor symptoms due to progressive generalized muscular weakness and muscle atrophy because of gradual impairment of the cranial and spinal motor neurons. Formerly, autonomic symptoms such as neurogenic bladder and constipation were considered to be negative features. However, various abnormalities of autonomic function, especially related to the cardiovascular system, have been identified in ALS by physiological and pharmacological studies [22, 23]. To assess autonomic function, quantitative analysis of MSNA along with BP and HR at rest and during head-up tilt was performed in ALS patients, and the results were compared with those obtained in healthy subjects by our colleagues in 1993 [24].

The results that they reported were as follows:

- In representative recordings, MSNA bursts in a patient with ALS are remarkably increased at rest compared with a healthy subject (Fig. 2.5).
- Resting MSNA is significantly higher in ALS patients than in the control group, although systolic BP, diastolic BP, and HR are similar in both groups at rest.
- In the control group, resting MSNA is positively correlated with age, while there are no correlations between the parameters of resting MSNA and age in the ALS group (Fig. 2.6).
- There is no consistent relationship between resting MSNA and disease duration, or between resting MSNA and the severity of muscular atrophy or spasticity.

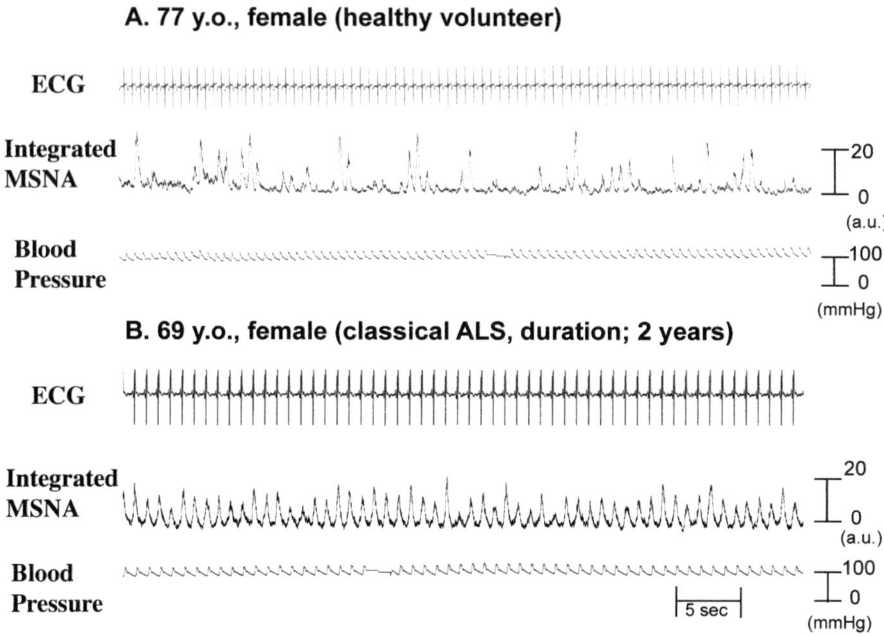

Fig. 2.5 Representative recordings (*top trace*, ECG; *middle trace*, integrated neurogram of muscle sympathetic nerve activity (MSNA); *bottom trace*, blood pressure) of a healthy volunteer (**a**) and a patient with amyotrophic lateral sclerosis (ALS, **b**) at rest

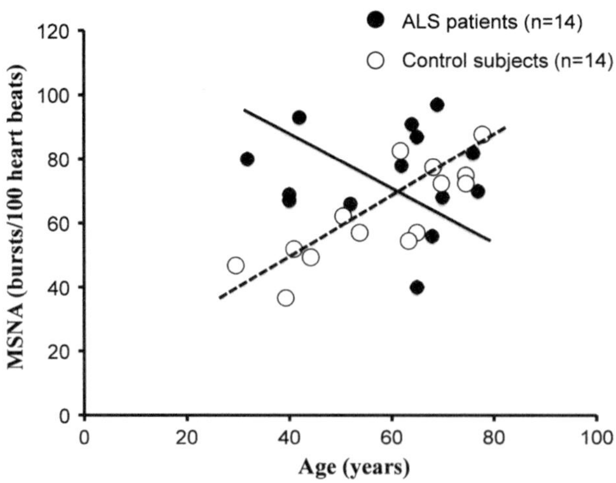

Fig. 2.6 Correlations between the burst incidence of muscle sympathetic nerve activity (*MSNA*) at rest and age in healthy controls and patients with amyotrophic lateral sclerosis (*ALS*)

- The response of MSNA to head-up tilt is weaker in the ALS group than in the controls, but the changes of blood pressure and heart rate are similar in both groups.

MSNA was reported to be markedly elevated in younger ALS patients, while MSNA is similar in elderly ALS patients and age-matched controls, suggesting that the lack of a difference in older subjects may be related to the increase of MSNA that normally occurs with aging [10, 11]. The changes of MSNA in ALS patients were reported to be smaller than in controls when they are placed in the head-up tilt position. Because ALS patients do not show postural hypotension, the muted response may result from increased baseline MSNA. The reason why MSNA is elevated is that since nerve conduction studies such as F-wave amplitudes suggest a marked increase in discharge from anterior horn cells secondary to reinnervation, the pathogenic agents responsible for ALS easily affect motor neurons, while sympathetic neurons are comparatively resistant to the same agents but are very excitable.

Most previous investigations have found that BP and HR are slightly increased in patients with ALS [22, 23]. Nogues et al. reported resting sinus tachycardia in ambulant ALS patients with normal breathing [25]. In the abovementioned study, resting MSNA was found to be increased in ALS patients, despite the absence of known cardiovascular abnormalities or significant respiratory impairment. Although a decreased PaO_2 or increased $PaCO_2$ is known to elevate MSNA [26], respiratory impairment is unlikely to account for the increase of MSNA in ALS patients.

2.3.2 *ALS and Other Neuromuscular Disorders*

It could not be ruled out that the increase of MSNA in ALS patients was influenced by a reduction of daily activities, bulbar symptoms, or subclinical respiratory distress. Furthermore, it was not clear whether this increase was specific to patients with ALS. To further clarify the influence of muscle wasting and limitation of daily activities on sympathetic outflow to the muscles, MSNA was compared between patients with ALS and patients with a similar level of disability due to other neuromuscular disorders by our colleagues in 1995 [1].

The results they obtained were as follows:

- There are no differences of disability scores, respiratory function, and resting HR or BP between the ALS patients and control patients (Fig. 2.7).
- The resting MSNA burst frequency is positively correlated with age in both the patients with ALS and the control patients and is significantly higher in the ALS patients compared with the controls (Fig. 2.7).
- Even in ALS patients aged more than 50 years, resting MSNA is slightly higher than in the control patients.
- In ALS patients, resting MSNA is not correlated with the disease duration, the prognosis, the disability score, or PaO_2 and $PaCO_2$.
- During the head-up tilt, changes of BP and the MSNA burst frequency are slightly smaller in ALS patients than in control patients.

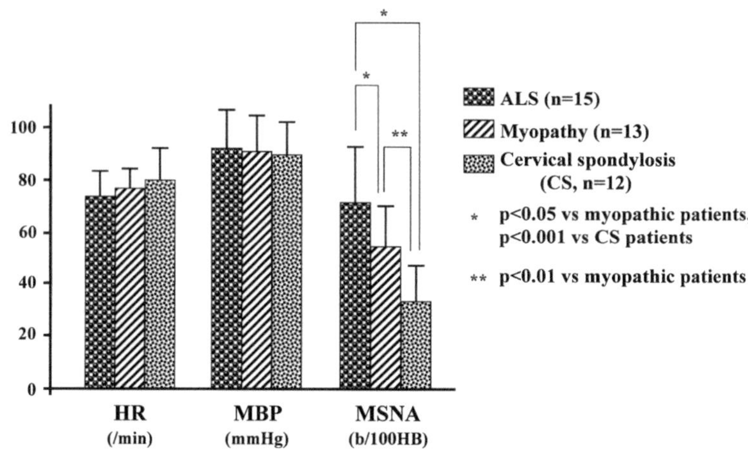

Fig. 2.7 Comparison between patients with amyotrophic lateral sclerosis (*ALS*) and patients with other neuromuscular disorders in each parameter at rest. *HR* heart rate, *MBP* mean blood pressure, *MSNA* muscle sympathetic nerve activity

None of the ALS patients had subjective dyspnea at rest, and MSNA is not correlated with PaO_2, $PaCO_2$, or forced vital capacity in either the ALS or control patients, suggesting that the mild to moderate hypoxemia present in ALS has little effect on resting MSNA. There were no correlations between MSNA and the duration of disease, the disability score, or the prognosis, probably because ALS patients show wide variation in the rate of progression and symptoms. These results suggest that some ALS patients have an increase of MSNA that is independent of the extent of muscle wasting or limitation of daily activities.

The muted response of MSNA in ALS patients may have resulted from higher resting activity. With regard to blood pressure, ALS patients show slight elevation of resting blood pressure [27], and it is possible that a continuous increase of sympathetic outflow to muscles causes downregulation of the vasoconstrictor response of intramuscular vessels. The cause of the increase of MSNA in ALS has yet to be clarified. Since recent investigations into the pathological or genetic background have revealed a link between ALS and frontotemporal dementia, pathological changes of the limbic system in medial temporal lobe (frequently affected in FTLD) might contribute to abnormalities in the cardiovascular system such as increased MSNA.

2.3.3 *Chronological Changes of MSNA in ALS*

There is a need for further studies, considering that sympathetic hyperactivity resembles overexcitability of motor neurons in the brain stem and spinal cord in ALS. Although various studies of autonomic function have been reported

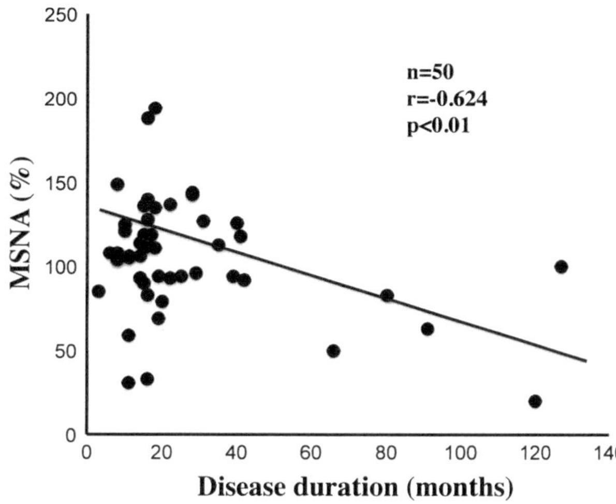

Fig. 2.8 A correlation between resting MSNA standardized from a control regression line and disease duration in patients with amyotrophic lateral sclerosis (ALS)

[1, 22–24], the chronological changes of sympathetic activity over the course of ALS was not evaluated. In 2004, correlations were analyzed between MSNA and disease duration, age, or disability [28]. Recordings were performed twice or more in the same patients at intervals of at least 6.

The following results were obtained:

- Standardized resting MSNA shows a negative correlation with disease duration, although several patients with a duration under 12 months had low MSNA (Fig. 2.8).
- There is a negative correlation between standardized MSNA and the level of disability.
- In patients who underwent recording of MSNA more than once, mean standardized MSNA shows a significant decrease at the second examination compared with the first.

Considering the negative correlation between MSNA and disease duration, it is possible that MSNA increases in the early stage of ALS and then gradually decreases with a longer disease duration. In relation to disability, most ALS patients tend to show a decrease of MSNA as motor function deteriorates, since patients with longer disease duration have more severe disability. Histopathological evidence of a slight decrease of intermediolateral column neurons in the spinal cord in ALS patients [29–31] is consistent with the chronological changes of sympathetic outflow to muscles shown in the abovementioned study. It is speculated that intermediolateral neurons may decrease as the result of continuous excessive stimulation, such as that related to increased resting MSNA.

2.3.4 Other Studies of MSNA in ALS

In 2002, Oey et al. reported that the resting heart rate was increased and that the response of MSNA to lower body negative pressure was reduced in ALS patients compared with age-matched healthy volunteers [32]. However, the baseline blood pressure and MSNA parameters did not differ between the two groups. According to the report by Nygren and Fagius in 2011 [3], patients with ALS showed a higher resting heart rate and higher resting MSNA than healthy controls. Although baroreflex inhibition was preserved, MSNA showed a smaller response to sympathoexcitatory maneuvers in ALS patients. They concluded that elevation of MSNA might be a primary feature of ALS. One reason why the MSNA findings vary among studies could be that ALS patients show considerable variation in disease progression and motor symptoms. Another reason would be the chronological changes of MSNA confirmed in the abovementioned study. Future investigations into the pathogenesis of ALS should not only focus on the motor system but also on changes in autonomic function.

2.4 MSNA in Other Neurodegenerative Disorders

2.4.1 Spinocerebellar Degeneration (SCD)

SCD, such as olivopontocerebellar atrophy or multiple system atrophy (MSA) with prominent autonomic symptoms (which used to be known as Shy-Drager syndrome), are often associated with symptoms of autonomic dysfunction [33, 34]. Although many investigators have previously reported that patients with various SCD subtypes have varying degrees of noradrenergic or cholinergic autonomic dysfunction [35, 36], few studies on sympathetic outflow to muscle and skin have been performed with microneurography [37–39]. In SCD patients with or without mild autonomic dysfunction, quantitative analysis of MSNA, BP, and HR at rest and during head-up tilt was performed in 1999 by our colleagues, and the results were compared with those obtained in healthy subjects [40].

The main findings were as follows:

- All SCD patients show lower resting MSNA than the healthy controls, while there are no significant differences of age, resting HR, or resting BP between the patients and controls (Fig. 2.9).
- There is no consistent relationship between MSNA and the severity of cerebellar ataxia or disease progression.
- During head-up tilt, the increase of BP and the mean amplitude of MSNA are significantly smaller in SCD patients than controls, but changes of HR and MSNA burst frequency are similar in both groups.

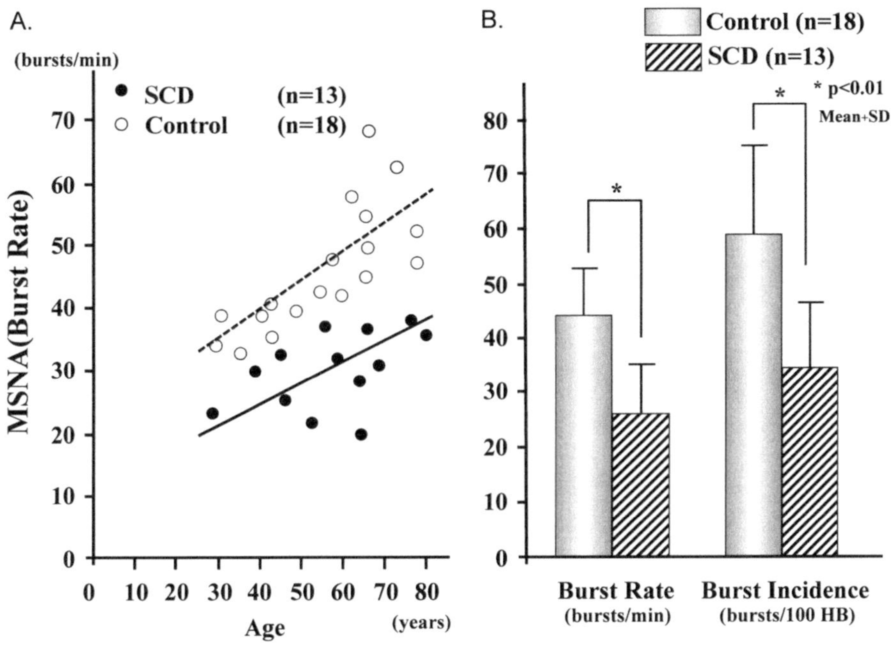

Fig. 2.9 (**a**) A correlation between the burst rate of resting MSNA and age in healthy controls and in patients with spinocerebellar degeneration (SCD). (**b**) Comparison of the burst rate (bursts/minute) and the burst incidence (bursts/100 heart beats) of muscle sympathetic nerve activity at rest between patients with spinocerebellar degeneration (SCD) and healthy controls

- Subclinical sympathetic dysfunction is observed in most cases of sporadic SCD without autonomic symptoms such as orthostatic dizziness or hypotension because of mild impairments of baroreflex modulation and vasomotor center.

In previous investigations, an impaired blood pressure response to cold immersion, reduced heart rate variability, sudomotor dysfunction, and neurogenic bladder were occasionally observed in patients with cortical cerebellar atrophy who only had mild autonomic symptoms [41, 42]. In the abovementioned study, it was confirmed that characteristic MSNA abnormalities can even be observed in patients who have SCD subtypes without autonomic symptoms, suggesting the existence of underlying noradrenergic sympathetic dysfunction.

2.4.2 Multiple System Atrophy (MSA)

A few authors have mentioned the characteristic features of MSNA in patients with MSA, which commonly show moderate or severe autonomic impairments. Kachi et al. reported that MSNA was markedly decreased at rest and there was poor increase during head-up tilt resulting in orthostatic hypotension in a patient with

Shy-Drager syndrome [37]. Hakusui et al. revealed that MSNA in five MSA patients with postprandial and orthostatic hypotension showed no increase after receiving 75 g glucose intake, while all healthy controls showed increased MSNA. They stated that postprandial hypotension is caused by the lack of sympathetic compensation for the systemic hypotensive stress of splanchnic blood pooling that occurs after food ingestion [38]. Donadio et al. found that MSNA could only be recorded in one of eight patients with MSA and that quantitative value at rest increases compared with controls' values with the expectation adversely [39]. When cardiovascular autonomic abnormalities are evident, it is difficult to obtain clear MSNA recordings in patients with MSA. Recordable sympathetic activity in patients with shorter disease duration suggests a progressive deterioration of peripheral sympathetic function positively correlated with disease duration, in addition to impairments of central autonomic neurons in MSA.

2.4.3 Cervical Spondylosis (CS) and Traumatic Spinal Cord Injury

In 1997, MSNA was recorded and correlations between the quantitative MSNA parameters and pyramidal tract symptoms were analyzed in patients with CS [43].

The following results were obtained:

- There are no differences between the CS patients and controls with respect to age, resting HR, or resting BP.
- The resting MSNA burst frequency is positively correlated with age in controls and resting MSNA is significantly lower in CS patients than controls.
- In CS patients, the resting MSNA burst frequency (expressed as a percentage of the predicted value) shows a significant negative correlation with the severity of pyramidal tract symptoms (Fig. 2.10).
- During head-up tilt, changes of HR and BP are similar in CS patients and controls, but changes of MSNA are slightly greater in the CS patients because of their lower baseline MSNA.

In the abovementioned study, there was an inverse relationship between MSNA and the severity of pyramidal tract symptoms, suggesting that resting MSNA bursts may be reduced by compression of descending sympathetic pathways in the cervical spinal cord. Since the response of MSNA to head-up tilt is preserved or exaggerated, cardiovascular neurologic symptoms such as vertigo or orthostatic hypotension might not be observed in CS patients. Further investigation of the autonomic nervous system may help to define the pathogenesis of cardiovascular neurologic symptoms in patients with early to advanced CS.

Patients with severe traumatic spinal cord lesions at the levels of C5 to Th8 exhibited sparse resting sympathetic activity on neurogram and showed only reflex

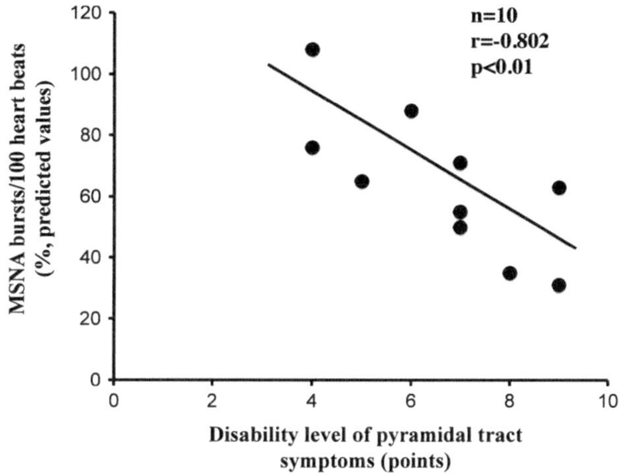

Fig. 2.10 A correlation between resting MSNA standardized from a control regression line and disability levels of pyramidal tract symptoms in patients with cervical spondylosis

bursts [44, 45]. This may suggest that excitability of decentralized spinal sympathetic neurons to muscles is decreased, resulting in weak MSNA in these patients.

2.4.4 *Polyneuropathy and Guillain-Barré Syndrome (GBS)*

Polyneuropathy, especially diabetic polyneuropathy, is frequently associated with autonomic impairments. However, when autonomic symptoms such as orthostatic hypotension are evident, the microneurographic recording of MSNA is often difficult. Even if MSNA can be recorded, it is reported that conduction velocities of MSNA determined as the conduction distance between the R-wave on electrocardiography and the peak of MSNA bursts are similar to normal values [46, 47]. The authors speculate that since postganglionic sympathetic fibers in polyneuropathy do not exhibit gradual changes in conduction velocities, normal or no conduction is observed in MSNA recordings according to the all or none law.

Guillain-Barré syndrome (GBS) is a representative disorder of post-infectious polyradiculoneuropathy with acute onset. Autonomic symptoms such as increased heart rate and blood pressure or elevated sweating are frequently observed in cases with severe motor impairments. In MSNA recordings in GBS patients with autonomic symptoms, it is reported that the burst frequency is increased in the acute phase and is decreased to normal values in the remission phase [48]. This increase in the acute phase is speculated to be due to demyelinated lesions of the afferent nerve fibers from arterial intrathoracic baroreceptors in glossopharyngeal and vagus nerves. Pure autonomic failure (PAF) is a subtype of idiopathic polyneuropathy showing various degrees of impairments of sympathetic and parasympathetic nerves. Although it is very difficult to obtain a clear recording of MSNA in patients with PAF [39, 49], Liguori et al. state that the burst frequency or conduction latency is normal in 14 cases with autonomic small fiber neuropathy [50].

References

1. Shindo K, Tsunoda S, Shiozawa Z. Increased sympathetic outflow to muscles in patients with amyotrophic lateral sclerosis: a comparison with other neuromuscular patients. J Neurol Sci. 1995;134:57–60.
2. Shindo K, Watanabe H, Tanaka H, Ohashi K, Nagasaka T, Tsunoda S, et al. Age and duration related changes in muscle sympathetic nerve activity in Parkinson's disease. J Neurol Neurosurg Psychiatry. 2003;74:1407–11.
3. Nygren I, Fagius J. High resting level and weak response of baroreflex-governed sympathetic outflow in amyotrophic lateral sclerosis. Muscle Nerve. 2011;43:432–40.
4. Aminoff MJ, Wilcox CS. Assessment of autonomic function in patients with a Parkinsonian syndrome. Br Med J. 1971;4:80–4.
5. Appenzeller O, Goss JE. Autonomic deficits in Parkinson's syndrome. Arch Neurol. 1971;24: 50–7.
6. Gross M, Bannister R, Godwin-Austen R. orthostatic hypotension in Parkinson's disease. Lancet. 1972;22:174–6.
7. Rajput AH, Rozdilsky B. Dysautonomia in Parkinsonism: a clinicopathological study. J Neurol Neurosurg Psychiatry. 1976;39:1092–100.
8. van Dijk JG, Haan J, Zwinderman K, Kremer B, van Hilten BJ, Roos RAC. Autonomic nervous system dysfunction in Parkinson's disease: relationships with age, medication, duration, and severity. J Neurol Neurosurg Psychiatry. 1993;56:1090–5.
9. Takeuchi S, Sugiyama Y, Mano T, Iwase S, Matsukawa T, Okada H, et al. Muscle sympathetic nerve activity in Parkinson's disease. Ann Res Inst Environ Med Nagoya Univ. 1994;45: 263–8.
10. Iwase S, Mano T, Watanabe T, Saito M, Kobayashi F. Age-related changes of sympathetic outflow to muscles in humans. J Gerontol Med Sci. 1991;46:M1–5.
11. Sundlof G, Wallin BG. Human muscle nerve sympathetic activity at rest. Relationship to blood pressure and age. J Physiol. 1978;274:621–37.
12. Turkka J, Suominen K, Tolonen U, Sotaniemi K, Myllyla VV. Selegiline diminishes cardiovascular autonomic responses in Parkinson's disease. Neurology. 1997;48:662–7.
13. Sachs C, Berglund B, Kaijser L. Autonomic cardiovascular responses in parkinsonism: effect of levodopa with dopa-decarboxylase inhibition. Acta Neurol Scand. 1985;71:37–42.
14. Ludin SM, Steiger UH, Ludin HP. Autonomic disturbances and cardiovascular reflexes in idiopathic Parkinson's disease. J Neurol. 1987;235:10–5.
15. Hakusui S, Yasuda T, Yanagi T, Tohyama J, Hasegawa Y, Koike Y, et al. A radiological analysis of heart sympathetic functions with *meta*-[^{123}I]iodobenzylguanidine in neurological patients with autonomic failure. J Auton Nerv Syst. 1994;49:81–4.
16. Orimo S, Ozawa E, Nakade S, Sugimoto T, Mizusawa H. ^{123}I-metaiodobenzylguanidine myocardial scintigraphy in Parkinson's disease. J Neurol Neurosurg Psychiatry. 1999;67: 189–94.
17. Braune S, Reinhardt M, Schnitzer R, Riedel A, Lücking CH. Cardiac uptake of [^{123}I] MIBG separates Parkinson's disease from multiple system atrophy. Neurology. 1999;53:1020–5.
18. Watanabe H, Ieda T, Katayama T, Takeda A, Aiba I, Doyu M, et al. Cardiac ^{123}I-metaiodobenzylguanidine (MIBG) uptake in dementia with Lewy bodies: comparison with Alzheimer's disease. J Neurol Neurosurg Psychiatry. 2001;70:781–3.
19. Shindo K, Kaneko E, Watanabe H, Sugimoto T, Ohta E, Ohashi K, et al. Analysis of the relationship between muscle sympathetic nerve activity and cardiac ^{123}I-metaiodobenzylguanidine uptake in patients with Parkinson's disease. Mov Disord. 2005;20: 1419–24.
20. Iriki M, Kozawa E. Factors controlling the regional differentiation of sympathetic outflow-Influence of the chemoreceptor reflex. Brain Res. 1975;87:281–91.

21. Sverrisdottir YB, Green AL, Aziz TZ, Bahuri NFA, Hyam J, Basnayake SD, et al. Differentiated baroreflex modulation of sympathetic nerve activity during deep brain stimulation in humans. Hypertension. 2014;63:1000–10.
22. Tamura N, Shimazu K, Oh-iwa K, Kim H, Hamaguchi K. Increased sympathetic nervous activity in motor neuron disease a hemodynamic study. Clin Neurol (Tokyo). 1983;23:152–8 (in Japanese).
23. Chida K, Sakamaki S, Takasu T. Alteration in autonomic function and cardiovascular regulation in amyotrophic lateral sclerosis. J Neurol. 1989;238:127–30.
24. Shindo K, Tsunoda S, Shiozawa Z. Microneurographic analysis of muscle sympathetic nerve activity in amyotrophic lateral sclerosis. Clin Auton Res. 1993;3:131–5.
25. Nogues MA, Stalberg EV. Automatic analysis of heart rate variation: II. Findings in patients attending an EMG laboratory. Muscle Nerve. 1989;12:1001–8.
26. Somers VK, Mark AL. Abboud FM Interaction of baroreceptor and chemoreceptor reflex control of sympathetic nerve activity in normal humans. J Clin Invest. 1991;87:1953–7.
27. Ziegler MG, Lake CR, Wood JH, Brooks BR. Relationship between cerebrospinal fluid norepinephrine and blood pressure in neurologic patients. Clin Exp Hypertens. 1980;2: 995–1008.
28. Shindo K, Shimokawa C, Watanabe H, Iida H, Ohashi K, Nitta K, et al. Chronological changes of sympathetic outflow to muscles in amyotrophic lateral sclerosis. J Neurol Sci. 2004;227: 79–84.
29. Brownell B, Oppenheimer DR, Hughes JT. The central nervous system in motor neurone disease. J Neurol Neurosurg Psychiatry. 1970;33:338–57.
30. Kennedy PGE, Duchen LW. A quantitative study of intermediolateral column cells in motor neuron disease and the Shy-Drager syndrome. J Neurol Neurosurg Psychiatry. 1985; 48:1103–6.
31. Takahashi H, Oyanagi K, Ikuta F. The intermediolateral nucleus in sporadic amyotrophic lateral sclerosis. Acta Neuropathol. 1993;86:190–2.
32. Oey PL, Vos PE, Wieneke GH, Wokke JHJ, Blankestijn PJ, Karemaker JM. Subtle involvement of the sympathetic nervous system in amyotrophic lateral sclerosis. Muscle Nerve. 2002; 25:402–8.
33. Bannister R, Oppenheimer DR. Degenerative diseases of the nervous system associated with autonomic failure. Brain. 1972;95:457–74.
34. Chokroverty S. Autonomic dysfunction in olivopontocerebellar atrophy. Adv Neurol. 1984;41: 105–41.
35. Kuroiwa Y, Shimada Y, Toyokura Y. Postural hypotension and low R-R interval variability in parkinsonism, spinocerebellar degeneration, and Shy-Drager syndrome. Neurology. 1983;33: 463–7.
36. Magalhaes M, Wenning GK, Daniel SE. Autonomic dysfunction in pathologically confirmed multiple system atrophy and idiopathic Parkinson's disease – a retrospective comparison. Acta Neurol Scand. 1995;91:98–102.
37. Kachi T, Iwase S, Mano T. Effect of L-threo- 3,4-dihydroxy-phenylserineon muscle sympathetic nerve activities in Shy-Drager syndrome. Neurology. 1988;38:1091–4.
38. Hakusui S, Sugiyama Y, Iwase S, Hasegawa Y, Koike Y, Mano T, et al. Postprandial hypotension: microneurographic analysis and treatment with vasopressin. Neurology. 1991;41: 712–5.
39. Donadio V, Cortelli P, Elam M, Di Stasi V, Montagna P, Holmberg B, et al. Autonomic innervation in multiple system atrophy and pure autonomic failure. J Neurol Neurosurg Psychiatry. 2010;81:1327–35.
40. Shindo K, Tsunoda S, Shiozawa Z. Muscle sympathetic activity in cerebellar degeneration with or without mild autonomic dysfunction. Acta Neurol Scand. 1999;100:244–8.
41. Tamura N, Shimazu K, Kim H. Autonomic nervous functions in various subtypes of spinocerebellar degenerations – a hemodynamic study. Clin Neurol (Tokyo). 1986;26: 284–90 (in Japanese).

42. Sandroni P, Ahlskog JE, Fealey RD. Autonomic involvement in extrapyramidal and cerebellar disorders. Clin Auton Res. 1991;1:147–55.
43. Shindo K, Tsunoda S, Shiozawa Z. Decreased sympathetic outflow to muscles in patients with cervical spondylosis. Acta Neurol Scand. 1997;96:241–6.
44. Stjernberg L, Blumberg H, Wallin BG. Sympathetic activity in man after spinal cord injury. Brain. 1986;109:695–715.
45. Shindo K, Watanabe H, Tanaka H, Takamura N, Tsunoda S, Shiozawa Z. A comparison of sympathetic outflow to muscles between cervical spondylotic amyotrophy and ALS. ALS Other Motor Neuron Disord. 2002;3:233–8.
46. Fagius J, Wallin BG. Sympathetic reflex latencies and conduction velocities in patients with polyneuropathy. J Neurol Sci. 1980;47:449–61.
47. Fagius J. Microneurographic findings in diabetic polyneuropathy with special reference to sympathetic nerve activity. Diabetologia. 1982;23:415–20.
48. Fagius J, Wallin BG. Microneurographic evidence of excessive sympathetic outflow in the Guillain-Barré syndrome. Brain. 1983;106:589–600.
49. Dotson R, Ochoa J, Marchettini P, Cline M. Sympathetic neural outflow directly recorded in patients with primary autonomic failure: clinical observations, microneurography, and histopathology. Neurology. 1990;40:1079–85.
50. Liguori R, Giannoccaro MP, Di Stasi V, Pizza F, Cortelli P, Baruzzi A, et al. Microneurographic evaluation of sympathetic activity in small fiber neuropathy. Clin Neurophysiol. 2011;122:1854–9.

Chapter 3
Muscle Sympathetic Nerve Activity and Cardiovascular Disease

Shuji Joho

Abstract The sympathetic nervous system can rapidly respond to the onset of a disruption in arterial pressure and plays essential roles in regulating cardiovascular function together with the renin-angiotensin-aldosterone system. On the other hand, chronic sympathetic overactivity could be associated with the onset and progression of hypertension, heart failure, or arrhythmia. The fact that sympathetic overactivity is an independent predictor of a poor outcome in patients with heart failure has been known for about 30 years. Accurate assessment of sympathetic nerve activity is required in clinical practice. Various methods including measuring neurotransmitters (norepinephrine), evaluating the responses of effectors (heart rate or blood pressure), neurotransmitter imaging (meta-iodobenzylguanidine), and so on have been applied to evaluate sympathetic nerve functions. However, quantifying the time-varying sympathetic activity which maintains body homeostasis over time is difficult. Microneurographic recording is the only way to directly evaluate sympathetic nerve activity from the human peripheral nerves. This chapter describes recent observations obtained by microneurographic recordings of muscle sympathetic nerve activity that have been adopted to reconsider the relationship between sympathetic nerve activity and several cardiovascular diseases.

Keywords Sympathetic nervous system • Ischemic heart disease • Resistant hypertension • Heart failure • Pulmonary hypertension

3.1 The Need to Evaluate Sympathetic Nerve Activity from Neural Activity

The sympathetic nervous system regulates involuntarily heart rate, blood pressure, and many target organs. Sympathetic stimulation increases heart rate via the sinoatrial node, increases myocardial contractility and vascular resistance via the contraction of peripheral blood vessels at arteriolar level, and consequently

S. Joho (✉)
Second Department of Internal Medicine, University of Toyama, 2630 Sugitani, Toyama 930-0194, Japan
e-mail: sjoho@med.u-toyama.ac.jp

S. Iwase et al. (eds.), *Clinical Assessment of the Autonomic Nervous System*,
DOI 10.1007/978-4-431-56012-8_3

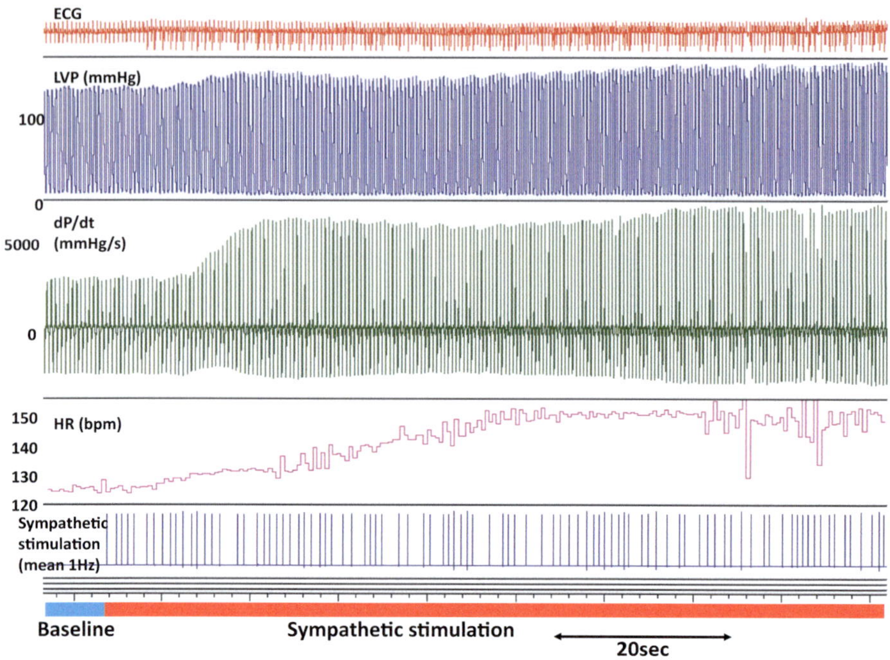

Fig. 3.1 Impact of sympathetic nerve stimulation on hemodynamics in a dog. Random electrical stimulation with an average of 1 Hz enhanced heart rate, left ventricular systolic pressure, and first derivative of left ventricular pressure (Modified from Ushijima R, Joho S. Evaluation for sympathetic nerve activity. Heart View. 2014;18(1):36)

increases blood pressure. Figure 3.1 shows an example of left cardiac sympathetic nerve stimulation in a dog with a non-failing heart. Electrical stimulation with random intervals (average 1 Hz) increased heart rate, as well as left ventricular systolic pressure and the first derivative of left ventricular pressure (LV dP/dt) after a few seconds. Thus, the sympathetic nervous system can regulate the heart rate and blood pressure within seconds.

The short-term activation of sympathetic nerve activity is known as the “fight or flight response” that prepares the body to either stay and fight or flee from a situation [1]. By contrast, the long-term excess activation of sympathetic nerves is pathological, being involved at the onset and progression of cardiovascular diseases such as hypertension, cardiac hypertrophy, myocardial ischemia, heart failure, and arrhythmias [2, 3]. The relationship between sympathetic nerve activity and cardiovascular diseases is important not only to understand autonomic status but also to consider its influence on prognosis and therapy. Therefore, sympathetic nerve activity should be accurately evaluated in each individual.

Sympathetic nerve activity in clinical practice can be evaluated by measuring blood or urinary levels of neurotransmitters (norepinephrine), measuring the responses of effectors (heart rate and blood pressure), neurotransmitter imaging (meta-iodobenzylguanidine (MIBG) scintigram), and measuring the electrical

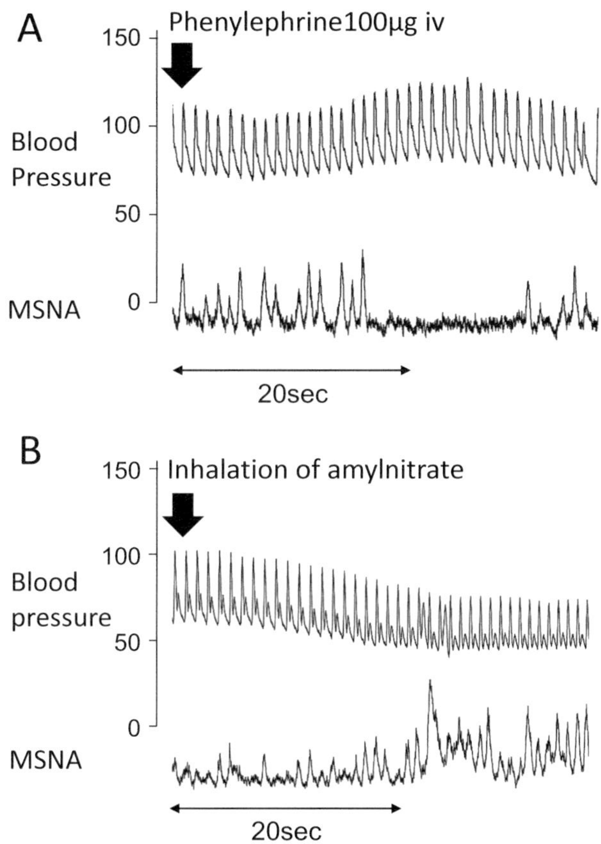

Fig. 3.2 Arterial pressure and muscle sympathetic nerve activity (*MSNA*). Control individual received phenylephrine injections and inhaled amyl nitrate (Modified from Ushijima R, Joho S. Evaluation for sympathetic nerve activity. Heart View. 2014;18(1):41)

activity of nerves. These methods should be selectively used depending on the nature of the studies.

Muscle sympathetic nerve activity (MSNA) is a vasomotor activity that governs the vascular smooth muscles distributed throughout the skeletal muscle. MSNA varies in conjunction with the instantaneous fluctuations of blood pressure via arterial baroreceptors to maintain constant blood pressure. A percutaneously inserted microelectrode is directly placed in muscle nerve fascicles of peripheral nerves. The major advantages of this method are that sympathetic neural activity can be directly and instantaneously recorded unlike any other indirect assessment of sympathetic activity. Thus, variations in sympathetic activity over time can be visualized using this method (Fig. 3.2).

Despite the advantages of MSNA, it is not frequent to apply this method in the clinical tests. The main reason might be mainly due to the difficulty of recording technique. However, to verify findings using MSNA is important for establishing evidence of the relationship between sympathetic nerve activity and cardiovascular diseases. We review recent clinical findings of the relationship between microneurographically recorded MSNA and various cardiovascular diseases as well as the pathophysiology underlying these diseases.

3.2 Ischemic Heart Disease

Animal experiments have shown that cardiac sympathetic nerve activity is enhanced by stimulating sympathetic afferent fibers with myocardial ischemia [4–6]. However, whether myocardial ischemia elicits sympathetic hyperactivity in patients with coronary artery disease has remained unclear. A recent study of MSNA during interventional coronary angioplasty in patients with coronary artery disease found that MSNA was increased by an average of 36 % from baseline during coronary artery occlusion [7] and in 10 of 12 patients regardless the site of such occlusion. Another study found that treating myocardial ischemia by coronary angioplasty can effectively reduce MSNA 1 month after balloon coronary angioplasty [8]. Neither of these studies found any significant changes in heart rate and blood pressure. These results imply that to determine changes in sympathetic activity related to myocardial ischemia from the response to the effector is difficult. Changes in MSNA without a change in blood pressure might occur through changes in input from sympathetic afferent fibers derived from the ischemic myocardium rather than by any alterations in blood pressure.

On the other hand, MSNA peaks during the acute phase (2–4 days after onset) of myocardial infarction and gradually decreases during the chronic stage (several months after onset) [9]. Furthermore, MSNA is significantly higher in patients with heart failure accompanied by ischemic than nonischemic cardiomyopathy [10]. Myocardial ischemia is frequently residual even after coronary angioplasty intervention in patients with coronary heart disease dependently upon the severity of residual coronary artery stenosis. Residual myocardial ischemia stimulates the sympathetic ascending fibers in the myocardium, which might increase the level of MSNA.

3.3 Resistant Hypertension

Resistant hypertension is defined as blood pressure above a target of 140/90 mmHg despite the concomitant administration of at least three antihypertensive agents from at least three classes of such agents including diuretics [11]. The level of MSNA was significantly higher in such patients, compared with those who had nonresistant hypertension even after excluding complications associated with sympathetic hyperactivity such as heart failure, obesity, obstructive sleep apnea, and renal failure [12]. Sympathetic hyperactivity reduces renal blood flow through the renal efferent sympathetic nerve, which could decrease sodium diuresis. Thus, renal sympathetic nerve denervation increases renal blood flow through a blockade of renal sympathetic efferents and probably increases natriuresis [13].

Catheter-based radiofrequency ablation has recently been used to disrupt the renal nerves of patients with resistant hypertension [14]. Renal sympathetic denervation results in an obvious reduction in renal norepinephrine spillover from the kidney, which indicates the effectiveness of the intervention. Importantly, MSNA, whole-body norepinephrine spillover, and blood pressure became decreased after this procedure [15]. Moreover, electrical autonomic nerve stimulation of the renal artery can increase blood pressure in the experimental setting, but this response is completely abolished after radiofrequency ablation of the renal artery [16]. These results could support the possibility that inhibition of afferent renal nerve activity might contribute to a reduction in central sympathetic drive.

So far, several studies examined the effect of renal denervation on blood pressure and MSNA in patients with resistant hypertension [17–20]. However, the findings were inconsistent. Renal denervation substantially reduced MSNA in two reports [18, 20], but did not change in other reports [17, 19].

This could be explained by a lack of sufficient renal sympathetic nerve ablation resulting in unsatisfactory blood pressure and sympathetic responses to renal denervation. Recently, it was investigated that the potential impact of various renal anatomies on blood pressure and MSNA responses to renal denervation in patients with resistant hypertension [20]. Interestingly, renal denervation significantly reduced blood pressure and MSNA in patients with single renal arteries bilaterally, but not in patients with dual renal arteries on either one or both sides [20].

Renal ischemia or injury might stimulate afferent signaling from the kidney to central integrative nuclei, with the consequence of exaggerated efferent sympathetic outflow to vasculature and the kidneys [21]. Besides sustained blood pressure reduction, the regression of left ventricular hypertrophy, a reduction in the size of the left atrium, an increase in left ventricular ejection fraction, and improved left ventricular diastolic function were evident at 6 months after renal denervation [22] (Fig. 3.3). Renal denervation inhibited the progression from cardiac hypertrophy to heart failure in animal models [23]. Therefore, renal denervation might help to interrupt this vicious cycle.

3.4 Chronic Heart Failure

3.4.1 Mechanisms of Sympathetic Hyperactivity in Chronic Heart Failure

Sympathetic overactivation is common in patients with chronic heart failure and it is recognized as a cause of exacerbated heart failure. Figure 3.4 shows representative traces of MSNA obtained from a healthy individual and from a patient with

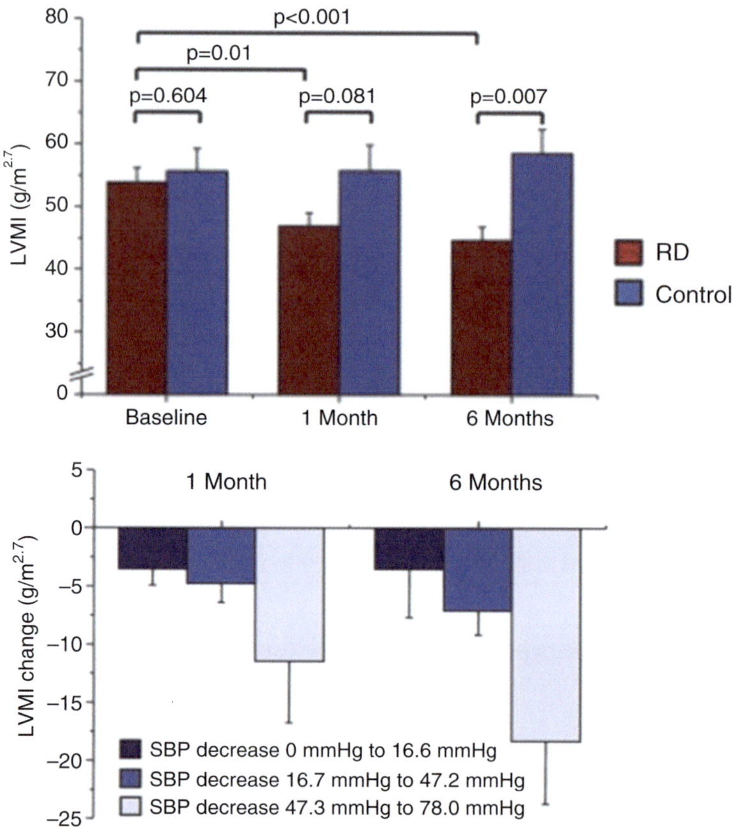

Fig. 3.3 Time course of left ventricular mass (*LVMI*) and changes in LVMI in group with renal sympathetic nerve ablation (*RD*) and control group (From reference [22]). LVMI was decreased in RD, but not in control group (*upper*). Effects of renal denervation on left ventricular hypertrophy were compared among three groups of patients whose blood pressure changed after renal denervation. Left ventricular hypertrophy regressed even in group with minimal antihypertensive effects

chronic heart failure. A heartbeat-synchronized MSNA burst occurs for every heartbeat in patients with heart failure, suggesting obviously increased sympathetic activity. Both plasma norepinephrine levels [24] and MSNA [25] can serve as independent prognosticators of chronic heart failure (Fig. 3.5).

The cause of central sympathetic hyperactivity in patients with chronic heart failure has not been established. Recent concepts include factors from the central nervous system that affect the sympathetic activity (Fig. 3.6) [26].

A tonic inhibitory restraint on the brain stem centers arises from arterial and cardiopulmonary baroreceptors.

Inactivation of the inhibitory baroreceptors through decrease in cardiac filling pressure or arterial pressure yields an decrease in afferent inhibition of the brain stem vasomotor center with an increase in efferent sympathetic tone. When the

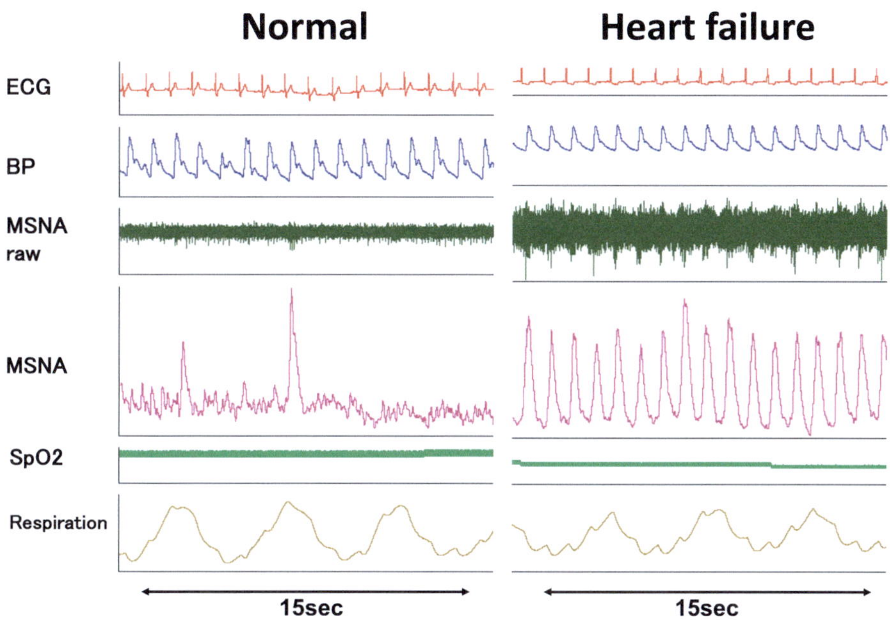

Fig. 3.4 Representative tracing of muscle sympathetic nerve activity (*MSNA*), electrocardiogram, blood pressure, oxygen saturation, and respiration obtained from healthy individual (*left*) and patient with NYHA class III chronic heart failure (*right*). Neural activity is synchronized with heart beat (burst) in patient

pump function of the heart fails, both decreased blood pressure and depressed contractility attenuate afferent impulses from aortic baroreceptors and result in sympathetic hyperactivity. Conversely, stimulating afferent nerves from the aortic baroreceptor using "baroreceptor activation therapy" inhibits MSNA [27].

Other mechanisms of sympathetic activation referred to as excitatory stimuli are independent of baroreceptor unloading and possibly cause a sympathetic reaction that is excessive for maintaining homeostasis through increasing the set point of the central sympathetic activity in the paraventricular nucleus [26]. Excitatory stimulus mechanisms through chemoreceptors located in the carotid sinus and aortic arch, skeletal muscle (muscle metaboreceptors), kidney, and lungs include enhanced mechanisms of sympathetic nerve activity [26]. Activation of these afferent mechanisms via ischemic metabolites in exercising muscle or via arterial hypoxemia, acidemia, and hypercapnia results in an increase in afferent excitatory influences on the brain stem vasomotor centers and causes an increase in efferent sympathetic neural outflow. These mechanisms are activated with the development of heart failure and then form a vicious cycle of exacerbating heart failure.

Thus, sympathetic hyperactivity that characterizes heart failure might reflect the sum of (1) a decrease in the afferent inhibitory restraint arising from cardiopulmonary and arterial baroreceptors (compensatory responses to reduced blood pressure

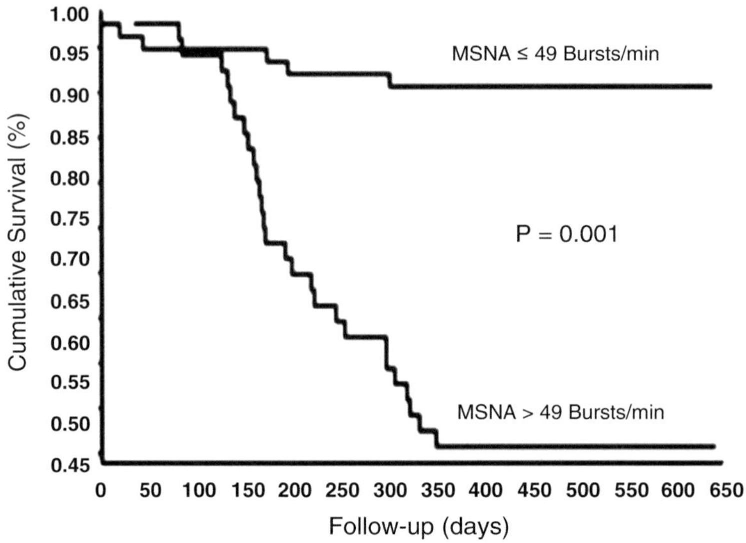

Fig. 3.5 Kaplan-Meier analysis of cumulative rates of survival in patients with heart failure stratified into two groups based on muscle sympathetic nerve activity (*MSNA*, bursts/min) (From reference [25]). Survival rate was significantly lower in patients with MSNA >49 bursts/min ($P = 0.001$)

or systolic function) and/or (2) an increase in the afferent excitatory influence of muscle metaboreceptors and/or arterial chemoreceptors.

3.4.2 Effects of Standard Pharmacological Therapy

Considerable improvements in the therapeutic approach to heart failure have led to a reduction in hospitalization and a significant reduction in mortality rates over the past 20 years [28, 29]. These improvements were due to the establishment of a pharmacological approach with beta-adrenergic blockers, angiotensin-converting enzyme (ACE) inhibitors, angiotensin receptor blockers (ARB), and mineralocorticoid receptor blockers. As a current standard treatment for chronic heart failure, β-blockers dose dependently block the action of endogenous epinephrine and norepinephrine on β-receptors and weaken the toxicity of catecholamines to heart muscle cells. On the other hand, β-blockers do not reduce MSNA in the acute phase [30], but do so over the long term [31]. Mineralocorticoid receptor blockers, ACE inhibitors, ARB, and digitalis are also standard treatments for heart failure. Either ACE inhibitors or ARB are usually administered to all patients, and mineralocorticoid receptor blockers are given to ≥50 % of patients. Statins are also administered to many patients with risk factors for coronary artery disease. Importantly, these drugs have a central sympathetic inhibitory effect [32–35]. An increase in

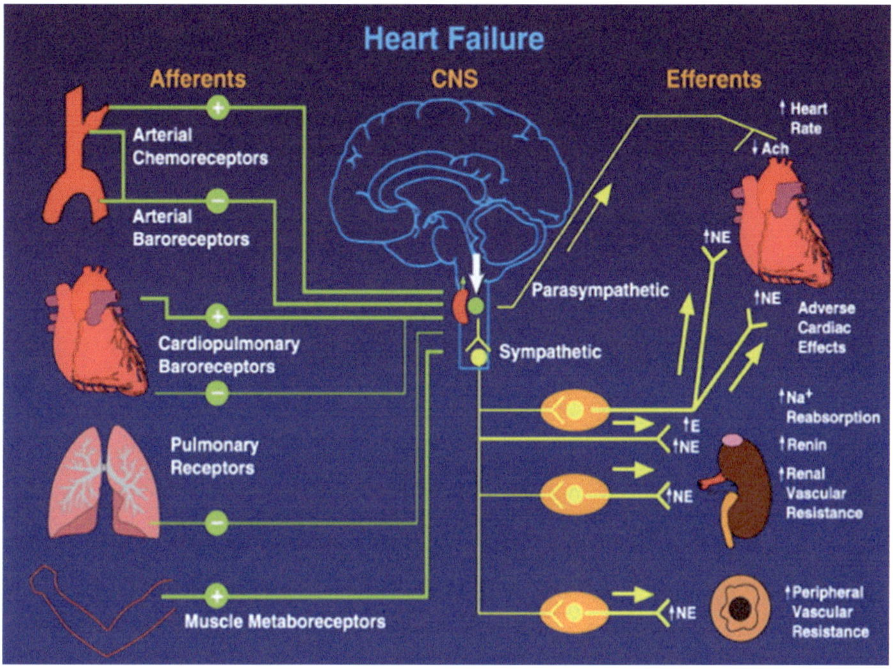

Fig. 3.6 Mechanisms of central sympathetic hyperactivity in chronic heart failure (From reference [26])

reactive oxygen species in the sympathetic central rostral ventrolateral medulla (RVLM) enhances the sympathetic activity [36]. For example, statins attenuate sympathetic activity by suppressing active oxygen [37].

Therefore, these medications should inhibit central sympathetic activity in addition to the original purpose.

3.4.3 Sympathetic Suppression by Treatment for Breathing Abnormalities

Breathing abnormalities such as rapid shallow breathing and periodic breathing during the daytime or sleep apnea during the nighttime are commonly found (50–60%) in patients with chronic heart failure [38–40]. Patients with breathing abnormalities have notably higher sympathetic nerve activity [40, 41]. Decreased lung stretch [42, 43], intermittent hypoxia [44, 45], and hypercapnia [46, 47] might all increase MSNA (Fig. 3.6).

Enhanced sympathetic activity can be suppressed by correcting breathing abnormalities in patients with heart failure. Device-guided slow and deep respiration suppresses steady-state sympathetic nerve activity in patients with heart failure and

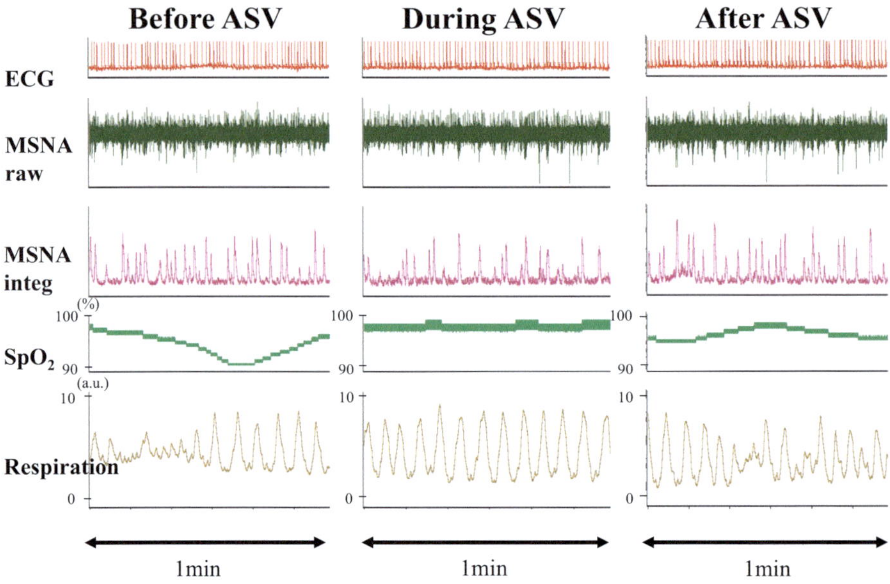

Fig. 3.7 Effect of adaptive servo-ventilation (*ASV*) on muscle sympathetic nerve activity (*MSNA*) and respiration (From reference [50]). Respiration became stabilized and MSNA became decreased during application of ASV

high levels of resting sympathetic tone [48]. Adaptive servo-ventilation (ASV), which is a novel method of ventilatory support designed for Cheyne-Stokes respiration (CSR) [49], improved respiratory instability and decreased MSNA in patients with heart failure (Fig. 3.7) [50]. The sympathoinhibitory action of ASV might be due mainly to respiratory normalization via a servo-ventilation effect [50]. Although long-term use of continuous positive airway pressure (CPAP) decreases MSNA in heart failure patients with obstructive sleep apnea [51], short-term use (30 min) of CPAP did not always decrease MSNA in patients with heart failure [52–55].

Patients with heart failure and obstructive sleep apnea have sympathetic hyperactivity during the daytime [56]. One month of CPAP reduced MSNA by 12 bursts/100 beats in patients with obstructive sleep apnea [51] and CPAP increased their cardiac function [57] while improved their prognosis [58]. On the other hand, sympathetic hyperactivity is also evident in patients with heart failure and central sleep apnea [56]. In patients with heart failure and moderate to severe central sleep apnea, ASV decreased MSNA by 9 bursts/100 beats in association with improved cardiac function [59] (Fig. 3.8). Furthermore, changes in left ventricular ejection fraction and MSNA closely correlated with the suppression of central sleep apnea using ASV. Thus, respiratory stabilization might be associated with improved sympathetic nerve activity and cardiac function [59].

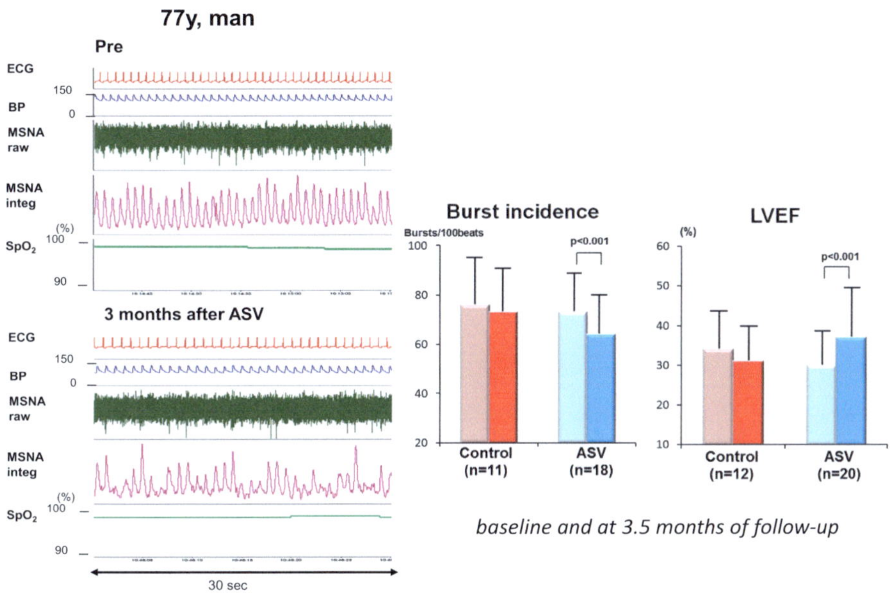

Fig. 3.8 Representative trace of electrocardiogram (*ECG*), blood pressure (*BP*), muscle sympathetic nerve activity (*MSNA*), and oxygen saturation (SpO_2) at baseline and at 3 months after adaptive servo-ventilation (*ASV*) in a patient with chronic heart failure (From reference [59]). Sympathetic nerve activity was reduced and left ventricular ejection fraction (*LVEF*) was increased in ASV group, whereas MSNA and cardiac function remained unaltered in control group

3.5 Pulmonary Hypertension

Like patients with chronic heart failure, sympathetic activity is elevated in patients with pulmonary hypertension complicated with right heart failure [60, 61] and the prognosis is poor for patients with pulmonary hypertension who have increased MSNA [62]. However, the mechanisms of sympathetic hyperactivity remain unclear in these patients. Balloon atrial septostomy, which is a palliative treatment for severe pulmonary hypertension that makes a short circuit from the right to the left atrium, can efficiently increase the systemic circulation and cardiac output [61]. This procedure notably decreased MSNA despite worsening oxygenation. Decreased MSNA was closely correlated with changes in right atrium pressure, but not with increased cardiac output.

Intravascular ablation of the sympathetic fibers distributed around the pulmonary artery trunk has recently been applied to patients with pulmonary arterial hypertension [63]. This procedure reduces mean pulmonary arterial pressure from 55 to 36 mmHg, indicating that efferent sympathetic fibers distributed into the lungs might have a specific effect on pulmonary circulation. This reduction was observed even under standard therapy for pulmonary hypertension, consisting of endothelin receptor blockers, phosphodiesterase inhibitors, and prostacyclin, suggesting that the pathway of sympathetic denervation into the lungs is independent of other established mechanisms of pulmonary vasodilation.

The effects of β-blockers on right heart failure due to pulmonary hypertension have also been investigated in animal models of pulmonary hypertension [64, 65]. Right ventricular function was maintained in this model, but this issue warrants future discussion.

3.6 Conclusion

Recent experiences highlight the need to understand the implication of sympathetic hyperactivity in the specific fundamental processes involved in cardiovascular diseases. Directly recording sympathetic nerve activity might deepen understanding of the pathophysiology of these diseases.

Acknowledgments This work was supported by JSPS KAKENHI Grant Number 15K01363.

References

1. Guyton AC, Hall JE. The autonomic nervous system; the adrenal medulla. In: Guyton AC, Hall JE, editors. Textbook of medical physiology. 9th ed. Philadelphia: W.B. Saunders Company; 1996. p. 769–81.
2. Katz AM, Konstam MA. The neurohumoral response in heart failure: functional signaling. In: Katz AM, Konstam MA, editors. Heart failure. Pathophysiology, molecular biology, and clinical management. 2nd ed. Philadelphia: Lippincott Williams & Wilkins, a Wolters Kluwer Business; 2009. p. 89–133.
3. Mann DL. Pathophysiology of heart failure. In: Bonow RO, Mann DL, Zipes DP, Libby P, editors. Braunwald's heart disease. A textbook of cardiovascular medicine. 9th ed. Philadelphia, PA: Elsevier Saunders; 2012. p. 487–504.
4. Malliani A, Schwartz JP, Zanchetti A. A sympathetic reflex elicited by experimental coronary occlusion. Am J Physiol. 1969;217:703–9.
5. Malliani A, Peterson DF, Bishop VS, Brown AM. Spinal sympathetic cardiocardiac reflex. Circ Res. 1972;30:158–66.
6. Minisi AJ, Thames MD. Activation of cardiac sympathetic afferents during coronary occlusion. Evidence for reflex activation of sympathetic nervous system during transmural myocardial ischemia in the dog. Circulation. 1991;84:357–67.
7. Ciećwierz D, Hering D, Somers VK, Wdowczyk-Szulc J, Kara T, Skarzyński P, et al. Sympathetic neural responses to coronary occlusion during balloon angioplasty. J Hypertens. 2007;25:1650–4.
8. Gomes ME, Aengevaeren WR, Lenders JW, Verheugt FW, Smits P, Tack CJ. Improving myocardial perfusion by percutaneous coronary intervention reduces central sympathetic activity in stable angina. Clin Cardiol. 2010;33:E16–21.
9. Graham LN, Smith PA, Stoker JB, Mackintosh AF, Mary DA. Time course of sympathetic neural hyperactivity after uncomplicated acute myocardial infarction. Circulation. 2002;106:793–7.
10. Notarius CF, Spaak J, Morris BL, Floras JS. Comparison of muscle sympathetic activity in ischemic and nonischemic heart failure. J Card Fail. 2007;13:470–5.

11. Calhoun DA, Jones D, Textor S, Goff DC, Murphy TP, Toto RD, et al. Resistant hypertension: diagnosis, evaluation, and treatment: a scientific statement from the American Heart Association Professional Education Committee of the Council for High Blood Pressure Research. Circulation. 2008;117:e510–26.
12. Grassi G, Seravalle G, Brambilla G, Pini C, Alimento M, Facchetti R, et al. Marked sympathetic activation and baroreflex dysfunction in true resistant hypertension. Int J Cardiol. 2014;177:1020–5.
13. Kompanowska-Jezierska E, Walkowska A, Johns EJ, Sadowski J. Early effects of renal denervation in the anaesthetised rat: natriuresis and increased cortical blood flow. J Physiol. 2001;531(Pt 2):527–34.
14. Krum H, Schlaich M, Whitbourn R, Sobotka PA, Sadowski J, Bartus K, et al. Catheter-based renal sympathetic denervation for resistant hypertension: a multicentre safety and proof-of-principle cohort study. Lancet. 2009;373:1275–81.
15. Schlaich MP, Sobotka PA, Krum H, Lambert E, Esler MD. Renal sympathetic-nerve ablation for uncontrolled hypertension. N Engl J Med. 2009;361:932–4.
16. Chinushi M, Izumi D, Iijima K, Suzuki K, Furushima H, Saitoh O, et al. Blood pressure and autonomic responses to electrical stimulation of the renal arterial nerves before and after ablation of the renal artery. Hypertension. 2013;61:450–6.
17. Brinkmann J, Heusser K, Schmidt BM, Menne J, Klein G, Bauersachs J, et al. Catheter-based renal nerve ablation and centrally generated sympathetic activity in difficult-to-control hypertensive patients: prospective case series. Hypertension. 2012;60:1485–90.
18. Hering D, Lambert EA, Marusic P, Walton AS, Krum H, Lambert GW, et al. Substantial reduction in single sympathetic nerve firing after renal denervation in patients with resistant hypertension. Hypertension. 2013;61:457–64.
19. Vink EE, Verloop WL, Siddiqi L, van Schelven LJ, Liam Oey P, Blankestijn PJ. The effect of percutaneous renal denervation on muscle sympathetic nerve activity in hypertensive patients. Int J Cardiol. 2014;176:8–12.
20. Hering D, Marusic P, Walton AS, Duval J, Lee R, Sata Y, et al. Renal artery anatomy affects the blood pressure response to renal denervation in patients with resistant hypertension. Int J Cardiol. 2016;202:388–93.
21. Ye S, Zhong H, Yanamadala V, Campese VM. Renal injury caused by intrarenal injection of phenol increases afferent and efferent renal sympathetic nerve activity. Am J Hypertens. 2002;15:717–24.
22. Brandt MC, Mahfoud F, Reda S, Schirmer SH, Erdmann E, Böhm M, et al. Renal sympathetic denervation reduces left ventricular hypertrophy and improves cardiac function in patients with resistant hypertension. J Am Coll Cardiol. 2012;59:901–9.
23. Nozawa T, Igawa A, Fujii N, Kato B, Yoshida N, Asanoi H, et al. Effects of long-term renal sympathetic denervation on heart failure after myocardial infarction in rats. Heart Vessels. 2002;16:51–6.
24. Cohn JN, Levine TB, Olivari MT, Garberg V, Lura D, Francis GS, et al. Plasma norepinephrine as a guide to prognosis in patients with chronic congestive heart failure. N Engl J Med. 1984;311:819–23.
25. Barretto AC, Santos AC, Munhoz R, Rondon MU, Franco FG, Trombetta IC, et al. Increased muscle sympathetic nerve activity predicts mortality in heart failure patients. Int J Cardiol. 2009;135:302–7.
26. Floras JS. Sympathetic nervous system activation in human heart failure: clinical implications of an updated model. J Am Coll Cardiol. 2009;54:375–85.
27. Heusser K, Tank J, Engeli S, Diedrich A, Menne J, Eckert S, et al. Carotid baroreceptor stimulation, sympathetic activity, baroreflex function, and blood pressure in hypertensive patients. Hypertension. 2010;55:619–26.
28. Swedberg K, Cleland J, Dargie H, Drexler H, Follath F, Komajda M, et al. Guidelines for the diagnosis and treatment of chronic heart failure: executive summary (update 2005): the task

force for the diagnosis and treatment of chronic heart failure of the European Society of Cardiology. Eur Heart J. 2005;26:1115–40.
29. Levy D, Kenchaiah S, Larson MG, Benjamin EJ, Kupka MJ, Ho KK, et al. Long-term trends in the incidence of and survival with heart failure. N Engl J Med. 2002;347:1397–402.
30. Azevedo ER, Kubo T, Mak S, Al-Hesayen A, Schofield A, Allan R, et al. Nonselective versus selective beta-adrenergic receptor blockade in congestive heart failure: differential effects on sympathetic activity. Circulation. 2001;104:2194–9.
31. De Matos LD, Gardenghi G, Rondon MU, Soufen HN, Tirone AP, Barretto AC, et al. Impact of 6 months of therapy with carvedilol on muscle sympathetic nerve activity in heart failure patients. J Card Fail. 2004;10:496–502.
32. Ferguson DW, Berg WJ, Sanders JS, Roach PJ, Kempf JS, Kienzle MG. Sympathoinhibitory responses to digitalis glycosides in heart failure patients. Direct evidence from sympathetic neural recordings. Circulation. 1989;80:65–77.
33. Grassi G, Cattaneo BM, Seravalle G, Lanfranchi A, Pozzi M, Morganti A, et al. Effects of chronic ACE inhibition on sympathetic nerve traffic and baroreflex control of circulation in heart failure. Circulation. 1997;96:1173–9.
34. Hikosaka M, Yuasa F, Yuyama R, Mimura J, Kawamura A, Motohiro M, et al. Candesartan and arterial baroreflex sensitivity and sympathetic nerve activity in patients with mild heart failure. J Cardiovasc Pharmacol. 2002;40:875–80.
35. Menon DV, Arbique D, Wang Z, Adams-Huet B, Auchus RJ, Vongpatanasin W. Differential effects of chlorthalidone versus spironolactone on muscle sympathetic nerve activity in hypertensive patients. J Clin Endocrinol Metab. 2009;94:1361–6.
36. Kishi T, Hirooka Y, Kimura Y, Ito K, Shimokawa H, Takeshita A. Increased reactive oxygen species in rostral ventrolateral medulla contribute to neural mechanisms of hypertension in stroke-prone spontaneously hypertensive rats. Circulation. 2004;109:2357–62.
37. Deo SH, Fisher JP, Vianna LC, Kim A, Chockalingam A, Zimmerman MC, et al. Statin therapy lowers muscle sympathetic nerve activity and oxidative stress in patients with heart failure. Am J Physiol Heart Circ Physiol. 2012;303:H377–85.
38. Mortara A, Sleight P, Pinna GD, Maestri R, Prpa A, La Rovere MT, et al. Abnormal awake respiratory patterns are common in chronic heart failure and may prevent evaluation of autonomic tone by measures of heart rate variability. Circulation. 1997;96:246–52.
39. Ponikowski P, Anker SD, Chua TP, Francis D, Banasiak W, Poole-Wilson PA, et al. Oscillatory breathing patterns during wakefulness in patients with chronic heart failure: clinical implications and role of augmented peripheral chemosensitivity. Circulation. 1999;100:2418–24.
40. Poletti R, Passino C, Giannoni A, Zyw L, Prontera C, Bramanti F, et al. Risk factors and prognostic value of daytime Cheyne-Stokes respiration in chronic heart failure patients. Int J Cardiol. 2009;137:47–53.
41. Van de Borne P, Oren R, Abouassaly C, Anderson E, Somers VK. Effect of Cheyne-Stokes respiration on muscle sympathetic nerve activity in severe congestive heart failure secondary to ischemic or idiopathic dilated cardiomyopathy. Am J Cardiol. 1998;8:432–6.
42. Naughton MT, Floras JS, Rahman MA, Jamal M, Bradley TD. Respiratory correlates of muscle sympathetic nerve activity in heart failure. Clin Sci (Lond). 1998;93:277–85.
43. Goso Y, Asanoi H, Ishise H, Kameyama T, Hirai T, Nozawa T, et al. Respiratory modulation of muscle sympathetic nerve activity in patients with chronic heart failure. Circulation. 2001;104:418–23.
44. Ueno H, Asanoi H, Yamada K, Oda Y, Takagawa J, Kameyama T, et al. Attenuated respiratory modulation of chemoreflex-mediated sympathoexcitation in patients with chronic heart failure. J Card Fail. 2004;10:236–43.
45. Gilmartin GS, Lynch M, Tamisier R, Weiss JW. Chronic intermittent hypoxia in humans during 28 nights results in blood pressure elevation and increased muscle sympathetic nerve activity. Am J Physiol Heart Circ Physiol. 2010;299:H925–31.

46. Yamamoto K, Eubank W, Franzke M, Mifflin S. Resetting of the sympathetic baroreflex is associated with the onset of hypertension during chronic intermittent hypoxia. Auton Neurosci. 2013;173:22–7.
47. Morgan BJ, Crabtree DC, Palta M, Skatrud JB. Combined hypoxia and hypercapnia evokes long-lasting sympathetic activation in humans. J Appl Physiol. 1995;79:205–13.
48. Harada D, Asanoi H, Takagawa J, Ishise H, Ueno H, Oda Y, et al. Slow and deep respiration suppresses steady-state sympathetic nerve activity in patients with chronic heart failure: from modeling to clinical application. Am J Physiol Heart Circ Physiol. 2014;307:H1159–68.
49. Teschler H, Döhring J, Wang YM, Berthon-Jones M. Adaptive pressure support servo-ventilation: a novel treatment for Cheyne-Stokes respiration in heart failure. Am J Respir Crit Care Med. 2001;164:614–9.
50. Harada D, Joho S, Oda Y, Hirai T, Asanoi H, Inoue H. Short term effect of adaptive servo-ventilation on muscle sympathetic nerve activity in patients with heart failure. Auton Neurosci. 2011;161:95–102.
51. Usui K, Bradley TD, Spaak J, Ryan CM, Kubo T, Kaneko Y, et al. Inhibition of awake sympathetic nerve activity of heart failure patients with obstructive sleep apnea by nocturnal continuous positive airway pressure. J Am Coll Cardiol. 2005;45:2008–11.
52. Naughton MT, Floras JS, Rahman MA, Jamal M, Bradley TD. Respiratory correlates of muscle sympathetic nerve activity in heart failure. Clin Sci (Lond). 1998;95:277–85.
53. Heindl S, Dodt C, Krahwinkel M, Hasenfuss G, Andreas S. Short term effect of continuous positive airway pressure on muscle sympathetic nerve activity in patients with chronic heart failure. Heart. 2001;85:185–90.
54. Zubin Maslov P, Breskovic T, Shoemaker JK, Olson TP, Johnson BD, Eterovic D, et al. Firing patterns of muscle sympathetic neurons during short-term use of continuous positive airway pressure in healthy subjects and in chronic heart failure patients. Respir Physiol Neurobiol. 2013;187:149–56.
55. Ushijima R, Joho S, Akabane T, Oda Y, Inoue H. Differing effects of adaptive servoventilation and continuous positive airway pressure on muscle sympathetic nerve activity in patients with heart failure. Circ J. 2014;78:1387–95.
56. Spaak J, Egri ZJ, Kubo T, Yu E, Ando S, Kaneko Y, et al. Muscle sympathetic nerve activity during wakefulness in heart failure patients with and without sleep apnea. Hypertension. 2005;46:1327–32.
57. Kaneko Y, Floras JS, Usui K, Plante J, Tkacova R, Kubo T, et al. Cardiovascular effects of continuous positive airway pressure in patients with heart failure and obstructive sleep apnea. N Engl J Med. 2003;348:1233–1241.
58. Kasai T, Narui K, Dohi T, Yanagisawa N, Ishiwata S, Ohno M, et al. Prognosis of patients with heart failure and obstructive sleep apnea treated with continuous positive airway pressure. Chest. 2008;133:690–6.
59. Joho S, Oda Y, Ushijima R, Hirai T, Inoue H. Effect of adaptive servoventilation on muscle sympathetic nerve activity in patients with chronic heart failure and central sleep apnea. J Card Fail. 2012;18:769–75.
60. Velez-Roa S, Ciarka A, Najem B, Vachiery JL, Naeije R, van de Borne P. Increased sympathetic nerve activity in pulmonary artery hypertension. Circulation. 2004;110:1308–12.
61. Ciarka A, Vachièry JL, Houssière A, Gujic M, Stoupel E, Velez-Roa S, et al. Atrial septostomy decreases sympathetic overactivity in pulmonary arterial hypertension. Chest. 2007;131:1831–7.
62. Ciarka A, Doan V, Velez-Roa S, Naeije R, van de Borne P. Prognostic significance of sympathetic nervous system activation in pulmonary arterial hypertension. Am J Respir Crit Care Med. 2010;181:1269–75.
63. Chen SL, Zhang FF, Xu J, Xie DJ, Zhou L, Nguyen T, et al. Pulmonary artery denervation to treat pulmonary arterial hypertension: the single-center, prospective, first-in-man PADN-1 study (first-in-man pulmonary artery denervation for treatment of pulmonary artery hypertension). J Am Coll Cardiol. 2013;62:1092–100.

64. Bogaard HJ, Natarajan R, Mizuno S, Abbate A, Chang PJ, Chau VQ, et al. Adrenergic receptor blockade reverses right heart remodeling and dysfunction in pulmonary hypertensive rats. Am J Respir Crit Care Med. 2010;182:652–60.
65. de Man FS, Handoko ML, van Ballegoij JJ, Schalij I, Bogaards SJ, Postmus PE, et al. Bisoprolol delays progression towards right heart failure in experimental pulmonary hypertension. Circ Heart Fail. 2012;5:97–105.

Chapter 4
Skin Sympathetic Nerve Activity and Thermoregulatory Control in Humans

Satoshi Iwase, Naoki Nishimura, Yuko Kuwahara, and Junichi Sugenoya

Abstract Skin sympathetic nerve activity (SSNA) is microneurographically recorded from the skin nerve fascicle in the peripheral nerves with properties of an irregular, pulse asynchronous burst with respiratory variation, followed by sweating and/or vasoconstriction, elicited by mental stress and arousal stimuli. It comprises sudomotor and vasoconstrictor nerve activities as well as vasodilator ones. SSNA function in thermoregulation in humans, however, is also elicited by mental stress or cognition. SSNA is lowest at thermoneutral temperature and enhanced in the presence of ambient warm and cool air. Burst amplitude is well correlated to sweat rate change or skin blood flow reduction rate. The clinical application of SSNA comprises the following: (1) clarification of sweating events, (2) clarification and diagnosis of anhidrosis or hyperhidrosis, (3) clarification of thermoregulatory function and diagnosis of thermoregulatory disorder, (4) clarification of pathophysiology and diagnosis of vascular disorders, (5) clarification of the relationship between cognitive function and SSNA, and (6) determination of pharmacological effects attributable to change in neuroeffector responses. Lastly, SSNA's significance as a feedforward thermoregulatory tool is discussed since SSNA contributes more to rapid thermoregulatory response than thermoregulation using convectional thermotransmission. Especially feedforward thermoregulatory sweating response is estimated to be highly phylogenetic than vasoconstrictive response.

Keywords Sweating • Vasoconstriction • Core temperature • Neural thermoregulatory pathway • Convectional thermoregulatory pathway • Feedforward thermoregulation

---

S. Iwase (✉) • N. Nishimura • Y. Kuwahara • J. Sugenoya
Department of Physiology, Aichi Medical University, 1-1 Yazako-Karimata, Nagakute, Aichi 480-1195, Japan
e-mail: s_iwase@nifty.com

S. Iwase et al. (eds.), *Clinical Assessment of the Autonomic Nervous System*,
DOI 10.1007/978-4-431-56012-8_4

4.1 Introduction

Skin sympathetic nerve activity (SSNA) is microneurographically recorded from the skin nerve fascicle in the peripheral nerves and contributes to thermoregulatory function in humans. It controls the skin blood flow by regulating the arteriolar sphincter, erects the hair by constricting the arrector pili muscles, and excretes sweat from the sweat glands. Activation of the SSNA constricts the skin vessels, thereby suppressing the heat flux from the skin, namely, vasoconstrictor activity [1, 2]. Contrarily, it accelerates the sweating to cool the skin by evaporation through sudomotor nerve activity. This chapter focuses on the thermoregulatory function of the skin sympathetic nerve activity in humans and discusses how humans developed their civilization using this thermoregulatory capability.

In the present chapter, the recording method, the properties, the difference from the muscle sympathetic nerve activity, and the identification are described. Changes in SSNA under various environmental conditions, as well as the pathophysiology of SSNA, are described in an overall review by Mano et al. [3–7].

4.2 Recording Skin Sympathetic Nerve Activity

SSNA is recorded from the skin nerve fascicles by inserting a tungsten microelectrode into the peripheral nerves subcutaneously without anesthesia. The electrode is insulated with epoxy resin, and the tip is exposed to record the nerve impulses as the voltage from the reference electrode. The shaft diameter of the recording electrode is usually 100 μm, tip diameter is 1 μm, and impedance is 3–5 MΩ. The recorded signals are fed into a high-impedance bioamplifier and band-pass filters of 500–5 kHz and are displayed on an oscilloscope with sound monitoring. The signal is usually recorded in a magnetic data recorder or digitized with a sampling frequency of >5 kHz and stored on the hard disk of a personal computer.

In order to record SSNA, a three-step procedure should be undertaken: (1) detection of the nerve tract using electric stimulation, (2) penetration of the nerve fascicle with an electrode, and (3) moving the electrode tip to record the required nerve activity and to enhance the S/N ratio. When the electrode tip penetrates into the nerve fascicle, the subject feels paresthesia or a grasped sensation. Gentle touches to the skin or taps of the muscle belly determine whether the tip is situated in the skin or muscle fascicles, respectively.

SSNA is identified by the following criteria [3–7]: (1) spontaneous nonrhythmic efferent burst discharge from the skin fascicle; (2) followed by sweating or vasoconstriction after the burst discharge with a certain latency; (3) eliciting burst discharge by mental stimuli, sound, pain, or electric stimulation with a certain (~1 s) latency; (4) longer burst duration of ~300 ms compared with MSNA (~200 ms); and (5) eliciting burst discharge by sudden inspiration with a certain latency. The criterion for good recording is that the burst amplitude is double the noise level.

4.3 Quantification and Evaluation of Skin Sympathetic Nerve Activity

Since SSNA is recorded as the burst activities, the quantification of SSNA is derived from [burst number per minute (burst rate)], from [burst number per minute] × [mean burst amplitude] (total SSNA), or from [total burst area when the maximum burst amplitude is regarded as 1000]. SSNA contains sudomotor nerve activity and vasoconstrictor activity and, furthermore, vasodilator and pilomotor activities, and the analysis of SSNA requires the differentiation of these activities since these activities run in the same nerve fascicle interminglingly by using laser Doppler flowmetry and skin potential change.

4.4 Properties of Skin Sympathetic Nerve Activity

The role of SSNA is the thermal and mental control of sweating and skin blood flow, and its goal is to maintain the body temperature through sweating/vasoconstriction against environmental change of temperature, as well as mental stress [1, 2, 8–11].

Pressure hemihidrosis is the phenomenon that occurs asymmetrically on the unilateral trunk when pressured on the contralateral side. It was firstly reported by Kuno [12], and the mechanism that skin pressure applied to one side of the body produces an ipsilateral suppression and contralateral facilitation was reported (Takagi and Kobayashi 1955) [13]. After the introduction of SSNA, Sugiyama et al. [14] reported that skin pressure reduced ipsilaterally the amplitude of sudomotor bursts. Furthermore, Okagawa et al. [15] concluded that the ipsilateral suppression of cutaneous blood flow was not mediated by the vasoconstrictor nerve system although ipsilateral suppression of sweating elicited by skin pressure was mediated by the sudomotor nervous system, indicating that the occurrence of the spinal reflex due to skin pressure is not uniform between the sudomotor and the vasoconstrictor nerve systems, which represent different organizations at the level of the spinal cord.

In humans, SSNAs innervating the glabrous skin including the palm and sole and those innervating the hairy skin in the generalized body surface behave differently. The former mainly contain mental sudomotor/vasomotor activities, while the latter include thermoregulatory ones [8–10]. Furthermore, since sudomotor nerve activity includes vasodilatory activity, a comparison of these activities is useful in analysis of the regional differentiation of sympathetic nerve activity in humans [16].

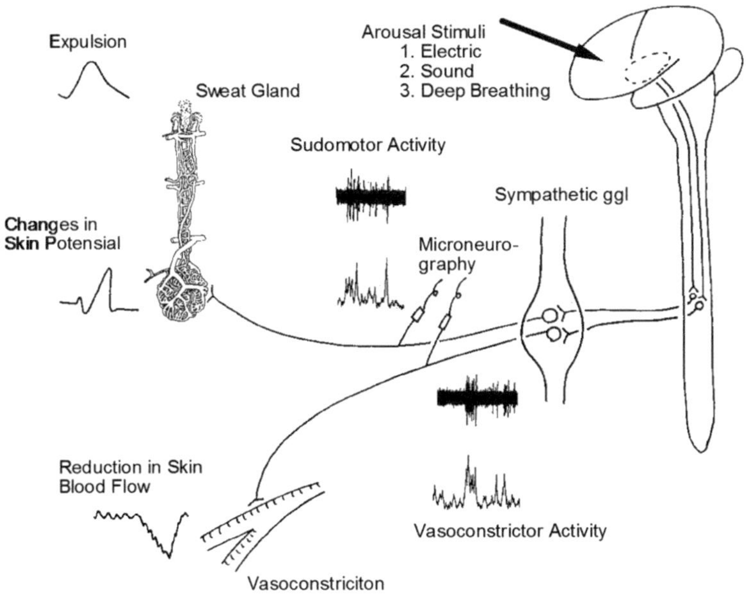

Fig. 4.1 Schematic illustration of skin sympathetic nerve activity. Skin sympathetic nerve activity mainly includes sudomotor nerve activity governing the sweat glands and vasoconstrictive nerve activity innervating the skin arteriolar presphincters of skin vessels. Sympathetic skin response is produced by the movement of sweat in the subcutaneous tissue, and output at the sweat gland orifice provides the sweat expulsion. In contrast, vasoconstriction by vasoconstrictor SSNA causes the reduction in skin blood flow measured by laser Doppler flowmetry

The sudomotor and vasoconstrictor activities are well correlated with the skin potential change and the rate of reduction of vasoconstriction, respectively; therefore, it is well documented that sympathetic skin response and sympathetic flow response measured by laser Doppler flowmetry are good reflections of sudomotor and vasoconstrictor nerve activity, respectively (Fig. 4.1).

The following describes the basic properties of SSNA.

4.4.1 The Relationship Between SSNA Bursts and Sweating Rate and Skin Blood Flow (Fig. 4.2)

The report by Bini et al. [1, 2] showed that laser Doppler flowmetry could enable the accurate measurement of skin blood flow, and sweat rate could be measured quantitatively by development of the ventilated capsule method. The full-wave rectified and integrated SSNA sudomotor and vasoconstrictor bursts were proved to

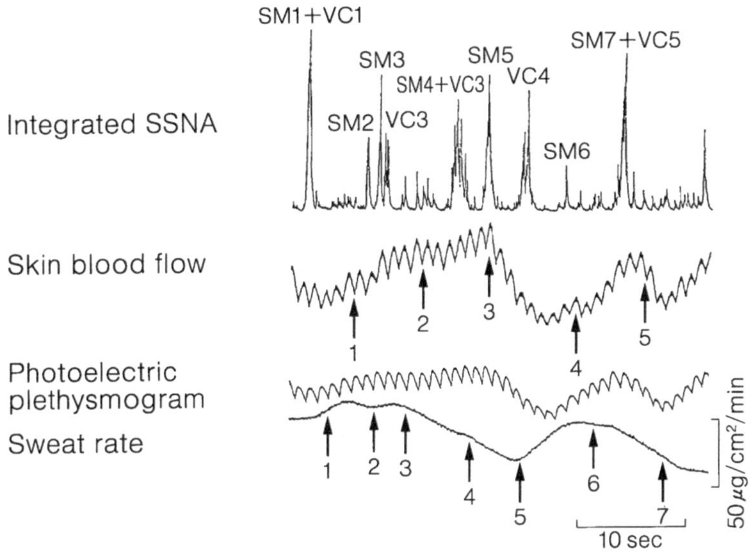

Fig. 4.2 Sudomotor and vasoconstrictor components in skin sympathetic nerve activity. Traces are full-wave rectified and integrated skin sympathetic nerve activity (SSNA), skin blood flow measured by laser Doppler flowmetry, photoelectric plethysmogram, and sweat rate measured by the ventilated capsule method. The latency from the SSNA burst to the inflection of sweat rate identifies the property of the SSNA burst as sudomotor, and that to the reduction point of laser Doppler skin blood flow identifies as vasoconstrictor. The burst indicators are *SM* sudomotor, *VC* vasoconstrictor, and *SM+VC* mixed. The numbers after the burst indicators correspond to the sweat rate inflections and the skin blood flow changes [17]

be correlated with the slope of the sweat rate and the rate of reduction of skin blood flow by laser Doppler flowmetry, respectively [17].

4.4.2 Effect of Thermal Change in the Environment (Fig. 4.3)

SSNA contributes to thermoregulation by enhancing or reducing its activity, which is lowest under thermoneutral conditions. This thermoneutral temperature depends on ethnic factors and is estimated to be 27 °C in the case of the Japanese [10].

SSNA changes depending on the recorded nerve when the environmental temperature changes. In the peroneal nerve innervating hairy skin, the lowest level was designated at the thermoneutral temperature, and vasoconstrictor SSNAs were predominant at 15 °C, while sudomotor SSNA is predominant at 35 °C. Therefore, the shape of the SSNA showed a hairpin curve, reaching its nadir at 27 °C. In contrast, changes in the tibial nerve SSNA showed the lowest value at 35 °C, with

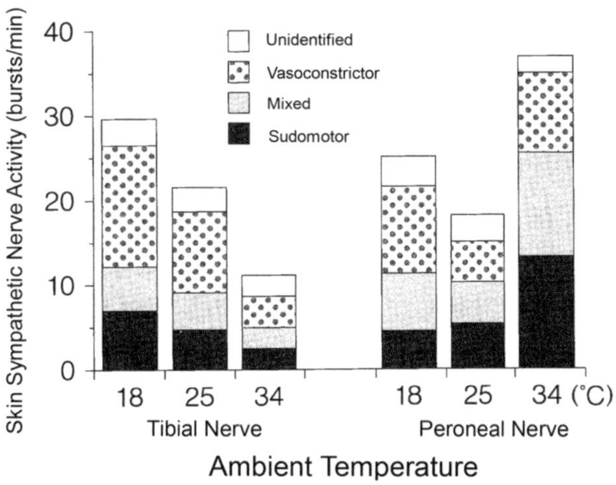

Fig. 4.3 Changes in sudomotor and vasoconstrictor components in skin sympathetic nerve activity depending on the ambient temperature from the tibial nerve and the peroneal nerve
The tibial SSNA innervating to the glabrous skin (sole) and the peroneal SSNA innervating to the hairy skin (dorsum pedis) change their components depending on the ambient temperature [10]

gradual enhancement as environmental temperature decreased. Since the tibial SSNA reflects the stress received from the environment, environmental cold stress enhances the vasoconstrictor SSNA [9, 10].

In this way, sudomotor and vasoconstrictor SSNAs contribute to human thermoregulatory function by altering its components depending on the environmental temperature.

4.4.3 Conduction Velocity of Skin Sympathetic Nerve Activity (Fig. 4.4)

A double recording technique [4] enabled us to measure the conduction velocity of SSNA. Two microelectrodes were inserted into the proximal and distal sites of one nerve, the time differences between burst rises or burst peaks were measured, and the velocity was calculated by dividing the distance between the electrodes by this time difference. The conduction velocities of MSNA thus measured were reported to be 0.72 m/s in the median nerve and 1.09 m/s in the peroneal nerve, while those of SSNA were 0.93 m/s in sudomotor, 0.76 m/s in vasoconstrictor at rest, 1.12 m/s in sudomotor, and 0.81 m/s in vasoconstrictor in elicited bursts by electrical stimulation [18].

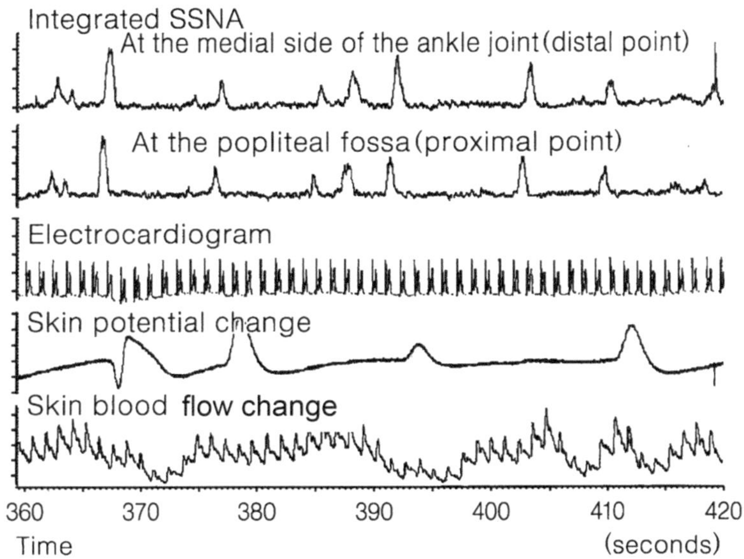

Fig. 4.4 Conduction velocity measurement in skin sympathetic nerve activity. Traces are integrated SSNA recorded at the medial malleolus, integrated SSNA recorded at the popliteal fossa, electrocardiogram, skin potential change, and skin blood flow change measured by laser Doppler flowmetry. The distance between the recording points at the medial malleolus and the popliteal fossa divided by the burst rise time differences provides the sudomotor and vasoconstrictor conduction velocities [18]

4.4.4 Change in Skin Sympathetic Nerve Activity During Sleep

In contrast to the MSNA change during sleep, which reduces its activity as sleep stage proceeds, SSNA changes were reported to be the component changes of sudomotor and vasoconstrictor. Generally, it is said that sympathetic nerve activity functions to cool the body during non-REM sleep, but to warm the body during REM sleep.

Takeuchi et al. [19] reported that SSNA reduces its activity until sleep stage 2 as the sleep stage proceeds, while in turn it increases during sleep stages 3 + 4. Kobayashi et al. [20] observed SSNA change during sleep by its effector responses and described that the sudomotor activity during NREM sleep is enhanced compared with that in REM sleep, while vasoconstrictor activity is enhanced during REM sleep compared with that in NREM sleep, the same tendency as reported by Noll et al. [21].

Thus, vasoconstrictor SSNA reduces its activity as the sleep stage proceeds, to cool the human body, while sudomotor SSNA is slightly enhanced to cool the human body during slow-wave NREM sleep (stages 3+4). Contrarily, vasoconstrictor is enhanced and sudomotor is reduced to warm the body during REM sleep.

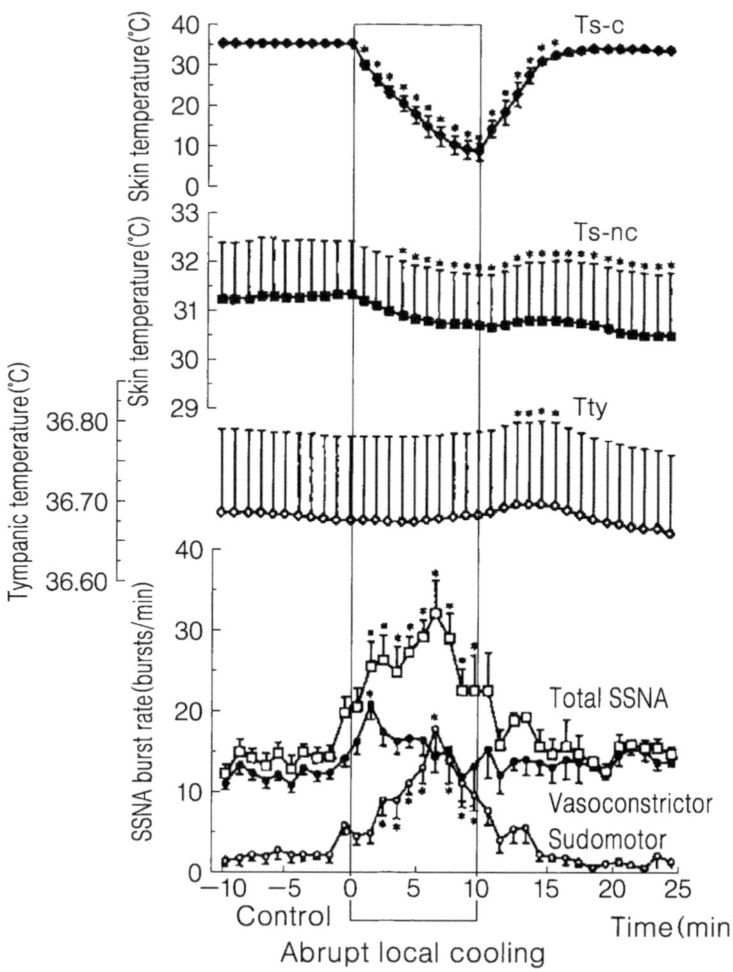

Fig. 4.5 Changes in body temperature and SSNA following acute local cooling. Traces are skin temperature of cooling site, skin temperature of non-cooling site, tympanic temperature, and SSNA burst rate with the sudomotor and vasoconstrictor components. Acute local cooling was provided by inserting both hands into a dry icebox [22]

4.4.5 Effect of Skin Sympathetic Nerve Activity on the Core Temperature Change

Upon exposure to a cold environment, vasoconstrictor SSNA is enhanced to reduce the heat dissipation from the skin surface, and thus, skin blood flow decreases. By this vasoconstriction, it has been observed that the core temperature, measured as the tympanic temperature (Tty), gradually increases with a certain latency, showing the usefulness of recording SSNA in studies of human thermoregulation [22] (Fig. 4.5).

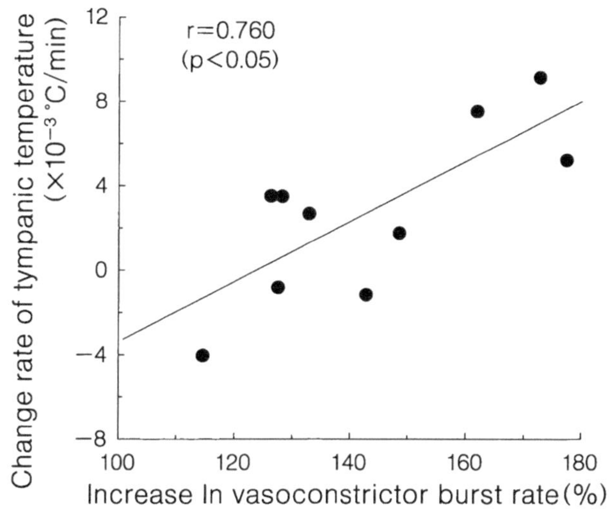

Fig. 4.6 The relationship between the increase in vasoconstrictor component of SSNA taking the baseline reading as 100 % and the change rate of tympanic temperature per min
A significant correlation is observed between the two, indicating that the larger the individual who is capable of SSNA, the higher the tympanic temperature, i.e., the core temperature [22]

When we analyzed the magnitude of increase in SSNA upon cold exposure and the increase in the core temperature, a significant correlation between increase in vasoconstrictor burst rate (%) and the rate of change of tympanic temperature was found (Fig. 4.6). This means that the greater the capability to change the SSNA upon cold exposure, the more excellent thermoregulatory function the individual possesses. The cross-correlation between the SSNA and the core temperature showed that the peak latency is approximately 10 min, meaning that it takes approximately 10 min to raise the core temperature upon cold exposure (Fig. 4.7) [22–25].

In contrast, we also reported that rewarming from a cold environment of ~15 °C suppresses the vasoconstrictor SSNA (Fig. 4.8), and the greater suppression, rate of vasoconstrictive SSNA is associated with the larger suppression of the core temperature (Fig. 4.9) [25]. Cross-correlogram between SSNA discharge and the core temperature (tympanic temperature) change designated that the time delay exhibited ~7 min (Fig. 4.10), indicating that not only the cooling but also warming takes approximately 7–10 min to raise the core temperature through the convectional pathway.

The height of integrated sudomotor SSNA is found to be proportional to the number of sweat glands observed with a video microscope [26]. This means that the height of integrated SSNA represents the driving force to expel sweating droplets, and it well correlates to the slope of the sweat rate.

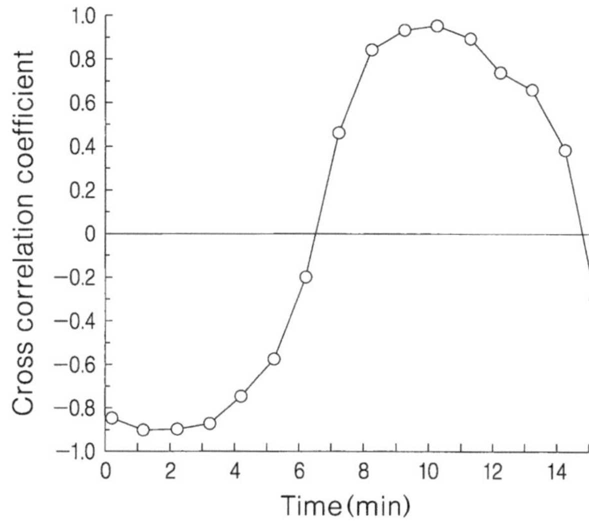

Fig. 4.7 Cross-correlation between SSNA and tympanic temperature during ambient temperature drop. The cross-correlogram exhibited at its maximum at 10 min indicating that the SSNA discharge provides the maximum tympanic temperature rise at 10 min after the discharge [22]

4.4.6 Effect of Aging on Skin Sympathetic Nerve Activity in Humans

Although it is known that MSNA increases with advancing age, age-related changes in resting SSNA have not been clarified. Attenuation of peripheral effectors has been reported by Watanabe et al. [27]; however, another report described that SSNA in the elderly is decreased compared with that in the young and middle aged, and responsiveness to the thermal environment is also reduced [28].

In a recent study by Greaney et al. [29], it was observed that SSNA response to whole-body cooling is blunted in the elderly, which is related to a marked impairment in reflex cutaneous vasoconstriction, although cutaneous vascular responsiveness to adrenergic stimuli is not reduced and that to a non-thermoregulatory stimulus is preserved. Nevertheless, interindividual differences of resting SSNA are very large, which makes this issue difficult to investigate.

4.5 Clinical Application of Skin Sympathetic Nerve Activity

4.5.1 Investigation of Sweating Physiology

Thermoregulatory sweating is only observed in some primates, including humans. The phenomenon of "sweat expulsion," which involves the emission of sweat from the sweat gland orifices, has been clarified to correspond to SSNA burst discharges [30]. The more impulses are included in sudomotor bursts, the more sweat glands

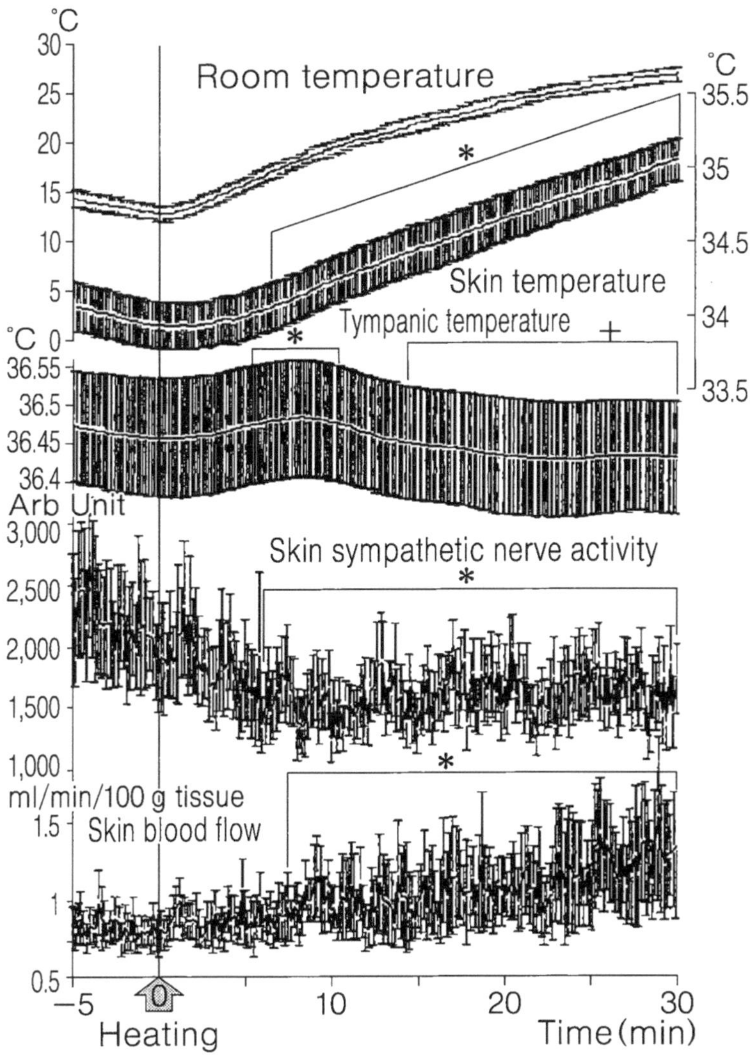

Fig. 4.8 SSNA suppression and tympanic temperature drop with ambient temperature decrease The traces are room, skin, tympanic temperatures, SSNA, and skin blood flow. The room temperature was decreased to 10 °C and rewarmed to 28 °C. With the room temperature rise, SSNA was suppressed, and skin blood flow was increased, which enhanced the heat dissipation from the skin surface and resulted in a tympanic temperature drop

are driven by SSNA bursts, and the more sweating is facilitated, which contributes to the thermoregulation of the human body.

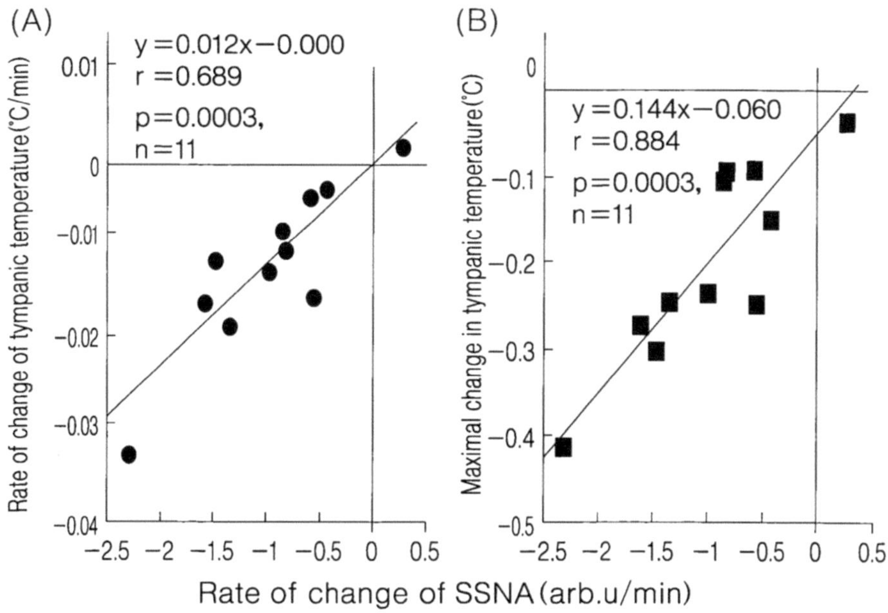

Fig. 4.9 The relationships between the change rate of SSNA and the tympanic temperature change rate (**a**) and maximal change in tympanic temperature (**b**) The established significant correlation indicates the larger the capacity to enhance SSNA, the larger the tympanic temperature to change [23]

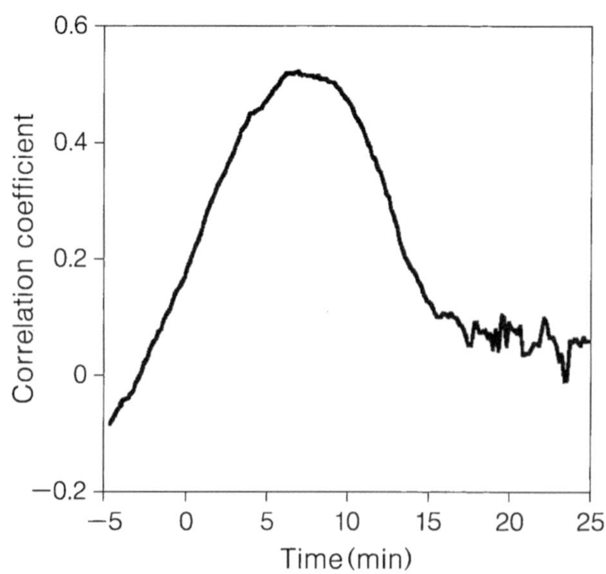

Fig. 4.10 Cross-correlation between SSNA and tympanic temperature during ambient temperature rewarming The cross-correlogram exhibited at its maximum at 7 min indicating that the SSNA discharge suppression provides the maximum tympanic temperature drop at 7 min after the discharge suppression [25]

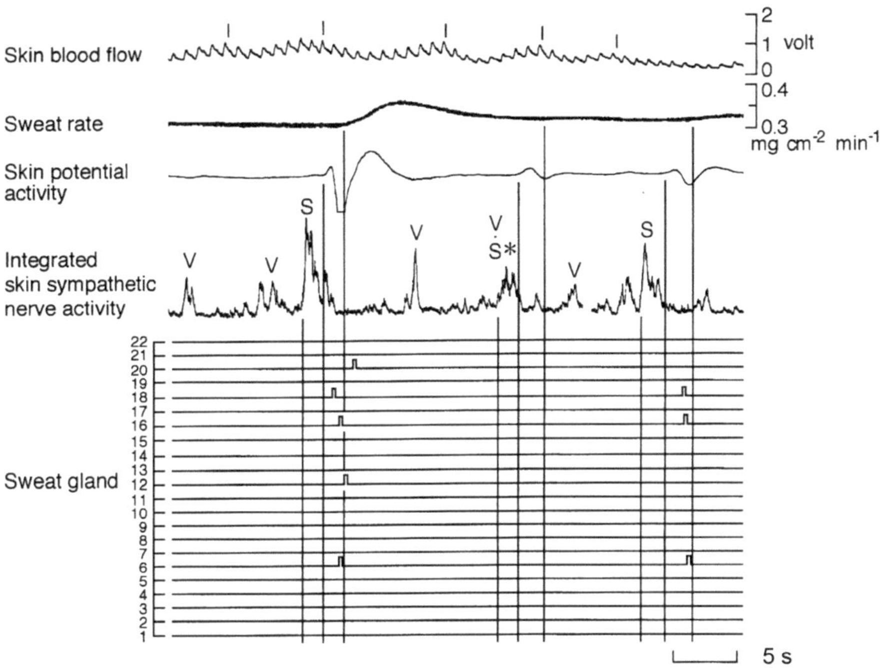

Fig. 4.11 Secretion of individual sweat glands and various effector responses to spontaneous sudomotor bursts during mild sweating. Skin blood flow, sweat rate, skin potential activity, and skin sympathetic nerve activity recorded from the tibial nerve are indicated. The *lower panel* indicates the timing of sweat secretion from individual sweat glands observed on a small area of the sole by a videomicroscope. The numerals 1–3 along the vertical bar represent individual glands. Two sudomotor bursts (S with no asterisk) elicited sweat secretion from three or five glands (Nos. 6, 12, 16, 18, 20). Another sudomotor burst (S with asterisk) did not elicit any sweat secretion, although the burst caused small responses in skin potential and local sweat rate. Skin vasoconstrictor bursts (shown as V) induced a transient reduction in skin blood flow. One of the sudomotor bursts contains the vasoconstrictor component. Three *vertical lines* show the onsets of sudomotor burst, skin potential response, and sweat response

4.5.2 Investigation of Anhidrosis/Hypohidrosis

Anhidrosis/hypohidrosis is defined as an absence of sweating. It may have multiple etiologies, but the recording of SSNA can identify the site of the lesion (Fig. 4.11) [31–36]. Attenuated SSNA represents neurogenic anhidrosis/hypohidrosis, and no change or activation, especially activation under heat exposure, indicates that a lesion is at a sweat gland. In neurogenic anhidrosis/hypohidrosis, a peripheral origin increases the burst rate, while a central origin does not exhibit an SSNA change (Fig. 4.12).

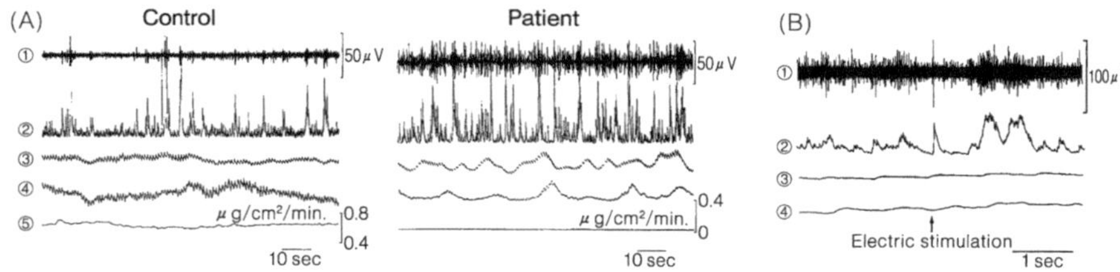

Fig. 4.12 SSNA from acquired idiopathic generalized anhidrosis (AIGA). Traces are ① raw trace of SSNA, ② integrated SSNA, ③ laser Doppler skin blood flow, ④ photoelectric plethysmogram, and ⑤ sweat rate (hidrogram) in the control subject, and in AIGA patient in baseline recording (**a**), and after electric stimulation (**b**). No sweat rate increase was observed after electric stimulation at the wrist joint [37]

Idiopathic and secondary anhidrosis and hypohidrosis exist, among which acquired idiopathic generalized anhidrosis (AIGA, 36, 32, 33, 34, 35) is characterized by a generalized absence of sweating without other autonomic and neurologic dysfunctions. AIGA is classified into three subgroups: (1) sudomotor neuropathy, (2) idiopathic pure sudomotor failure (IPSF), and (3) sweat gland failure (SGF), with each subgroup presenting a different pathogenesis. Sudomotor neuropathy is neurogenic and may sometimes be accompanied by vasoconstrictor neuropathy (Fig. 4.13). IPSF is the most common presentation and is characterized by loss of sweating in the absence of any neurological features or destruction of sweat glands by skin biopsy. IPSF is associated with cholinergic urticaria and estimated to be of autoimmune origin. In SGF, skin biopsy demonstrates the destruction and devastation of the sweat glands, which suggests the possibility of the resultant devastation of IPSF. SSNA is normal or even enhanced in IPSF and SGF without neurological deficits but with a generalized absence of sweating, sometimes associated with heat intolerance, especially upon heat exposure [37–39].

Since IPSF has been proved to be improved by steroid pulse therapy, early diagnosis of the disease by microneurography and skin biopsy is recommend [40].

4.5.3 *Investigation of Hyperhidrosis*

Hyperhidrosis is defined as an excess of sweating beyond the amount required to return an elevated body temperature to normal, with a distinction between primary and secondary forms. Among primary forms, idiopathic palmoplantar hyperhidrosis is frequently found in Mongoloids, with profuse sweating in their palms and soles. The condition may become socially and professionally debilitating. Idiopathic palmoplantar hyperhidrosis begins in childhood and frequently runs in families. Epidemiologically, it has been estimated to occur 0.6–1 % of the population, while its rate has been reported to be 2.8 % in the United States [41].

The definition of palmoplantar hyperhidrosis is not quantitatively determined in terms of sweating more than a certain amount, but rather that the patient feels that

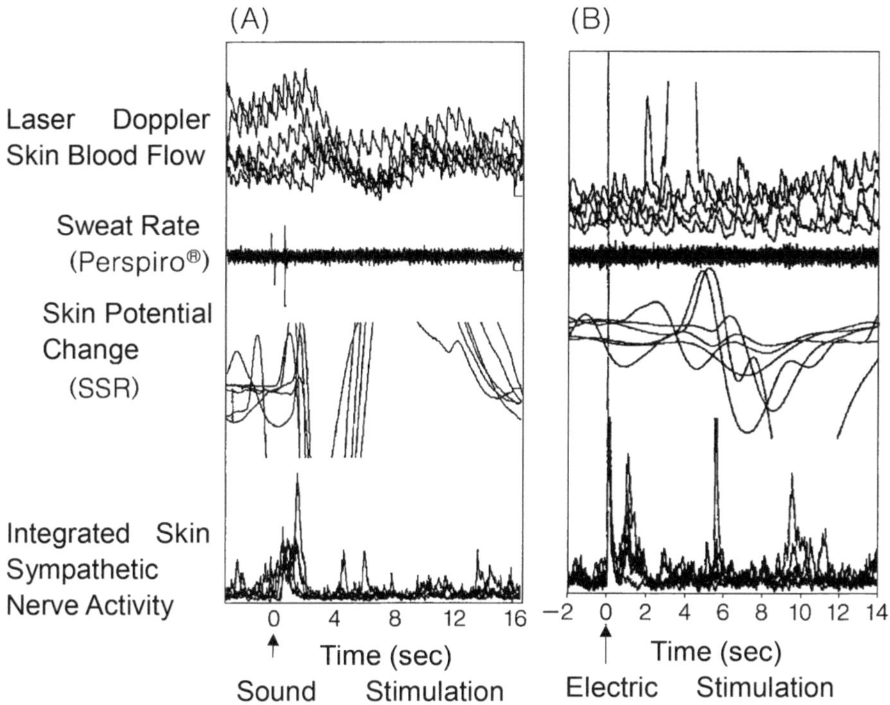

Fig. 4.13 Changes in skin blood flow, sweat rate, skin potential change, and integrated skin sympathetic nerve activity followed by sound/electric stimulations in acquired idiopathic generalized anhidrosis (**a**) and autonomic neuropathy (**b**) (**a**) In acquired idiopathic generalized anhidrosis, resting SSNA is enhanced, reflex response to sudden sound is favorable, and skin potential change exhibits fair response, although no sweating response is observed (**b**) In autonomic neuropathy, since both sudomotor and vasoconstrictor SSNA are disordered, reduced resting SSNA is observed, and electric stimulation elicits lowered reflex bursts, reduced skin potential change, as well as no sweat response, and no skin blood flow change

the hyperhidrotic state is troublesome. It affects men and women equally and most commonly occurs among people aged 25–64 years, some of whom may have been affected since early childhood; ~30–50 % have another family member afflicted, implying a genetic predisposition [40]. In 2006, a primary palmar hyperhidrosis locus was reported to be mapped to 14q11.2–q13 [42].

The activated sudomotor as well as vasoconstrictor SSNA governing the palms (median nerve) and soles (tibial nerve) was first reported in 1997, while that governing hairy skin (peroneal nerve) is not enhanced as such (Figs. 4.13 and 4.14). The intermingling of sudomotor and vasoconstrictor in the same nerve fascicle causes not only palmar hyperhidrosis but also vasoconstriction-caused hypothermia in the affected skin areas [32]. Hyperhidrosis observed in amyotrophic lateral sclerosis [43] or Guillain-Barré syndrome [30] is also caused by this enhanced SSNA.

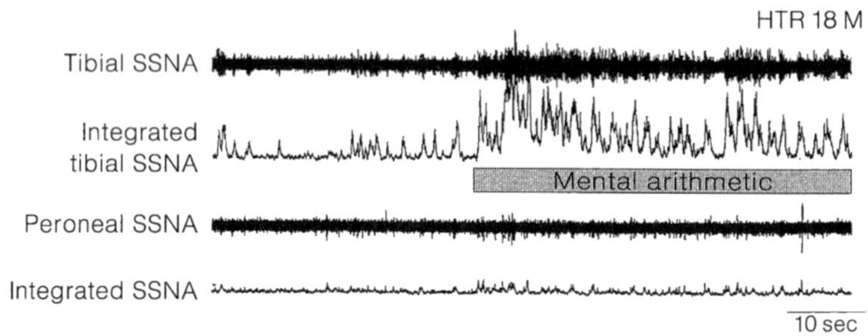

Fig 4.14 Tibial SSNA and peroneal SSNA in palmoplantar hyperhidrosis and their responses to mental arithmetic (psychological load). Traces are raw tibial SSNA and its full-wave rectified and integrated trace, innervating the sole, and peroneal SSNA and its full-wave rectified and integrated trace, innervating the dorsum pedis. The resting tibial SSNA is enhanced compared with peroneal SSNA, and the response to the mental arithmetic was markedly enhanced. The resting peroneal SSNA is not so activated, and the response is not so prominent. These traces indicate that the SSNA response to mental stimuli only is markedly enhanced [32]

4.5.4 Investigation of Thermoregulatory Function and Thermoregulatory Disorders

There are large interindividual differences in human thermoregulatory function, and we have reported that the activation capacity of vasoconstrictive SSNA determines individual thermoregulatory function [22, 23, 25]. Individuals who can suppress SSNA can minimize the change in core temperature. Thus, human cold tolerance and heat tolerance are acquirable by acclimatization by altering the sympathetic nerve activity toward environmental thermal change [24].

In hypothalamic thermodysregulation, microneurographic SSNA recording revealed the hypofunction of the hypothalamic warm neurons, whereas the cold neurons worked effectively, which suppressed sweating and resulted in thermodysregulation [27].

4.5.5 Investigation of Pathophysiology and Diagnosis of Vasomotor Disease

Diseases due to aberrant vasoconstriction, for example, Raynaud's disease and Buerger's disease, are sometimes treated and improved by sympathectomy. SSNA from these disorders with a certain perturbation presents different responses from that in normal subjects [24, 44–47]. These recordings are beneficial to assess the effectiveness of nerve block or sympathectomy.

4.5.6 *Investigation of Association Between Cognitive Function and Sympathetic Nerve Activity*

Studies on the association between cognitive function and sympathetic nerve activity were initiated by the analysis of reflex latency in Parkinson's disease [48], in which delay of reflex latency was reported.

One of the event-related potentials, P300, has a close relationship with cognitive function. Simultaneous recordings of event-related potential and SSNA with an oddball paradigm have revealed that cognition of the target elucidates a significantly strong sympathetic response compared with that upon no cognition, which indicates that the cognitive process strongly influences central sympathetic outflow in humans. Furthermore, analysis of the presence of reflex bursts and summation of event-related potentials clarified that the early component of P300 is closely related to the origin of SSNA [49].

4.5.7 *Evaluation of Drug Effects on Nerve-Effector Responses*

The relationships between the height of integrated burst discharge and sweat acceleration, and between that and reduction rate of skin blood flow, have been established (Fig. 4.15). Analysis of the slope of the regression lines can enable evaluation of drug effects in cases when such effects cannot be clarified by analysis of the changes in effector organs.

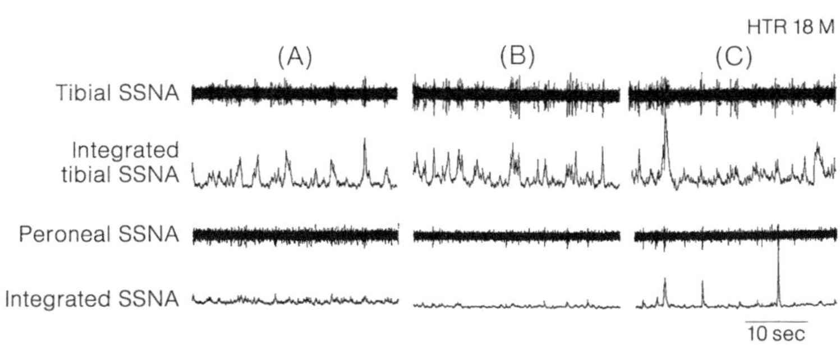

Fig. 4.15 Tibial SSNA and peroneal SSNA in palmoplantar hyperhidrosis and their responses to ambient temperature rise (psychological and physical load). Traces are, from the top to the bottom, tibial SSNA, its full-wave rectified and integrated trace, peroneal SSNA, and its integration at the ambient temperature of 25 °C in resting (**a**), during heating to ambient temperature of 30 °C (**b**), and 30 °C in resting (**c**). Resting SSNA was enhanced in the tibial and gradually augmented with an ambient temperature rise in both, but the augmentation was more prominent in the tibial SSNA. The degree of enhancement was greater in the mental load than in the physical [17]

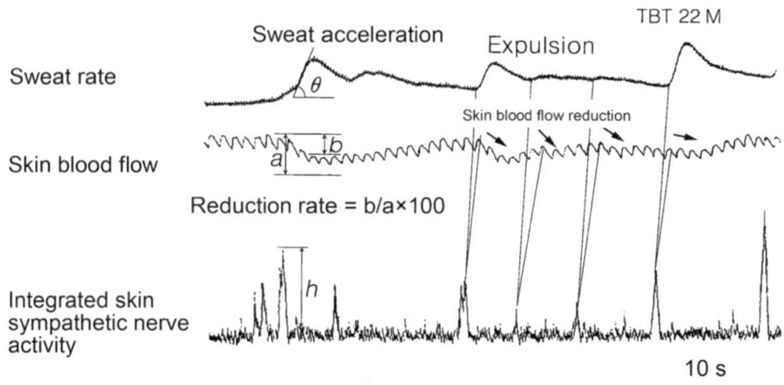

Fig. 4.16 The relationship between SSNA and laser Doppler skin blood flow. There established a significant correlation between the amplitude of integrated SSNA burst (h) and the slope of sweat rate rise (tan θ, sweat acceleration); the logarithm of reduction rate of skin blood flow log (b/a ×100) [17]

For example, sweat acceleration and vasoconstrictive effects of prostaglandin E_1 can be clarified by analyzing the relationship between sudomotor SSNA and sweat expulsion and vasoconstrictor SSNA and the rate of reduction of laser Doppler skin blood flow, respectively, even if the baseline measurement of both sweat rate and skin blood flow indicated no changes. The attenuating effect of prostaglandin E_1 on sudomotor and vasoconstrictive effects could be confirmed by the changes in the slope of the regression line from a steep gradient to a gradual gradient by the intravenous administration of prostaglandin E_1 (Figs. 4.16 and 4.17) [50].

4.6 Significance of SSNA as a Feedforward Tool

4.6.1 Environmental Temperature and Thermal Conduction

There are two ways to conduct information on the environmental temperature to the hypothalamus. One is a neural pathway that conducts the thermal information through a neural network, while the other is a convectional pathway that provides information through circulation by warmed or cooled blood to the hypothalamus.

The thermal sensation is transmitted to the thermoregulatory center in the hypothalamus, situated at the preoptic area and anterior hypothalamus. Warming this area facilitates heat dissipation, through vasodilatation and sweating. Inversely, cooling this area suppresses heat dissipation, through vasoconstriction or shivering. Thus, activation of these neurons facilitates sympathoactivation. After thermoreception at the preoptic area and anterior hypothalamus, the generation of comfortableness and uncomfortableness is involved with amygdala, orbitofrontal cortex, or cingulate cortex. This central command descends through the periaqueductal gray and raphe nucleus, connects to the rostral ventrolateral medulla

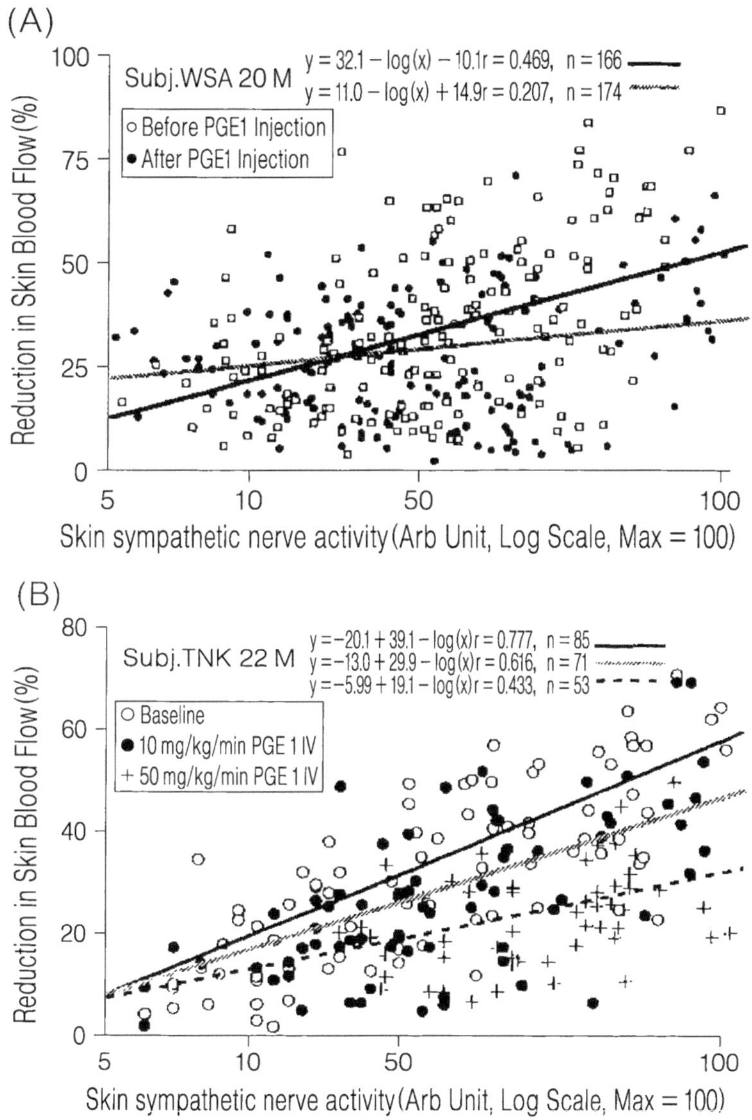

Fig. 4.17 Effect of prostaglandin E_1 administration on the relationships between logarithm of integrated SSNA and reduction rate of skin blood flow (**a**) Administration of prostaglandin E_1 reduces the slope of the regression line between the logarithm of integrated SSNA and the rate of reduction of skin blood flow. Prostaglandin E_1 administration reduces the regression line between log (integrated SSNA amplitude) and reduction rate of laser Doppler skin blood flow (**b**) Administration of prostaglandin E_1 reduces the slope of the regression line between the logarithm of integrated SSNA and the rate of reduction of laser Doppler skin blood flow dose dependently [50]

as the premotor neuron, and exchanges the neurons at the intermediolateral nucleus and sympathetic ganglion. The descending sympathetic outflow to the skin thus governs the sweat glands and skin vasculature to control the body temperature.

4.6.2 Chronological Delay in Thermoregulation Using Convectional and Neural Pathway

The time delay from vasoconstrictive SSNA discharge to the core temperature rise should be ~7–14 min, from the correlative analysis of SSNA and core temperature. In the same fashion, the time delay from the stimulation to the warm neurons to the core temperature drop through vasodilatation or sweating should be also ~7–14 min. Thus, thermoregulation using a convectional pathway, that is, negative feedback, requires approximately 10 min depending on the individual; contrarily, the time delay from the environmental temperature change through the neural pathway to thermoregulatory SSNA discharge is within 1 min. Therefore, employing this rapid neural pathway, namely, the feedforward network, is efficient to control the human body temperature.

In this way, feedforward control by this neural pathway has the merit of rapid initiation of thermoregulatory control compared with the convectional pathway which takes approximately 10 min, through the activation or suppression of SSNA. This capacity to activate/suppress the SSNA determines the thermoregulatory capacity through the feedforward neural pathway.

The reason why two thermo-informative pathways have developed in humans is that the convectional pathway, possessed at phylogenetically older levels, takes too long to transmit the thermoregulatory information to the central nervous system. This accelerated the development of the feedforward and neural pathway, which takes only one tenth of the time of the convectional pathway. In particular, thermoregulation by sweating is recognized in the primates, which is assumed to be phylogenetically high level of thermoregulation.

4.6.3 Phylogenetic Considerations

Differentiation of sudomotor and vasoconstrictor components of SSNA and correlative study with evoked potentials have revealed that the cognitive potential, P300, has close associations with sudomotor SSNA (Fig. 4.18) and somatosensory potential, N140, has close associations with vasoconstrictor SSNA (Fig. 4.19) [51]. These findings suggest that the sweating responding to neural transmission is a higher level of thermoregulatory function associated with cognition and behavior, which are phylogenetically higher-level cortical functions than skin vasoconstriction induced by stimulation of the skin or cold exposure.

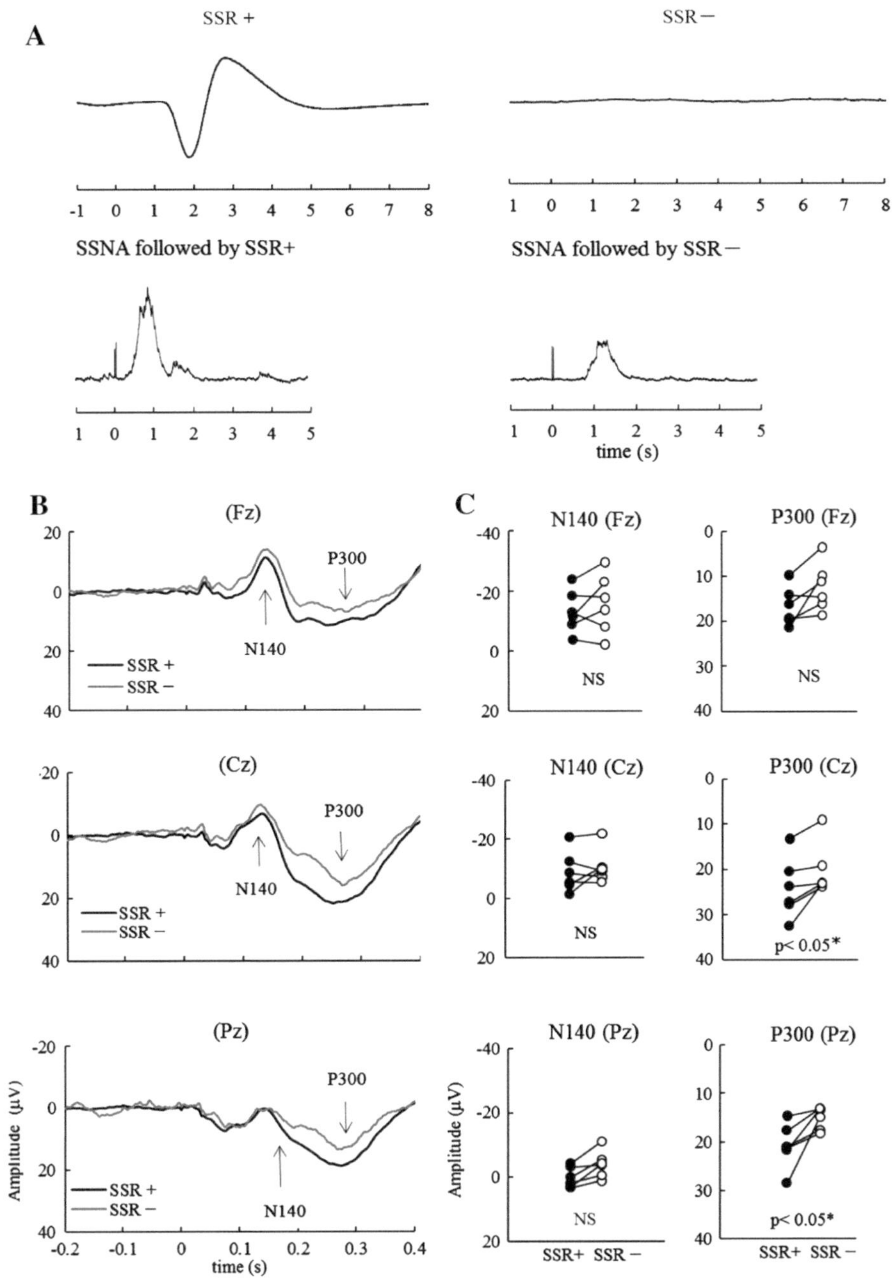

Fig. 4.18 Sudomotor SSNA is closely associated with cognition. Integrated SSNA with skin potential change (*SSR* sympathetic skin response) is assumed as sudomotor SSNA (SSR+), and SSNA without skin potential change is assumed as SSR− SSNA (**a**). The amplitude of integrated SSNA burst with SSR+ response accompanies with a significantly large P300 amplitude (**b**) in the evoked potential (in Cz and Pz,), whereas there is no significant difference in N140 (**c**)

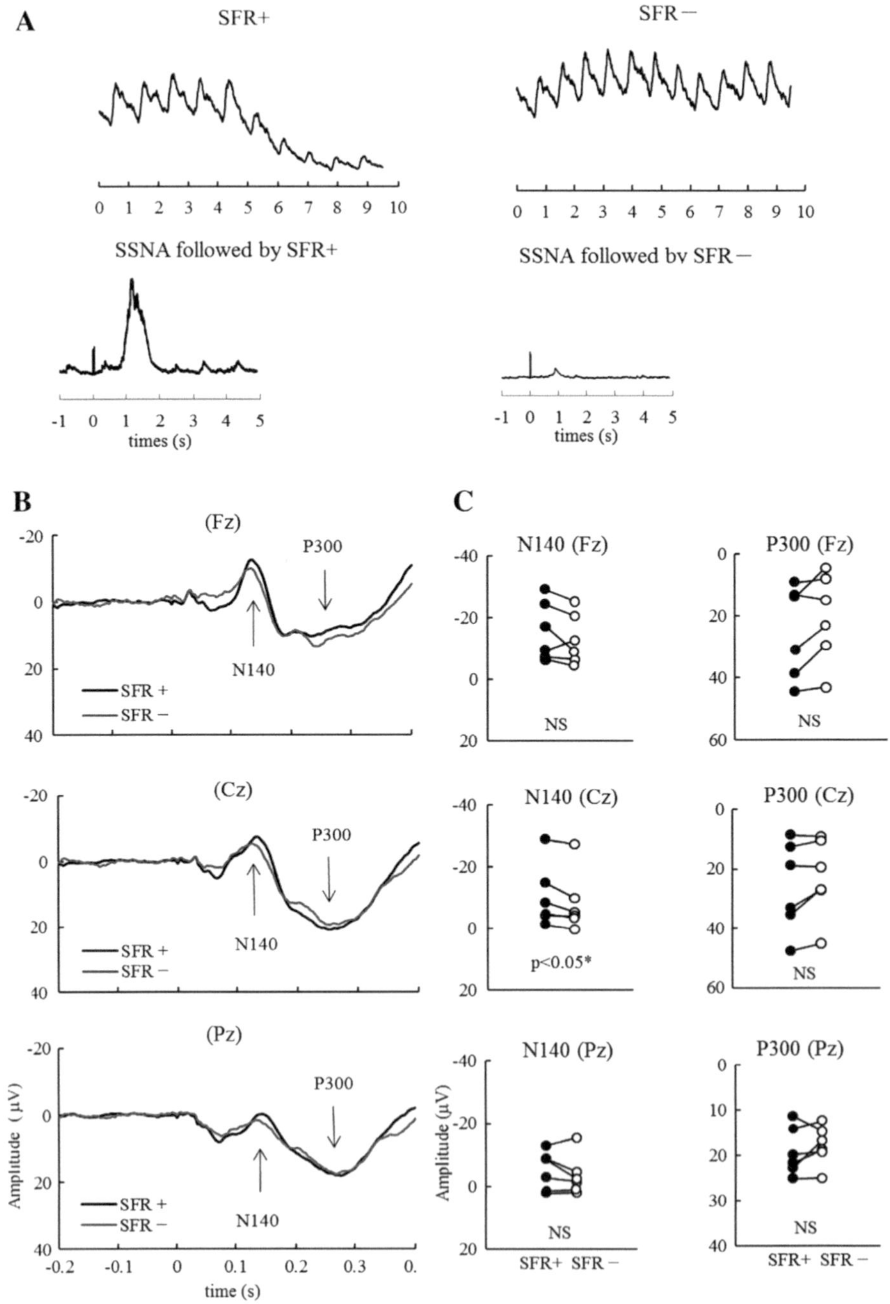

Fig. 4.19 Vasoconstrictor SSNA is closely associated with somatosensory sensation. Integrated SSNA with skin blood flow change (SFR, sympathetic flow response) is assumed as vasoconstrictor SSNA (SFR+) and without skin blood flow change is assumed as SSR− SSNA (**a**). The amplitude of integrated SSNA burst with SSR+ response accompanies with significantly large N140 amplitude (**b**) in the evoked potential (in Cz), whereas there is no significant difference in P300 (**c**)

This hypothesis that "mental sweating necessitates recognition whereas somatosensory stimulation is important for vasoconstriction" is supported by the fact that "mental sweating disappears but vasoconstriction continues under general anesthesia" and that "vasoconstriction is induced by the cold sensation to sensory loss skin."

In this way, the development of humans is considered to have occurred by them widening their activity under various thermal environments, preventing extreme body temperature rises or drops through more rapid thermoregulation using feedforward control of the body temperature via neural thermoregulatory control.

4.7 Conclusion

The recording techniques, properties, clinical applications, and significance of sympathetic nerve activity to the skin have been described. Time series analysis with other variables would clarify the various aspects of SSNA in humans. The significance of SSNA as a phylogenetic consideration should be clarified in future studies.

References

1. Bini G, Hagbarth K-E, Hynninen P, Wallin BG. Thermoregulatory and rhythm-generating mechanisms governing the sudomotor and vasoconstrictor outflow in human cutaneous nerves. J Physiol. 1980;306:537–52.
2. Bini G, Hagbarth K-E, Hynninen P, Wallin BG. Regional similarities and differences in thermoregulatory vaso- and sudomotor tone. J Physiol. 1980;306:553–65.
3. Mano T. Sympathetic nerve mechanisms of human adaptation to environments – findings obtained by recent microneurographic studies. Environ Med. 1990;34:1–35.
4. Mano T, Iwase S, Sugiyama Y. Double recording technique of sympathetic nerve traffic in microneurography. In: Shibasaki H, Kimura J, editors. Recent advances in clinical neurophysiology. Amsterdam: Elsevier Science; 1996. p. 416–9.
5. Mano T. Microneurographic research on sympathetic nerve response to environmental stimuli in humans. Jpn J Physiol. 1998;48:99–114.
6. Mano T. Muscular and cutaneous sympathetic nerve activity. In: Appenzeller O, editor. The autonomic nervous system (1), handbook of clinical neurology, vol. 174. Amsterdam: Elsevier Science; 1999. p. 649–65.
7. Mano T, Iwase S, Toma S. Microneurography as a tool in clinical neurophysiology to investigate peripheral neural traffic in humans. Clin Neurophysiol. 2006;117:2357–84.
8. Iwase S, Sugenoya J, Sugiyama Y, Mano T, Yamamoto K. Skin sympathetic nerve activity in the peroneal and the tibial nerves. Ann Rep Res Inst Environ Med Nagoya Univ. 1990;40:308–13.
9. Iwase S, Sugenoya J, Mano T, Sugiyama Y, Hakusui S, Yamamoto K. Reflex burst of skin sympathetic nerve activity and sudomotor/vasoconstrictor activity in humans. Ann Rep Res Inst Environ Med Nagoya Univ. 1991;41:1990.

10. Okamoto T, Iwase S, Sugenoya J, Mano T, Sugiyama Y, Yamamoto K. Different thermal dependency of cutaneous sympathetic outflow to glabrous and hairy skin in humans. Eur J Appl Physiol Occup Physiol. 1994;68:460–4.
11. Sugenoya J, Iwase S, Mano T, Ogawa T. Identification of sudomotor activity in cutaneous sympathetic nerves using sweat expulsion as the effector response. Eur J Appl Physiol Occup Physiol. 1990;61:302–8.
12. Kuno Y. The physiology of human perspiration. London: Churchill; 1934.
13. Takagi K, Kobayashi S. Skin pressure-vegetative reflex. Acta Med Biol. 1955;4:31–57.
14. Sugiyama Y, Watanabe T, Takeuchi S, Iwase S, Mano T. Asymmetric discharges of skin sympathetic nerve in humans—with special reference to pressure hemihidrosis. Environ Med. 1992;36:207–10.
15. Okagawa T, Sugenoya J, Iwase S, Mano T, Suzumura A, Matsumoto T, et al. Occurrence of the spinal reflex due to skin pressure in sudomotor and cutaneous vasoconstrictor nerve system of humans. Auton Neurosci. 2003;105:62–70.
16. Sugenoya J, Iwase S, Mano T, Sugiyama Y, Ogawa T, Nishiyama T, et al. Vasodilator component in sympathetic nerve activity destined for the skin of the dorsal foot of mildly heated humans. J Physiol. 1998;507(Pt 2):603–10.
17. Iwase S, Mano T, Sugenoya J, Sugiyama Y, Matsukawa T, Yamamoto K. Microneurographic study on sympathetic control of sweating and skin blood flow. In: Mabuchi K, editor. Advanced techniques and clinical applications in biomedical thermology. Chur: Harwood; 1994. p. 147–62.
18. Kondo M, Iwase S, Mano T, Kuzuhara S. Direct measurement of human sympathetic nerve conduction velocity. Muscle Nerve. 2004;29:128–33.
19. Takeuchi S, Iwase S, Mano T, Okada H, Sugiyama Y, Watanabe T. Sleep-related changes in human muscle and skin sympathetic nerve activities. J Auton Nerv Syst. 1994;47:121–9.
20. Kobayashi R, Koike Y, Hirayama M, Ito H, Sobue G. Skin sympathetic nerve function during sleep—a study with effector responses. Auton Neurosci. 2003;103:121–6.
21. Noll G, Elam M, Kunimoto M, Karlsson T, Wallin BG. Skin sympathetic nerve activity and effector function during sleep in humans. Acta Physiol Scand. 1994;151:319–29.
22. Sawasaki N, Iwase S, Mano T. Effect of skin sympathetic response to local or systemic cold exposure on thermoregulatory functions in humans. Auton Neurosci. 2001;87:274–81.
23. Michikami D, Iwase S, Kamiya A, Fu Q, Mano T, Suzumura A. Interrelations of vasoconstrictor sympathetic outflow to skin core temperature during unilateral sole heating in humans. Auton Neurosci. 2001;91:55–61.
24. Iwase S, Sawasaki N, Michikami D, Cui J, Kamiya A, Niimi Y, et al. Effect of activation and suppression of vasoconstrictive skin sympathetic nerve activity on the core temperature in humans. Auton Nerv Syst. 2001;38:7–10.
25. Iwase S, Cui J, Wallin BG, Kamiya A, Mano T. Effects of increased ambient temperature on skin sympathetic nerve activity and core temperature in humans. Neurosci Lett. 2002;327:37–40.
26. Nishiyama T, Sugenoya J, Matsumoto T, Iwase S, Mano T. Irregular activation of individual sweat glands in human sole observed by a videomicroscopy. Auton Neurosci. 2001;88:117–26.
27. Watanabe H, Shindo K, Ida H, Tanaka H, Nagasaka T, Shiozawa Z. Aging effects of sympathetic reflex activities on skin nerves. Gerontology. 2003;49:366–73.
28. Grassi G, Seravalle G, Turri C, Bertinieri G, Dell'Oro R, Mancia G. Impairment of thermoregulatory control of skin sympathetic nerve traffic in the elderly. Circulation. 2003;108:729–35.
29. Greaney JL, Stanhewics AE, Kenney WL, Alexander LM. Impaired increases in skin sympathetic nerve activity contribute to age-related decrements in reflex cutaneous vasoconstriction. J Physiol. 2015. doi:10.1113/JP270062 [Epub ahead of print].
30. Yamamoto K, Sobue G, Iwase S, Nagamatsu M, Mano T, Mitsuma T. Skin sympathetic nerve activity in Guillain-Barré syndrome: a microneurographic study. J Neurol Neurosurg Psychiatry. 1997;63:537–41.

31. Donadio V, Montagna P, Nolano M, Cortelli P, Misciali C, Pierangeli G, et al. Generalised anhidrosis: different lesion sites demonstrated by microneurography and skin biopsy. J Neurol Neurosurg Psychiatry. 2005;76:588–91.
32. Iwase S, Ikeda T, Kitazawa H, Hakusui S, Sugenoya J, Mano T. Altered response in cutaneous sympathetic outflow to mental and thermal stimuli in primary palmoplantar hyperhidrosis. J Auton Nerv Syst. 1997;64:65–73.
33. Iwase S. Acquired idiopathic generalized anhidrosis. Nihon Iji Shimpo. 2003;4136:1–10.
34. Iwase S. Acquired idiopathic generalized anhidrosis. Auton Nerv Syst. 2003;40:335–42.
35. Miyazoe S, Matsuo H, Ohnishi A, Tajima F, Fujishita S, Ichinose K, et al. Acquired idiopathic generalized anhidrosis with isolated sudomotor neuropathy. Ann Neurol. 1998;44:378–81.
36. Nakazato Y, Tamura N, Ohkuma A, Yoshimaru K, Shimazu K. Idiopathic pure sudomotor failure: anhidrosis due to deficits in cholinergic transmission. Neurology. 2004;63:1476–80.
37. Murakami K, Sobue G, Iwase S, Mitsuma T, Mano T. Skin sympathetic nerve activity in acquired idiopathic generalized anhidrosis. Neurology. 1993;43:1137–40.
38. Hakusui S, Yasuda T, Yanagi T, Yamamoto K, Iwase S, Ikeda T, et al. Pathophysiology of 2 cases of acquired anhidrosis. Auton Nerv Syst. 1996;33:366–70.
39. Yamamoto K, Sobue G, Iwase S, Kumazawa K, Mitsuma T, Mano T. Possible mechanism of anhidrosis in a symptomatic female carrier of Fabry's disease: an assessment by skin sympathetic nerve activity and sympathetic skin response. Clin Auton Res. 1996;6:107–10.
40. Chen YC, Wu CS, Chen GS, Khor GT, Chen CH, Huang P. Identification of subgroups of acquired idiopathic generalized anhidrosis. Neurologist. 2008;14:318–20. doi:10.1097/NRL.0b013e318173e818.
41. Haider A, Solish N. Focal hyperhidrosis: diagnosis and management. Can Med Assoc J. 2005;172(1):69–75. doi:10.1503/cmaj.1040708.PMC543948.
42. Higashimoto I, Yoshiura K, Hirakawa N, Higashimoto K, Soejima H, Otoki T, et al. Primary palmar hyperhidrosis locus maps to 14q11.2-q13. Am J Med Genet A. 2006;140A(6):567–72. doi:10.1002/ajmg.a.31127.PMID16470694.
43. Shindo K, Watanabe H, Ohta E, Nagasaka T, Shiozawa Z, Takiyama Y. Sympathetic sudomotor neural function in amyotrophic lateral sclerosis. Amyotroph Lateral Scler. 2011;12:39–44. doi:10.3109/17482968.2010.508529. Epub 2010 Aug 25.
44. Iwase S, Sawasaki N, Michikami D, Cui J, Kamiya A, Niimi Y, et al. Effects of activation and suppression of vasoconstrictor nerve activity in skin sympathetic nerve activity on core temperature change in humans. Auton Nerv Syst. 2001;38:641–4.
45. Iwase S, Okamoto T, Mano T, Kamiya A, Niimi Y, Qi F, et al. Skin sympathetic outflow in Buerger's disease. Auton Neurosci. 2001;87:286–92.
46. Sakakibara H, Iwase S, Mano T, Watanabe T, Kobayashi F, Furuta M, et al. Skin sympathetic activity in the tibial nerve triggered by vibration applied to the hand. Int Arch Occup Environ Health. 1990;62:455–8.
47. Hakusui S, Iwase S, Sugenoya J, Mano T. Microneurographic analysis of skin sympathetic nerve activity under cold environment in humans —with special reference to Raynaud's phenomenon—. Environ Med. 1989;33:103–6.
48. Ishida G, Nakashima K, Takahashi K. Skin nerve sympathetic activity reflex latency in Parkinson's disease. Acta Physiol Scand. 1990;81:121–4.
49. Ito H, Sugiyama Y, Mano T, Okada H, Matsukawa T, Iwase S. Skin sympathetic nerve activity and event-related potentials during auditory oddball paradigms. J Auton Nerv Syst. 1996;60:129–35.
50. Iwase S, Yamamoto K, Miwa C, Kamiya A, Niimi Y, Fu Q, et al. Skin sympathetic neuroeffector response is attenuated dose-dependently by systemic prostaglandin E1injection in humans. Neurosci Lett. 2000;292:191–4.
51. Kuwahara Y, Tsukahara R, Iwase S, Shimizu Y, Nishimura N, Sugenoya J, et al. Arousal electrical stimuli evoke sudomotor activity related to P300, and skin vasoconstrictor activity related to N140 in humans. Clin Neurophysiol. 2015;126:933–42.

Chapter 5
Muscle Sympathetic Nerve Activity and Syncope

Satoshi Iwase, Naoki Nishimura, and Tadaaki Mano

Abstract Syncope is defined as a transient loss of consciousness and postural tone, characterized by rapid onset, short duration, and spontaneous recovery; the process of syncope progression is here described with two types of sympathetic change. Simultaneous recording of microneurographically recorded muscle sympathetic nerve activity (MSNA) and continuous and noninvasive blood pressure measurement have disclosed what is going on during the course of syncope progression. For vasovagal or neurally mediated syncope, three stages are identified in the course of syncope onset, oscillation, imbalance, and catastrophe phases. Vasovagal syncope is characterized by sympathoexcitation, followed by vagal overcoming via the Bezold-Jarisch reflex. Orthostatic syncope is caused by response failure or a lack of sympathetic nerve activity to orthostatic challenge, followed by fluid shift and subsequent low cerebral perfusion. Four causes of the compensatory failure that trigger orthostatic syncope are considered: hypovolemia, increased pooling in the lower body, failure to activate sympathetic activity, and failure of vasoconstriction against sympathetic vasoconstrictive stimulation. Many pathophysiological conditions have been described from the perspectives of (1) exaggerated sympathoexcitation and (2) failure to activate the sympathetic nerve. We conclude that the sympathetic nervous system can control cardiovascular function, and its failure results in syncope; however, responses of the system obtained by microneurographically recorded MSNA would determine the pathophysiology of the onset and progression of syncope, explaining the treatment effect that could be achieved by the analysis of this mechanism.

Keywords Syncope • Vasovagal syncope • Orthostatic hypotension • Muscle sympathetic nerve activity • Neurally mediated syncope

S. Iwase, M.D., Ph.D. (✉) • N. Nishimura, Ph.D.
Department of Physiology, School of Medicine, Aichi Medical University, 1-1Yazako-Karimata, Nagakute, Aichi 480-1195, Japan
e-mail: s_iwase@nifty.com

T. Mano, M.D., Ph.D.
Gifu University of Medical Sciences, Gifu, Japan

S. Iwase et al. (eds.), *Clinical Assessment of the Autonomic Nervous System*,
DOI 10.1007/978-4-431-56012-8_5

5.1 Introduction

Syncope is defined as a transient loss of *consciousness* and postural tone, characterized by rapid onset, short duration, and spontaneous recovery, due to global cerebral *hypoperfusion* that most often results from *hypotension*. The accurate definition does not include types of unconsciousness not caused by cerebral hypoperfusion, for example, epileptic seizures, concussion, and cerebrovascular accidents, which are persistent structural impairments, and persistent states of unconsciousness, for example, coma and cerebrovascular diseases [26].

The pronunciation of "syncope" can differ from person to person; however [síŋkəpi] seems to be generally accepted, meaning "strike" or "cut-off" in ancient Greek.

The human autonomic nervous system (ANS) regulates systemic arterial pressure in order to maintain constant cerebral perfusion under conditions of uneven fluid shift caused by postural change. Syncope is the state of failure of this regulating function.

Many forms of syncope are preceded by a *prodromal* state that often includes *dizziness* and loss of vision, loss of hearing, loss of pain and feeling, *nausea* and abdominal discomfort, *weakness*, cold *sweating*, a feeling of hotness, *palpitations*, and other phenomena, which are often called "*presyncope*" [25].

In this chapter the process of syncope progression is described with two types of sympathetic change.

5.2 Autonomic Nervous System Relating to the Maintenance of Orthostasis

The autonomic nervous system (ANS) is the part of the peripheral nervous system that acts as a control system that functions largely below the level of consciousness to control visceral functions, including heart rate and contractility, digestion, respiratory rate, salivation, sweating, pupillary dilation, urination, sexual arousal, breathing, and swallowing, most of which are involuntary, while only the respiratory rate is subjected to the combination of voluntary and involuntary control [21].

ANS has two limbs: sympathetic and parasympathetic (sometimes including enteric). The general action of the sympathetic nervous system is to mobilize the body's nervous system fight-or-flight response, while the parasympathetic system is responsible for conserving energy. Via this system, animals feed, sleep, and rest. In other words, it facilitates "rest-and-digest" or "feed-and-breed" activities that occur when the body is at rest, especially after eating, including sexual arousal, salivation, lacrimation (the secretion of tears), urination, digestion, and defecation. The response can be summarized as "eat, sleep, and rest nerves."

The terminologies of the *sympathetic nervous system* and the *parasympathetic nervous system* are primarily anatomical; efferent autonomic nerves from the

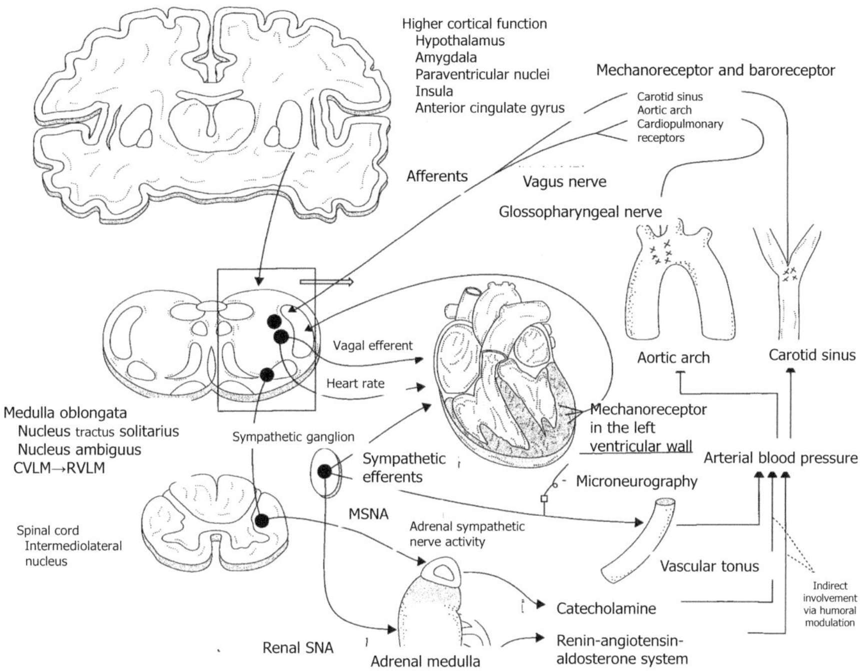

Fig. 5.1 Pathway of arterial baroreflex. The *square area* is illustrated in Fig. 5.2 in detail (See the text for the mechanism)

thoracic and lumbar spinal cord are named as *sympathetic*, and those from the brain stem and sacral region are named as *parasympathetic*.

5.3 Sympathetic Nervous System (Figs. 5.1 and 5.2)

The sympathetic nervous system is also called the thoracolumbar nervous system. The cardiovascular center is situated at the rostral ventrolateral medulla (RVLM), and the axons of the premotor neurons descend to the intermediolateral nucleus (ILM), where synapse to preganglionic neurons, which are myelinated B-fibers. The sympathetic preganglionic neurons exit the spinal cord as the white rami from the first thoracic nerve (T_1) to the second/third lumbar nerve ($L_{2/3}$) and enter the sympathetic trunk or chain, which is also known as the paravertebral ganglion. The sympathetic trunk forms the superior, middle, and inferior cervical ganglia, while the inferior ganglion often fuses with the first thoracic ganglion to form the stellate ganglion (in 80 % of cases) and travels downward to the fifth sacral ganglion, with which the stellate ganglion, the cardiac sympathetic innervation of T_1–T_3, and the

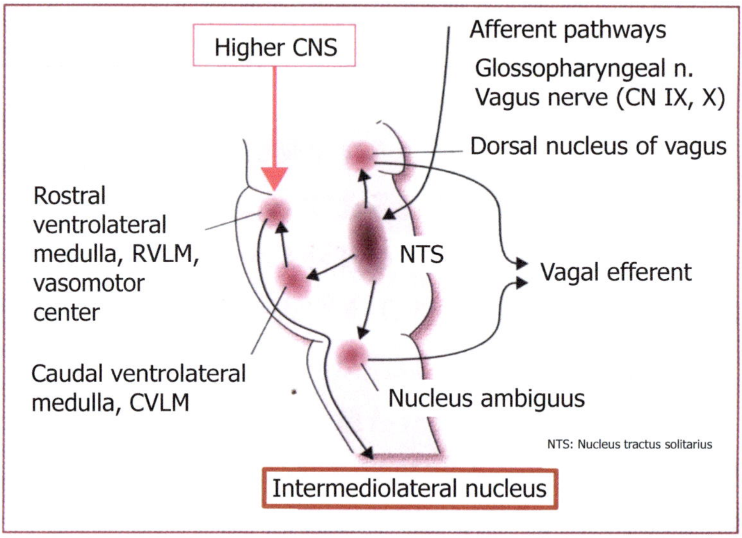

Fig. 5.2 Pathway of arterial baroreflex in the brain stem (See the text for the mechanism and abbreviations)

systemic vasoconstrictive innervation to the arterioles of the resistant vessels are directly involved [11, 22, 24].

5.4 Parasympathetic Nervous System (Figs. 5.1 and 5.2)

The parasympathetic nervous system is also called the craniosacral nervous system, which exits from the brain stem as the oculomotor (CN III), facial (CN VII), and glossopharyngeal (CN IX) nerves, innervating the ciliary, pterygopalatine, and otic ganglia, respectively, as well as the vagus nerve. The vagus nerve shares 85 % of the entire parasympathetic nervous system, and its innervation is distributed to the dura mater, auricle, upper pharynx, lung, heart, liver, kidney, stomach, small intestine, and transverse colon down to the splenic flexure. The parasympathetic nerve fibers exiting from the S_2 to S_4 innervate defecation and urogenital organs including the descending colon, rectum, kidney, bladder, prostate, and genital organs. Among these parasympathetic innervations, direct blood pressure control was shown to involve the cardiac branch of the vagus [11, 22, 24].

5.5 Autonomic Control of the Cardiovascular System

Baroreceptors are situated at the carotid sinus and the aortic arch; they monitor the arterial pressure and transmit information on this pressure to the central nervous system. Upon extension of the arterial wall by raising of the blood pressure, the baroreceptors (stretch receptors) generate an impulse depending on the arterial pressure via the glossopharyngeal nerve (CN IX) from the carotid sinus and via the vagus nerve (CN X) from the aortic arch and transmit the signals to the nucleus tractus solitarius (NTS) in the medulla [13–16].

From the NTS, excitatory information is sent via the glutamatergic neurons to the caudal ventrolateral medulla (CVLM), where the inversion of positive and negative signs is executed. The CVLM then transmits GABAergic (using γ-amino butyric acid as a neurotransmitter) neurons to the rostral ventrolateral medulla (RVLM). The topographical regional differentiation has been established [4].

The RVLM neurons are considered to be premotor neurons for the preganglionic efferent neurons whose cell bodies are situated in the intermediolateral nucleus (IML), with synapses to the preganglionic B-fiber neurons, and which exit from the spinal cord as white rami.

In contrast, input from NTS to the nucleus ambiguus and the dorsal nucleus of the vagus modulates the activity of the vagal nerve, innervating the heart and presphincters of the resistant vessels and causing decreases in heart rate and cardiac contractility.

As for the sympathetic efferent pathway, the cardiac sympathetic activity and the activity of muscle sympathetic nerves innervating skeletal muscles play a role in blood pressure regulation. Muscle sympathetic nerve activity (MSNA) can be recorded microneurographically from human peripheral nerves in situ, which is the only baromodulatory sympathetic nerve activity directly recordable in humans (Fig. 5.1). The techniques to record MSNA are described elsewhere [20].

These pathways provide a negative feedback mechanism to suppress the sympathetic nerve activity, while the vagus nerve activity is activated when blood pressure rises, and facilitate the sympathetic nerve activity and suppress the vagus nerve activity when blood pressure drops.

5.6 The Autonomic Nervous System and Syncope

How is the autonomic nervous system involved in the onset of syncope? It is primarily responsible for sympathetic responses toward environmental challenges. The most frequent challenge triggering syncope might be a postural change (Fig. 5.3). Syncope or fainting associated with orthostatic challenge includes vasovagal syncope (or in a wider sense, neurally mediated syncope) and orthostatic syncope (or orthostatic intolerance). The symptoms of these two situations seem to

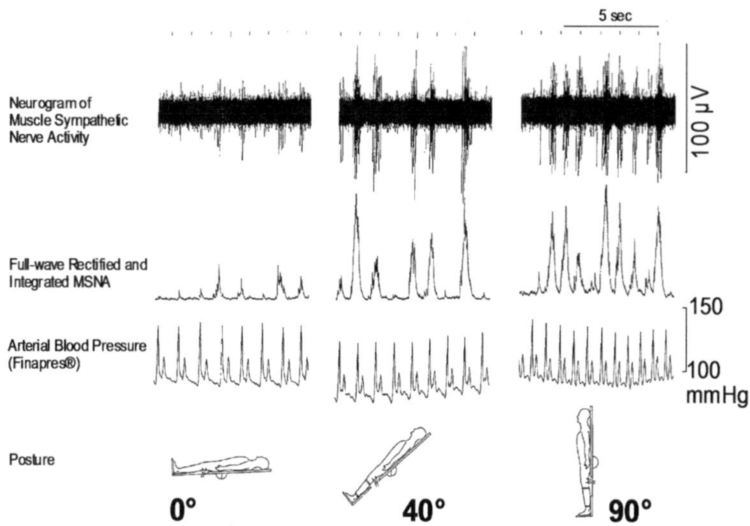

Fig. 5.3 Responses of microneurographically recorded muscle sympathetic nerve activity during orthostatic challenge. MSNA is sparsely discharged in a supine position, but is remarkably enhanced as the tilt angle becomes large

be the same, but the pathophysiology is different; they involve opposite directions of pathways of the autonomic nervous system.

In the following sections, syncope from the perspective of the autonomic nervous system is described, based on the premise that no structural or conduction disorder is present in the heart.

5.7 Vasovagal Syncope (or in a Wider Sense, Neurally Mediated Syncope)

The concept of vasovagal syncope was first proposed by Lewis in 1932, with primary symptoms of hypotension and bradycardia. Vasovagal syncope is triggered by an exaggerated sympathetic response to a situation; in other words, sympathoexcitation precedes the onset.

The Bezold-Jarisch reflex plays an important role in the onset of this type of syncope. This reflex is generated by the cerebral hypoperfusion due to vagal-activation-mediated sympathosuppression for the protection of the myocardium. This vagal activation may be caused by hyperactivity of the left ventricular wall, which activates the stretch receptors in the wall and trabeculae, and in turn the C-fiber afferents to NTS, which triggers bradycardia and decreased myocardial contractility [27].

Several factors are known to exacerbate and accelerate vasovagal syncope, including (1) fatigue, dehydration, and hypovolemia due to hemorrhage, followed

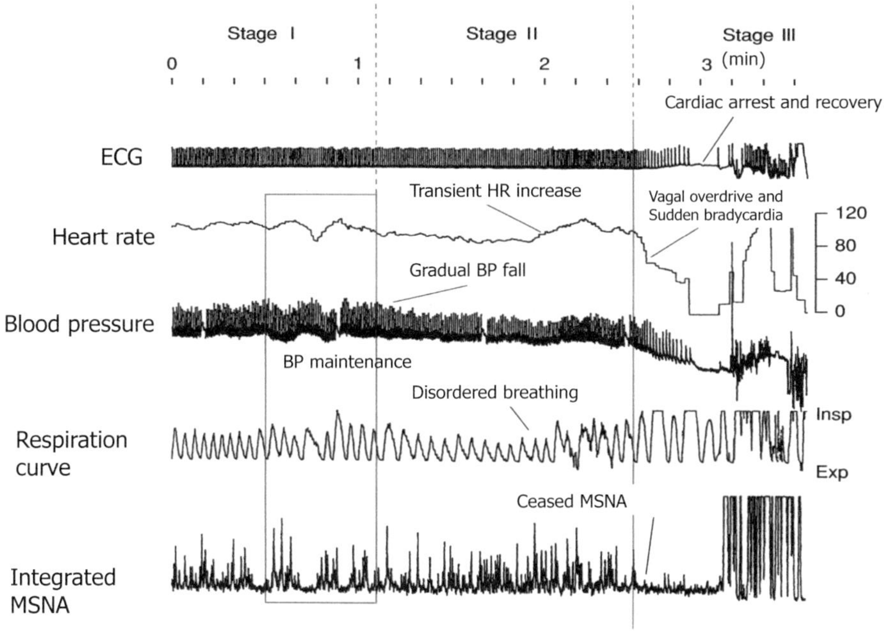

Fig. 5.4 Polygraph of vasovagal syncope. Over the course of syncope progression in three stages, syncope onset develops at stage III

by reduced venous return; (2) blood shift and pooling in the lower body; (3) hypersensitivity of the stretch receptors in the left ventricular wall; and (4) fear, emotional stress, and reaction to pain [6]. Among these, (3) is the trigger that is considered to be essential for the Bezold-Jarisch reflex.

An analysis of the polygraph of vasovagal syncope indicated that there are three phases in the onset of vasovagal syncope. The changes during 60° head-up tilt and polygraph trace (Fig. 5.4) and flow charts are described below (Fig. 5.5).

5.7.1 Stage I: Oscillation Phase [14, 17, 28, 29]

In stage I, the sympathetic nerve activity (MSNA) and blood pressure display an oscillation pattern with the levels being well maintained. The time lag from MSNA discharge to blood pressure rise by the constriction of arterioles is approximately 4–5 s, while the time lag from the blood pressure rise to vasodilatation by the suppression of sympathetic nerve activity is approximately 1 s from the descending pathway in the medulla oblongata to the peripheral vasodilating sphincter muscles via the synapse in the sympathetic chain to the postganglionic C-fiber, the time delay from the baroreceptor to the recording site of microneurography. The sympathetic discharge induces a blood pressure rise. This double feedback mechanism

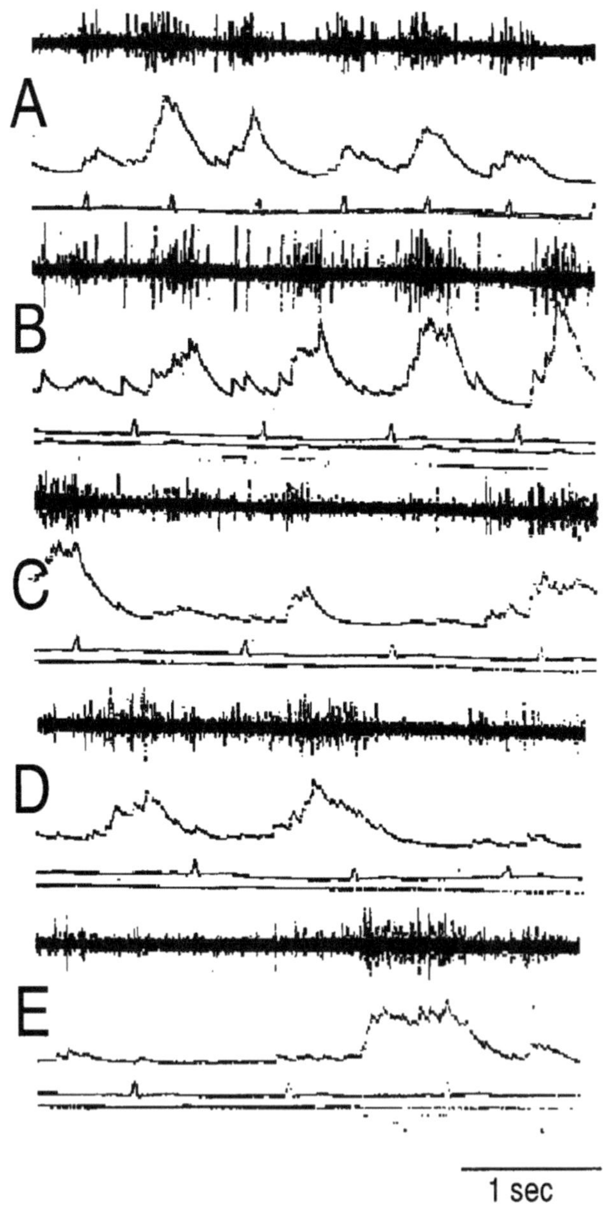

Fig. 5.5 Progression of vasovagal syncope from stage I to stage III

oscillates to maintain the blood pressure. In stage I, blood pressure wave and heart rate oscillate with a frequency of 0.1 Hz (10 s period), and the low frequency component (LF) of heart rate variability is enhanced. This phase includes mutual increases in both sympathetic and vagal activities.

5.7.2 *Stage II: Imbalance Phase [14, 15, 17, 19, 28]*

At this stage, venous return is reduced due to the failure to compensate for the enhanced venous pooling, which makes blood pressure maintenance difficult. The tidal volume is raised to increase the venous return. Failure of the double feedback mechanism arises with the enhanced power of the high-frequency (HF) component of the blood pressure wave and heart rate variability. The falling blood pressure due to decreased stroke volume cannot be compensated for by only peripheral vasoconstriction, inducing an extremely elevated heart rate. This tachycardia compensates for the reduced stroke volume, which could not be compensated for by decreased venous return or circulatory blood volume. In this state, cardiac echography sometimes reveals parallel movement of the interventricular septum and the left ventricular wall, which is called "paradoxical movement." It is estimated that this state involves the sympathetic nerve activity being overwhelmed by the parasympathetic nerve activity, and the reciprocal relationship between the diastolic blood pressure and MSNA, denoting the baroreflex function, disappears late in this phase.

5.7.3 *Stage III: Catastrophe Phase [14, 15, 17]*

In spite of the effort to maintain the blood pressure by an increased heart rate, the stage transitions to stage III, when a sudden heart rate drop with the cessation of MSNA is observed, resulting in an immediate fall in blood pressure. Some patients experience cardiac arrest simultaneously with cold sweat, nausea, blackout, and loss of consciousness. Dominant activation of parasympathetic nerve activity occurs concomitantly with sympathosuppression. In "neurally mediated syncope," this stage corresponds to the classification of cardiosuppressive, vasosuppresive, and mixed types, which are based on whether the suppressive drive is dominant over cardiac or vasomotor sympathetic drives or both.

Therefore, sympathoactivation is observed, followed by sudden tachycardia, which triggers the onset of vasovagal syncope or neurally mediated syncope.

5.8 Burst Property Change During the Progression of Syncope [14]

As the hypotensive attack began, the subject complained of nausea and exhibited pallor with sweating, and there was a disappearance of pulse synchrony of MSNA and an elongated burst duration (Fig. 5.6). After complete recovery, the electrode was reinserted to obtain an MSNA burst at a satisfactory level.

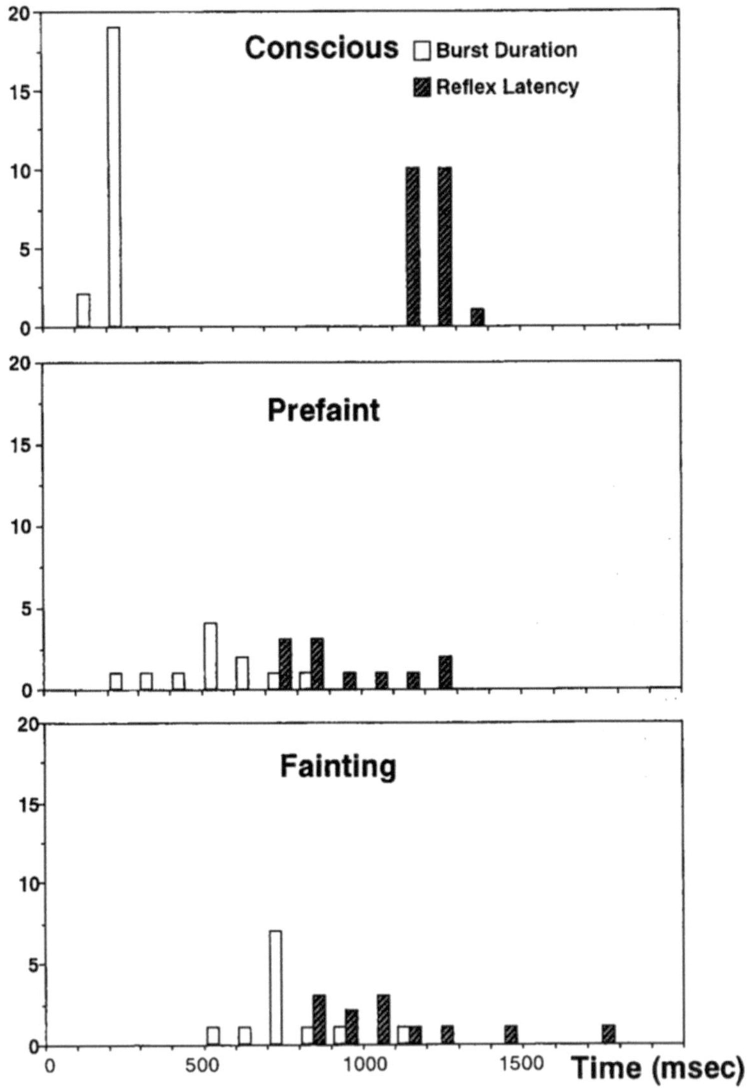

Fig. 5.6 Changes in MSNA in fainting: (**a**) when the subject is conscious, (**b**) when the subject complains of a fainting sensation, (**c**) when the subject complains of nausea and becomes pale with sweating, (**d**) when the subject loses consciousness, and (**e**) when the subject is laid down after the reinsertion. From (**a** to **e**), *upper trace* is original MSNA, *middle trace* is full-wave rectified and integrated MSNA, and *lower trace* is electrocardiogram. (**b**, **c**) Are prefaint states, and (**d**, **e**) are fainting states. In (**b**) burst duration became elongated, although pulse synchrony is still maintained. In (**c**–**e**), pulse synchrony disappears and burst duration becomes elongated more and more as the fainting proceeds

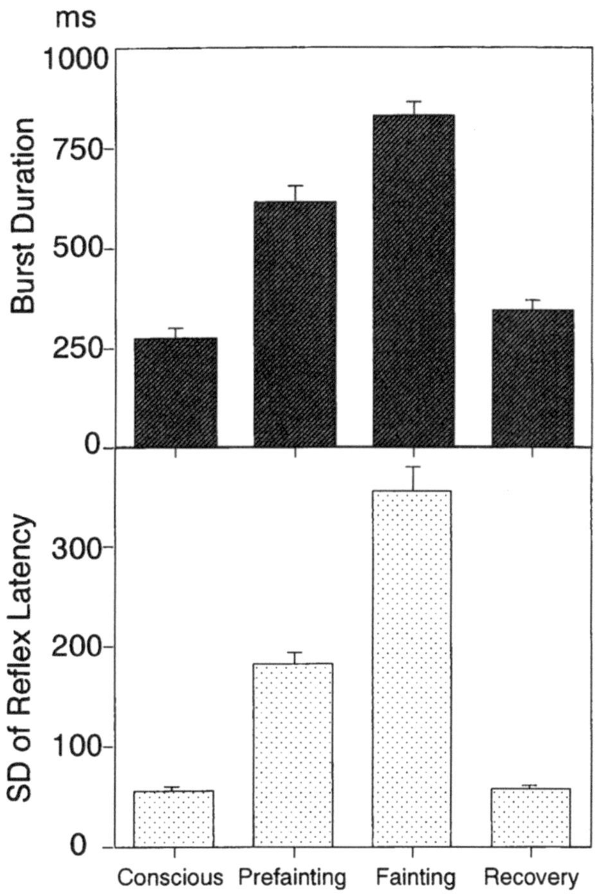

Fig. 5.7 Histogram of burst duration and reflex latency in fainting. Conscious, prefaint, and fainting correspond to (**a**–**e**), respectively. The reflex latency becomes more scattered, and burst duration becomes elongated as the fainting proceeds

The burst duration of a typical subject in a conscious state (stage I) was 246 ± 32 ms, whereas it was elongated to 556 ± 157 ms in prefainting (stage II) and further elongated in the fainting state (stage III) to 771 ± 123 ms (Fig. 5.7). After complete recovery of consciousness, it recovered to a fairly constant value of 303 ± 51 ms. The averaged burst duration in the six subjects was significantly reduced by the fainting event from 279 ± 52 ms in a conscious state to 619 ± 92 ms in prefainting, 834 ± 78 ms in fainting, and recovered to 349 ± 59 ms (Fig. 5.8, $F = 75.793$, $p < 0.0001$). The reflex latency in a conscious state in a typical subject was $1{,}062 \pm 41$ ms. When the subject complained of a fainting sensation and nausea (presyncopal state), it became shortened but scattered to 740 ± 139 ms, becoming further shortened and scattered to 723 ± 289 ms in the fainting state (Fig. 5.7). The standard deviations of reflex latencies in the six subjects were significantly increased by the fainting event from 57 ± 12 to 182 ± 33 ms in prefainting, 357 ± 58 ms in fainting, and recovered to 59 ± 8 ms (Fig. 5.8, $F = 102.329$, $p < 0.0001$).

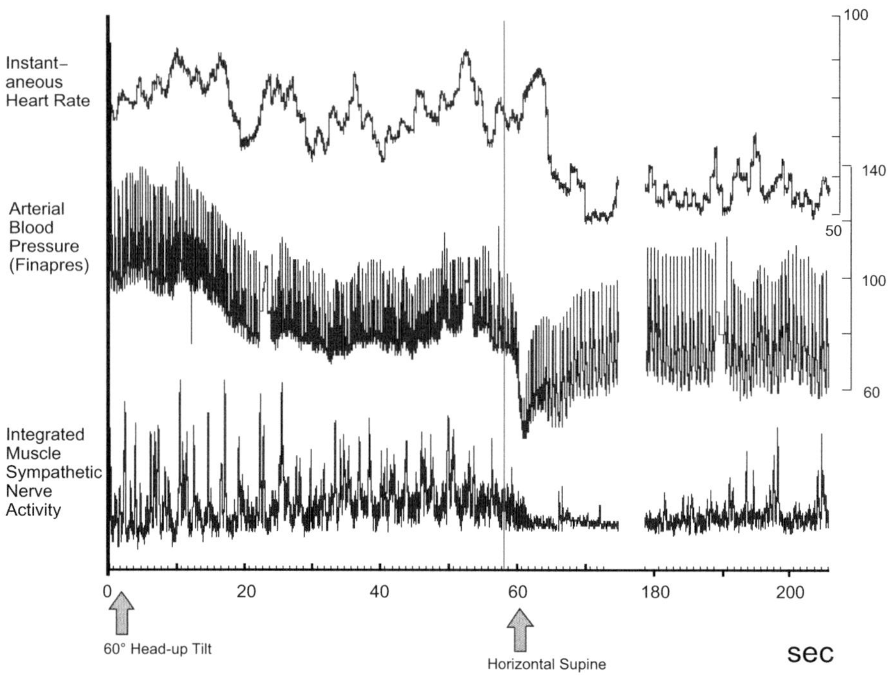

Fig. 5.8 Changes in burst duration (*upper plate*) and standard deviation of reflex latency (*lower plate*). As the fainting attack develops progressively, burst duration is more elongated and reflex latency is more scattered

During the syncopal attack, we called the subject's name. By this arousal stimulus, the subject discharged an MSNA burst during prefainting and fainting states with elongated burst duration ~1 s after calling. This phenomenon disappeared during name calling in the recovered state.

During fainting reaction, the MSNA-specific burst properties were altered into those of SSNA, which are as follows: (1) a rhythmic pulse synchronous pattern was disorganized to scattered reflex latency, (2) burst duration of 150–300 ms was elongated to 400–700 ms, and (3) no response to arousal stimuli in a conscious state became responsive with a latency of ~1 s.

The R-wave locked burst reflex latencies were scattered during the presyncopal period, and more scattered in the course of syncopal attack, findings that suggest the disappearance of pulse synchrony of the MSNA burst. This means the diminished input of the inhibitory rhythmic and pulse synchronous properties to the RVLM. Prolongation of MSNA burst duration and responsive properties of MSNA to arousal stimuli might be attributed to deafferentation from the baroreceptors [5] to NTS. These alterations of MSNA-specific properties indicate that diminished inhibitory input from the baroreceptors is suppressed in the course of fainting. In spite of this diminished inhibitory input to the CVLM, the central sympathetic outflow was incapable of enhancing MSNA that was inadequate to compensate for the blood pressure drop. This was then followed by progressive MSNA inhibition until its complete disappearance and syncope [34]. Similar phenomena were observed in MSNA during glossopharyngeal and vagal anesthesia by lidocaine [5] and during non-REM sleep [30].

The deafferentation from the baroreceptors during fainting might be attributable to suppression of the central sympathetic outflow proximal to the RVLM [34]. The abrupt sympathetic silence is proposed to be due to initial exaggerated sympathetic activation, in combination with relative ventricular hypovolemia and stimulated myocardial ventricular afferents [23, 32]. In tilt-induced fainting, MSNA bursts are gradually but not immediately diminished during fainting. It is probable that the inhibitory vagal suppression on the sympathetic volley is not sufficiently strong to suppress the MSNA bursts completely, but is rather mild and altered only the specific features of MSNA, which are the elongation of burst duration, disappearance of pulse synchrony, and burst elicitation by arousal stimuli. Greater suppression of the central sympathetic outflow suppressed the burst discharge and then it progressed to a hypotensive state with a cardiac standstill [33]. These findings indicate that this sympathetic silence was evoked by overactivation of sympathetic activity.

5.9 Orthostatic Hypotension or Orthostatic Intolerance

The orthostatic hypotension or orthostatic syncope or orthostatic intolerance associated with postural change is caused by response failure or a lack of sympathetic nerve activity to an orthostatic challenge. A change in posture from a supine to an upright position in humans induces a fluid shift of 300–800 mL from the thoracic cavity to the lower body, including the abdominal cavity or legs, resulting in a reduction in venous return. This causes a stroke volume decrease, which in turn activates the sympathetic nervous system via the arterial baroreflex and cardiopulmonary reflex. This sympathoactivation induces increases in heart rate, cardiac contractility, and peripheral resistance, while arterial blood pressure is maintained and cerebral blood flow is maintained constant.

In the case of activation failure, hypotension develops just after postural change, sometimes progressing to fainting or syncope. The essence of orthostatic hypotension is failure of the compensation of the fluid shift that follows postural change. The largest difference from the neurally mediated syncope or vasovagal syncope is the latency of syncope onset after orthostatic challenge. Neurally mediated syncope or vasovagal syncope is usually induced when individuals maintain an upright position for a comparatively long duration, possibly >10 min, whereas orthostatic hypotension develops in a short period. In tilt studies on 66 subjects, the onset latency was reported to be >1 min for 58 and >2 min for the remaining eight individuals [8]. One more large difference from neurally mediated or vasovagal syncope is that there are no prodromal symptoms before the onset, namely, vagal symptoms, including cold sweating or nausea. Therefore, syncope without vagal symptoms usually progresses with onset 2 min after a change of posture.

There are four causes of compensatory failure: (1) hypovolemia, (2) increased pooling in the lower body, (3) failure to activate the sympathetic activity, and (4) failure of vasoconstriction upon sympathetic vasoconstrictive stimulation.

5.9.1 Hypovolemia

Hypovolemia is the most prevalent cause of orthostatic hypotension, having associations with a hot environment, dehydration, hemorrhage, diarrhea, Addison's disease, and the administration of diuretics. Hypovolemia results in a fluid shift to the lower body, causing insufficient venous return, stroke volume, and cardiac output, which lowers cerebral perfusion and induces syncope. A large fluid shift increases the heart rate, but usually insufficiently. If the increase is exaggerated, this might be a cause of neurally mediated or vasovagal syncope. In this case, the blood pressure might be maintained for several minutes, while gradual blood pressure decreases might be a cause. Here, the difference between neurally mediated or vasovagal syncope and orthostatic hypotension might be reduced responsiveness of the sympathetic nerve activity to fluid shift.

5.9.2 Increased Pooling in the Lower Body

This occurs due to (3) or (4) below, or due to the increased pooling in the lower body in pathological conditions including vascular disease that reduces the elasticity of the vessels, as seen in varices or pregnancy. The mechanism of development is the same as in (1).

5.9.3 Failure to Activate the Sympathetic Activity

Any disorder that impairs the baroreflex function is included in this category, including carotid deafferentation, carotid sinus neuropathy, primary autonomic failure (Shy-Drager type of multiple system atrophy, MSA-P), pure autonomic failure, Parkinson's disease with autonomic failure, and acute pandysautonomia. In addition to the primary autonomic failures, secondary ones include diabetic neuropathy, amyloid neuropathy, chronic renal failure, alcoholic neuropathy, spinal cord injury, brain tumors, and postoperative symptoms of lumbar sympathectomy for the treatment of Buerger's disease or Raynaud's syndrome all have an impaired efferent sympathetic pathway, which induces marked pooling of the body fluid. One more state that causes orthostatic hypotension is depression. Structural brain abnormalities are associated with depression in the elderly, suggesting a link between vascular diseases and white matter atrophy. Gordon et al. [9] reported the association between the degree of orthostatic hypotension and white matter hyperintensity volume in a depression group, indicating that systolic blood pressure drops may be a factor contributing to white matter lesions in depression in the elderly.

5.9.4 *Attenuation of Vasoconstrictor Response to Sympathetic Stimulation*

Resistant vessels contract to enhance peripheral resistance, resulting in a rise in arterial blood pressure, while lack or attenuation of the vasoconstrictor response to sympathetic stimulation might be the cause of orthostatic hypotension. This attenuated response is primarily caused by the administration of vasodilative agents including α-adrenoreceptor blockers (e.g., prazosin), calcium antagonists (nifedipine), prostaglandin E_1 (limaprost), or nitroglycerin. We previously reported that although the microneurographically recorded sympathetic nerve activity is not altered, the effectiveness of the sympathetic activity on blood flow reduction is attenuated [13]. Orthostatic intolerance that is sometimes observed after long exposure to microgravity or a long period of bed rest is partly caused by this attenuated vasoconstrictive response to sympathoexcitation. In this case, the heart rate is not usually reduced or is in fact elevated by blood pressure reduction although MSNA gradually ceases (Fig. 5.9).

5.10 Sympathetic Nerve Activity Recording During Syncope

Several examinations have been proposed to investigate the pathophysiology of syncope, among which we propose recording the muscle sympathetic nerve activity from the tibial nerve microneurographically simultaneously with ECG, noninvasive continuous blood pressure wave (Finapres, Finomonitor, Jentow, etc.), and fluid shift measurements by the bioimpedance method. Several investigations related to syncope with polygraphic recording have been reported by the recording of microneurographically recorded MSNA, among which Kamiya et al. reported the disappearance of 10 s oscillation in MSNA and blood pressure in tilt-induced syncope [19], Ichinose et al. reported gradual blood pressure drop with the disappearance of 10 s oscillation in LBNP (lower body negative pressure)-induced syncope [12], and Cooke et al. reported oscillation disappearance in LBNP-induced syncope [3]. Vagal suppression seemed to be classified into cardiosuppressive and vasosuppressive types, as well as a mixed type [1, 2]. These two types were also

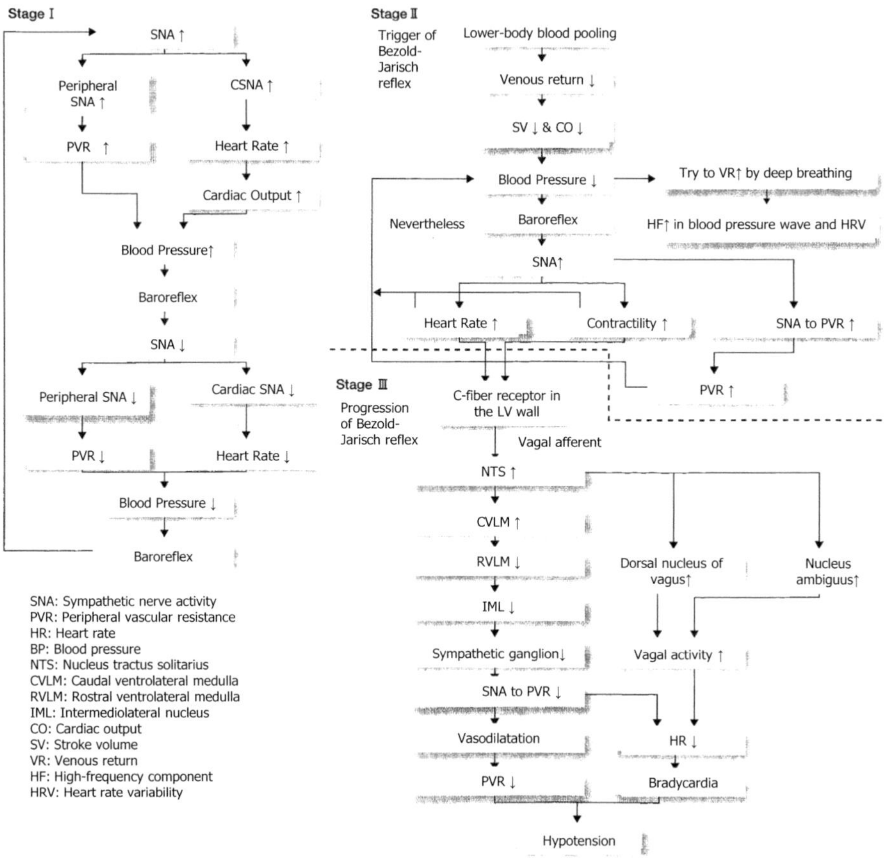

Fig. 5.9 Progression of orthostatic hypotension after 3 days of simulated microgravity exposure and dry immersion. MSNA was gradually decreased, and blood pressure drop was also observed progressively. At 60s after the tilt, gradually blood pressure fell, and MSNA was suppressed, while heart rate was elevated

evidenced by the MSNA recording by Vaddadi et al. [31] and Fu et al. [7], which showed suppression of MSNA, sudden bradycardia, or both. These reports clarified the pathophysiology of the progress of syncope by recording microneurographic MSNA and continuous blood pressure. In other words, in the cardiosuppressive type, sudden bradycardia occurs without MSNA disappearance, whereas MSNA ceases in the vasosuppressive type. These two types of vasovagal syncope should be registered.

In contrast to the activation-suppression process of MSNA in vasovagal syncope, the inability to follow the sympathetic nerve activity to hypotension is considered to cause orthostatic intolerance. Clarification of the pathophysiology is difficult because MSNA is seldom recorded from patients with such neurological impairments who are likely to have orthostatic hypotension. We have recorded

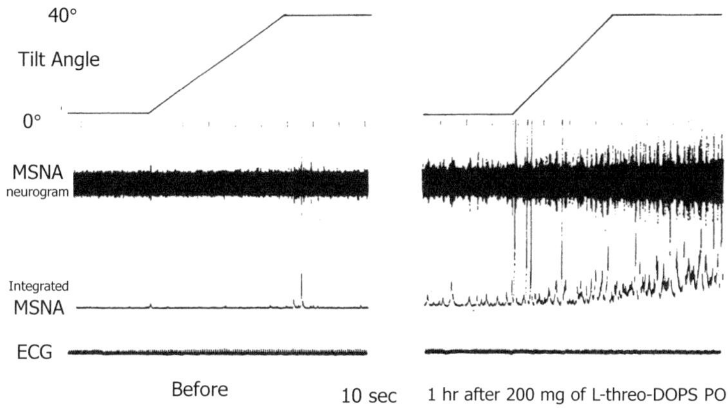

Fig. 5.10 Changes in MSNA discharges before and after administration of L-threo-3, 4-dihydroxyphenylserine (L-DOPS). Before the administration, there were few MSNA discharges, and enhancement due to tilting of the bed at 40° was small (*left plate*). The discharges became apparent 1 h after L-DOPS administration and were further enhanced by 40° head-up tilting (*right plate*)

Shy-Drager syndrome (multiple system atrophy, autonomic failure type) and reported that although MSNA appeared not to be discharged from the tibial nerve, administration of L-*threo*-DOPS (L-*threo*-3,4-dihydroxyphenylserine), a precursor of noradrenaline, ameliorated the orthostatic intolerance and induced MSNA discharge from the tibial nerve [18] (Fig. 5.10) in that the appearance of MSNA improved the symptom by the use of a nose drop of vasopressin [10].

These changes recorded by polygraphy during the fluid shift maneuver provided measures for diagnosing orthostatic hypotension or vasovagal syncope (Fig. 5.11). By the response of the sympathetic nervous system, we can differentially describe the underlying pathophysiology of orthostatic hypotension and vasovagal (or neutrally mediated) syncope.

5.11 Conclusion

The sympathetic nervous system can control cardiovascular function, and its failure results in syncope; however, responses of the system by microneurographically recorded MSNA would determine the pathophysiology of the onset and progression of syncope. An effective treatment could be achieved by the analysis of this mechanism.

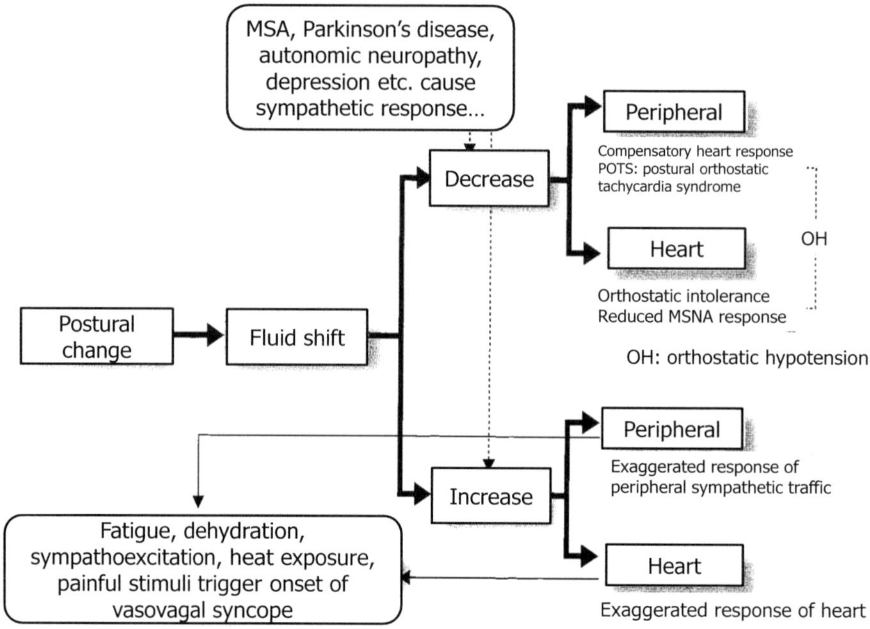

Fig. 5.11 Responses of sympathetic nerve activity at the onset of syncope

References

1. Brignole M, Alboni P, Benditt D, Bergfeldt L, Blanc JJ, Bloch Thomsen PE, Task Force on Syncope, European Society of Cardiology, et al. Guidelines on management (diagnosis and treatment) of syncope. Eur Heart J. 2001;22(15):1256–306.
2. Brignole M, Alboni P, Benditt DG, Bergfeldt L, Blanc JJ, Thomsen PE, Task Force on Syncope, European Society of Cardiology, et al. Guidelines on management (diagnosis and treatment) of syncope-update 2004. Executive summary. Eur Heart J. 2004;25(22):2054–72.
3. Cooke WH, Rickards CA, Ryan KL, Kuusela TA, Convertino VA. Muscle sympathetic nerve activity during intense lower body negative pressure to presyncope in humans. J Physiol. 2009;587(Pt 20):4987–99. doi:10.1113/jphysiol.2009.177352. Epub 2009 Aug 24.
4. Dampney R. Medullary pathways regulating sympathetic outflow: the need for more lateral thinking. Am J Physiol Regul Integr Physiol. 2004;286:R446–8.
5. Fagius J, Wallin BG, Sundröf G, Nerhed C, Englesson S. Sympathetic outflow in man after anaesthesia of the glossopharyngeal and vagus nerves. Brain. 1985;108:423–38.
6. Fenton AM, Hammill SC, Rea RF, Low PA, Shen WK. Vasovagal syncope. Ann Intern Med. 2000;133:714–25.
7. Fu Q, Verheyden B, Wieling W, Levine BD. Cardiac output and sympathetic vasoconstrictor responses during upright tilt to presyncope in healthy humans. J Physiol. 2012;590 (Pt 8):1839–48. doi:10.1113/jphysiol.2011.224998. Epub 2012 Feb 13.
8. Gehrking JA, Hines SM, Benrud-Larson LM, Opher-Gehrking TL, Low PA. What is the minimum duration of head-up tilt necessary to detect orthostatic hypotension? Clin Auton Res. 2005;15:71–5.
9. Gordon J, Ditto B, Lavoie K, Pelletier R, Campbell T, Arsenault A, et al. The effect of major depression on postexercise cardiovascular recovery. Psychophysiology. 2011;48(11):1605–10. doi:10.1111/j.1469-8986.2011.01232.

10. Hakusui S, Sugiyama Y, Iwase S, Hasegawa Y, Koike Y, Mano T, et al. Postprandial hypotension: microneurographic analysis and treatment with vasopressin. Neurology. 1991;41(5):712–5.
11. Hasegawa Y, Koike Y. Postpradial hypotension, supervised by Takahashi A, Nanzando, Tokyo. 2004. p. 1–264.
12. Ichinose M, Saito M, Fujii N, Kondo N, Nishiyasu T. Modulation of the control of muscle sympathetic nerve activity during severe orthostatic stress. J Physiol. 2006;576(Pt 3):947–58.
13. Iwase S, Yamamoto K, Miwa C, Kamiya A, Niimi Y, Fu Q, et al. Skin sympathetic neuroeffector response is attenuated dose-dependently by systemic prostaglandin E1 injection in humans. Neurosci Lett. 2000;292:191–4.
14. Iwase S, Mano T, Kamiya A, Niimi Y, Fu Q, Suzumura A. Syncopal attack alters the burst properties of muscle sympathetic nerve activity in humans. Auton Neurosci. 2002;95:141–5.
15. Iwase S. Pathophysiology of vasovagal syncope. Heart View. 2002;6:1142–9.
16. Iwase S. Ch. II-1 structure and function. In: Goto Y, Hongo M, editors. Basic and clinical autonomic nervous system. Osaka: Iyaku J; 2006. p. 1–340.
17. Iwase S. Ch. 2-2-1 vasovagal syncope. In: Nohara R, editor. Mastering syncope. Tokyo: Medical View; 2009. p. 1–256.
18. Kachi T, Iwase S, Mano T, Saito M, Kunimoto M, Sobue I. Effect of L-threo-3,4-dihydroxyphenylserine on muscle sympathetic nerve activities in Shy-Drager syndrome. Neurology. 1988;38:1091–4.
19. Kamiya A, Hayano J, Kawada T, Michikami D, Yamamoto K, Ariumi H, et al. Low-frequency oscillation of sympathetic nerve activity decreases during development of tilt-induced syncope preceding sympathetic withdrawal and bradycardia. Am J Physiol Heart Circ Physiol. 2005;289:H1758–69. Epub 2005 Jun 3.
20. Mano T, Iwase S, Toma S. Microneurography as a tool in clinical neurophysiology to investigate peripheral neural traffic in humans. Clin Neurophysiol. 2006;117:2357–84. Epub 2006 Aug 10.
21. Mathias CJ, Bannister SR. Autonomic failure "A Textbook of Clinical Disorders of the Autonomic Nervous system". 4th ed. Oxford and New York: Oxford University Press; 2002.
22. Netter FH. The Ciba collection of medical illustrations vol. 1, nervous system part I, anatomy and physiology 2. West Cadwell: Ciba Pharmaceutical Company; 1983.
23. Öberg B, Thorén P. Increased activity in left ventricular receptors during hemorrhage or occlusion of caval veins in the cat —a possible cause of the vasovagal reaction. Acta Physiol Scand. 1972;85:164–73.
24. Patten J. Neurological differential diagnosis. 2nd ed. London: Springer-Verlag; 1996.
25. Reeves AG, Swenson RS. Chapter 14: Evaluation of the dizzy patient. Disorders of the nervous system: a primer. Dartmouth Medical School. Retrieved 2012-01-06.
26. Ross RT. Syncope. London: WB Saunders; 1988.
27. Salo LM, Woods RL, Anderson CR, McAllen RM. Nonuniformity in the von Bezold-Jarisch reflex. Am J Physiol Regul Integr Comp Physiol. 2007;293:R714–20.
28. Schwartz CE, Lambert E, Medow MS, Stewart JM. Disruption of phase synchronization between blood pressure and muscle sympathetic nerve activity in postural vasovagal syncope. Am J Physiol Heart Circ Physiol. 2013;305(8):H1238–45. doi:10.1152/ajpheart.00415.2013. Epub 2013 Aug 9.
29. Sugiyama Y, Matsukawa T, Shamsuzzaman AS, Okada H, Watanabe T, Mano T. Delayed and diminished pressor response to muscle sympathetic nerve activity in the elderly. J Appl Physiol. 1996;80(3):869–75.
30. Takeuchi S, Iwase S, Mano T, Okada H, Sugiyama Y, Watanabe T. Sleep-related changes in human muscle and skin sympathetic nerve activities. J Auton Nerv Syst. 1994;47:121–9.
31. Vaddadi G, Esler MD, Dawood T, Lambert E. Persistence of muscle sympathetic nerve activity during vasovagal syncope. Eur Heart J. 2010;31:2027–33. doi:10.1093/eurheartj/ehq071. Epub 2010 Mar 19.

32. Walker JL, Thames MD, Abboud FM, Mark AL, Kopfenstein HS. Preferential distribution of inhibitory cardiac receptors in left ventricle of the dog. Am J Physiol. 1978;235:H188–92.
33. Wallin BG, Sundlöf G. Sympathetic outflow to muscles during vasovagal syncope. J Auton Nerv Syst. 1982;6:287–91.
34. Ziegler MG, Ehon C, Wilner KD, Specho P, Lake CR, McCutchen JA. Sympathetic nervous withdrawal in the vasodepressor (vasovagal) reaction. J Auton Nerv Syst. 1986;17:273–8.

Chapter 6
Single Fiber Analysis of Muscle Sympathetic Nerve Activity

Hisayoshi Murai, Shuichi Kaneko, and Masayuki Takamura

Abstract Direct recording of multiunit efferent muscle sympathetic nerve activity (MSNA) by microneurography is the best method for quantifying sympathetic nerve activity in humans. It has been still recognized as a gold standard method for evaluation of sympathetic nerve activity in human. Recently, single-unit MSNA analysis was developed in humans. Single-unit MSNA reveals (1) firing frequency of single-unit MSNA, (2) multiple firing of single-unit MSNA within one cardiac interval, and (3) functionally different neuron activities, which could not be obtained by multiunit MSNA analysis. Single-unit MSNA provides additional information regarding actual sympathetic neuron firing to peripheral. Several studies have already demonstrated that single-unit MSNA analysis shed insight into the mechanism of actual central sympathetic firing to peripheral in several cardiovascular diseases. In this chapter, we describe the differences between analysis of multiunit and single-unit MSNA and discuss the advantages of single-unit MSNA recording.

Keywords Arrhythmia • Muscle sympathetic nerve activity • Physiological stress

6.1 Introduction

Although it has been 40 years since Vallbo and Wallin [1, 2] developed microneurography to record multiunit efferent muscle sympathetic nerve activity (MSNA) directly, it is still considered as the best method for quantifying sympathetic nerve activity in healthy human subjects and in those with diseases associated with cardiovascular risk (see Chaps. 1, 2, 3, 4, and 5). Recently, analysis of single-unit MSNA was refined in humans by Macefield et al. [3]. Although the method of recording single-unit MSNA is similar to that of multiunit MSNA, obtaining single-unit MSNA spike is technically demanding. Single-unit MSNA reveals (1) firing frequency of single-unit MSNA, (2) multiple firing of single-unit MSNA within one

H. Murai (✉) • S. Kaneko • M. Takamura
Disease Control and Homeostasis, Graduate School of Medical Science, Kanazawa University, 13-1 Takara-machi, Kanazawa, Ishikawa 920-8641, Japan
e-mail: sakurasoma1209@yahoo.co.jp

S. Iwase et al. (eds.), *Clinical Assessment of the Autonomic Nervous System*,
DOI 10.1007/978-4-431-56012-8_6

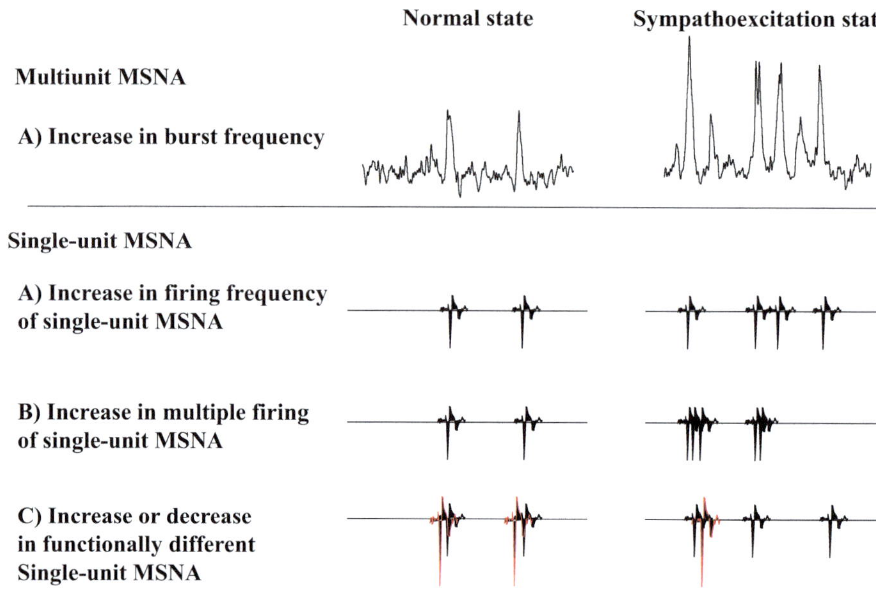

Fig. 6.1 Increases in burst intensity: potential mechanisms. Schematic representation of the three mechanisms that can bring about an increase in sympathetic burst intensity

cardiac interval, and (3) functionally different neuron activities (Fig. 6.1), which could not be observed by analysis of multiunit MSNA. The analysis of single-unit nerve is applied in not only muscle but also in skin sympathetic nerve activity, which can distinguish between sudomotor and skin vasoconstrictor unit activity in multiunit skin sympathetic nerve activity [4]. Using these assessments, analysis of single-unit neuron provides more detailed information regarding actual central neuron firing to peripheral in healthy human as well as cardiovascular diseases. In this chapter, we mainly describe the single-unit MSNA and the differences between multiunit and single-unit MSNA and discuss the advantages of single-unit MSNA recording.

6.2 Assessment of Multiunit MSNA

6.2.1 Limitation of Multiunit MSNA

Traditionally, multiunit MSNA is quantified by counting the number of bursts during a specified period of time and 100 heart beats [2]. As multiunit MSNA is mostly regulated by arterial baroreceptors, burst occurrence is synchronized with the cardiac interval. Pulse synchronous bursts are a specific feature of MSNA, which is different from skin sympathetic nerve activity and sudomotor or vasomotor nerve activity.

First limitation of analysis of multiunit MSNA is that counting multiunit MSNA does not increased over heart rate. Namely, under conditions of augmented sympathetic excitation, including heart failure, essential hypertension, or obstructive sleep apnea, multiunit MSNA is near the maximum response level to sympathoexcitatory stimulation (i.e., 100 bursts per 100 heart beats). To break through this limitation, total MSNA and/or normalized amplitude measurements have been utilized. Total MSNA is calculated as the product of the burst rate and the burst amplitude per minute, with all amplitudes normalized to the maximum amplitude [2]. The normalization process of absolute burst amplitudes has been shown to be a reproducible variable [4]. This approach assumes that the burst of the greatest amplitude reflects maximal recruitment of active neurons for that particular recording site. However, this approach cannot be used to compare subjects or to compare the same subject on different occasions, when the intervention changes the burst amplitude (i.e., shifts the distribution), because the normalization procedure will eliminate the change [5]. In addition, total multiunit MSNA cannot distinguish between changes in sympathetic nerve firing due to the recruitment of additional single-unit vasoconstrictor neurons and that due to an increase in firing rate of already active single-unit fibers.

The other limitation of multiunit MSNA is that multiunit MSNA is composed of many single-unit firing action potentials that might include functionally different firing responses to physiological stress. In the analysis of multiunit MSNA recordings during physiological stress, different firing responses cannot be observed because multiunit MSNA is calculated with all action potentials by integrating the mean of all voltages within 0.1 s. In an animal model using efferent renal sympathetic preparations, DiBona [6] showed that specific nerve firing frequencies and discrete afferent inputs exert important electrolyte, humoral, and vascular regulatory responses that are functionally distinct.

6.3 Analysis of Single-Unit MSNA

6.3.1 Recording of Single-Unit MSNA

Macefield et al. [3] refined single-unit MSNA analysis. This technically demanding method requires obtaining high signal to noise ratio and adjustment of the tungsten electrode until a large unitary discharge can be observed in a raw nerve action potential signal to discriminate a single-unit action potential. The two different points in recording single-unit MSNA compared with multiunit MSNA are that (1) data sampling from raw action potentials is needed to be over 10 Hz because spike interval is near 1 ms and (2) using high-impedance microelectrode (>10 mΩ) because of the reduction in electrical noise levels.

In offline analysis, spike detection software is necessary to analyze the data of single-unit MSNA from raw nerve action potentials. Several software are utilized in the recent year. We usually use Spike2 software (ver.5, Cambridge Electronics Design, Cambridge, UK) and/or Power Lab recording system (Model ML 785/8SP; ADInstruments, Bella Vista, Australia).

To identify single-unit MSNA, candidate single units were selected by isolating large identifiable unitary spikes in the raw neurogram within a distinct discharge amplitude range. Confirmation that these action potentials originated from a single fiber were made by established criteria: (1) spike synchronization with a multiunit burst, (2) triphasic spike morphology with the main phase being negative, and (3) superimposition of candidate action potentials with minimal variation using abovementioned software.

6.3.2 Advantage of Single-Unit MSNA

Several points were needed in obtaining excellent single-unit spikes. However, the technique provides an estimate of single-unit firing properties in relation to the number of active firings and/or the recruitment of fibers from central or reflex effects. Using this technique, additional measurements can be obtained with regard to the mean firing frequency, the firing probability (the percentage of cardiac intervals in which a unit fires), and the percentage of spikes a unit generates per cardiac interval.

Single-unit MSNA describes three possible scenarios to explain an increase in sympathetic outflow: (1) an increase in overall mean spike firing frequency without an increase in the rate of multiple firings per cardiac interval, (2) an increase in the firing frequency by multiple spike firing within one cardiac interval, and (3) an increase in the previous silent neuron active [7–9].

We demonstrated that single-unit MSNA can also be recorded during periods of physiological stress (e.g., handgrip [HG] exercise and the Valsalva maneuver) and that reflex sympathoexcitation could be attributed to changes in the frequency of single-unit spike firing within each multiunit sympathetic burst in healthy subjects. In particular, the firing of multiple spikes within one cardiac interval was significantly augmented during the Valsalva maneuver [10]. Multiunit MSNA is near the maximum response level to sympathoexcitatory stimulation (i.e., 100 bursts per 100 heart beats). However, single-unit MSNA can increase firing frequency even within one cardiac interval. In Fig. 6.2, during mild sympathetic activation (40 spikes/100 heartbeats), results for single-unit MSNA were similar to those for multiunit MSNA. However, during intense sympathetic activation, single-unit MSNA was greater than multiunit MSNA. This finding indicates that single-unit MSNA may overcome multiunit MSNA with regard to quantifying sympathetic nerve activity especially during intense sympathoexcitation state. Macefield et al. [11] demonstrated that single-unit MSNA tends to fire only once per cardiac interval even under conditions associated with elevated sympathetic nerve activity

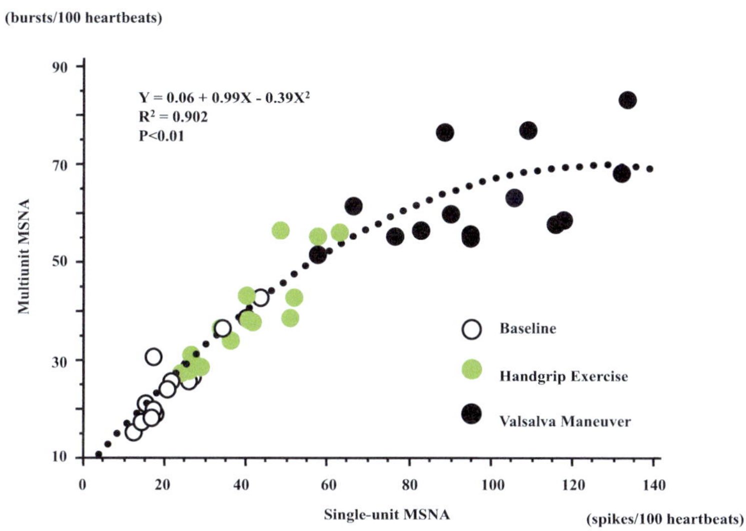

Fig. 6.2 Relationship between single-unit MSNA and multiunit MSNA. During mild sympathetic activation (40 spikes/100 heartbeats), results for single-unit MSNA were similar to those for multiunit MSNA. However, during intense sympathetic activation, single-unit MSNA was greater than multiunit MSNA

such as heart failure. These results suggest that single-unit MSNA has the capacity to increase multiple spike firing within one cardiac interval in a state of intense sympathoexcitation.

6.3.3 Assessment of Recording Single-Unit MSNA in Several Diseases and Sympathoexcitation State

Multiple spike firing of single-unit MSNA during one cardiac interval is thought as one of the important observations that is obtained from the analysis of single-unit MSNA. In a previous study in animal model, acute irregular and rapid nerve stimulation have been shown to evoke a greater effector organ response than regular stimulation through increased norepinephrine release in anesthetized rats [12]. In humans, Lambert et al. [13] reported that the incidence of multiple firing was associated with cardiac norepinephrine spillover. These results suggest that the firing of these instantaneous multiple spikes are thought to influence strong effector organ responses. Recent research has shown that a resting high heart failure firing frequency and/or incidence (percentage) of multiple spikes is related to cardiovascular risk factors, including hypertension [14], type 2 diabetes mellitus [15], obstructive sleep apnea [16], panic disorder [17], myocardial infarction [18], and congestive heart failure [10, 19]. The multiple firing frequency of single-unit MSNA within one cardiac interval may be involved in disease progression.

6.4 Assessment of Increase in Single-Unit MSNA Frequency During Physiological Stress and Arrhythmia

6.4.1 Advantage of Recording Single-Unit MSNA During Static Exercise in Heart Failure

Augmented sympathetic nerve activity is a characteristic feature of heart failure. Excessive sympathetic activation under resting conditions has been shown to increase from the early stages of the disease, and elevated levels of sympathetic nerve activity are correlated with a poor prognosis [20–23]. Sympathetic activity plays an essential role in maintaining blood pressure in acute heart failure, but excessive sympathetic activity in chronic heart failure has deleterious effects on the heart, including beta receptor downregulation [24], cardiac myocyte apoptosis [25], and calcium overload [26].

Although sympathetic nerve activity is difficult to assess in clinical settings, the assessment of sympathetic nerve activity is considered important in human heart failure. Since the development of microneurography, the increased response of multiunit MSNA was considered to indicate elevated central sympathetic nerve activity to the peripheral vascular bed. Augmented multiunit MSNA and its mechanism in healthy human subjects were reported previously [27]; however, there was a controversy with regard to the differences in reflex response of multiunit MSNA to handgrip exercise between heart failure patients and age-matched healthy controls [28–30].

It is likely that the responses of multiunit MSNA are dependent on disease severity, race, sex, and intensity of exercise. However, these controversial results may be attributable to the limitations of multiunit MSNA. In chronic heart failure, the level of multiunit MSNA is nearly maximal at rest, so multiunit MSNA cannot increase further. That is, multiunit MSNA cannot increase above 100 bursts/100 heartbeats because of pulse synchrony.

Thus, as mentioned above, single-unit MSNA analysis is useful for determining actual sympathetic neural firing within one cardiac interval. We demonstrated that the percentage of multiple single-unit spikes within one cardiac interval was increased during handgrip exercise in chronic heart failure patients compared to healthy subjects, although the response of multiunit MSNA was not significantly different between the two groups [19] (Fig. 6.3). These results suggest that single-unit MSNA responses contributed to the exaggerated sympathoexcitation measured during exercise in chronic heart failure patients. The instantaneous firing frequency within one cardiac interval may increase the peripheral vascular tone, which might contribute to exercise intolerance.

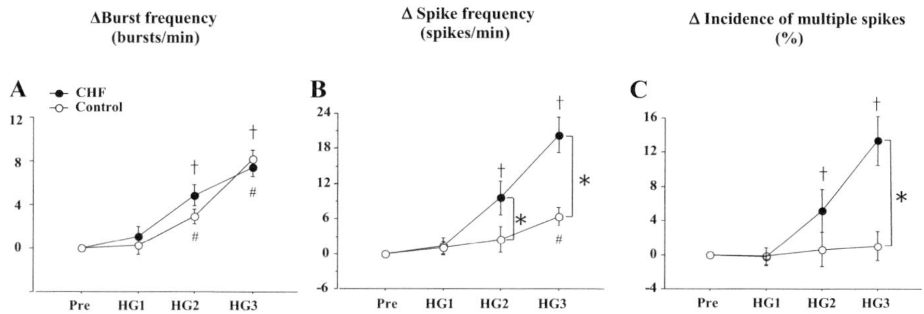

Fig. 6.3 HG exercise-mediated changes in multiunit MSNA parameters in control subjects (○) and heart failure patients (●). The changes in MSNA burst frequency (**a**) were similar between the two groups. However, the changes in single-unit MSNA spike frequency (**b**) and the incidence of multiple firings (**c**) during HG exercise were significantly increased in heart failure patients compared to healthy controls. The values are expressed as means ± SEM. $^{+}P < 0.05$ compared to heart failure at baseline. $^{\#}P < 0.05$ compared to control subjects at baseline. $^{*}P < 0.05$ compared to control subjects at the same time point

6.4.2 Assessment of Sympathetic Nerve Activity in Arrhythmia

Arrhythmia is a common complication of chronic heart failure caused by arrhythmogenic substrates [31]. Several studies indicated that a large multiunit MSNA burst occurred during premature ventricular contraction (PVC) [32] and atrial fibrillation (AF) [33, 34] in heart failure patients. The low diastolic pressure induced by these arrhythmic conditions unloads arterial baroreceptors and evokes a larger and longer multiunit MSNA burst [35, 36]. However, only counting multiunit MSNA could cause the actual level of sympathetic nerve activity to be underestimated, because a large sympathetic activity burst could produce prolonged sympathetic inhibition. In a human study, frequent PVC and AF were recognized as exclusion criteria for evaluating sympathetic outflow by multiunit MSNA analysis.

6.4.3 Recording of Single-Unit MSNA in Heart Failure with PVC

The mechanism underlying augmentation of the sympathetic nerve activity in heart failure has been assumed to involve a disorder of arterial baroreceptors. However, recent observations indicated that arterial baroreceptor function is preserved, maintaining appropriate blood pressure in heart failure [37]. Elam et al. [38] demonstrated the instantaneous augmentation of multiple single-unit firings following premature ventricular contraction (PVC) in heart failure patients. Frequent PVC is thought to induce moderate and severe left ventricular dysfunction [39]. Elimination of PVC with catheter ablation has been reported to improve cardiac dysfunction [40].

Cardiac norepinephrine spillover was reported to be related to multiple spike firing of single-unit MSNA within one cardiac interval [13]. Although the mechanisms of reduced left ventricular contraction induced by PVC remain unclear, multiple spike firing of single-unit MSNA is considered to cause the progression of heart dysfunction by instantaneous norepinephrine release to the heart.

6.4.4 Recording of Single-Unit MSNA in Heart Failure with Atrial Fibrillation

As mentioned above, the analysis of multiunit MSNA in AF is not without limitations in that a prolonged irregular ventricular response would cause a large burst followed by prolonged sympathetic inhibition. The results of previous assessments of sympathetic nerve activity in acute paroxysmal AF patients using multiunit MSNA are controversial. Grassi et al. [33] used multiunit MSNA to assess sympathetic nerve activity during AF and sinus rhythm (SR) in patients with paroxysmal AF and observed a reduction in multiunit MSNA during AF. In contrast, Wasmund et al. [34] found significant augmentation of multiunit MSNA during AF, which was induced by right atrial pacing. Recently, we analyzed the single-unit MSNA frequency in heart failure patients with AF [41]. Multiunit MSNA in heart failure patients with AF was decreased compared to that in heart failure patients with SR. However, the single-unit MSNA in heart failure patients with AF was significantly greater than that in heart failure patients with SR. Moreover, the incidence of multiple firing of single-unit MSNA within one cardiac interval was augmented in heart failure + AF patients compared to heart failure + SR patients, and it shifted toward multiple firing spikes of single-unit MSNA in heart failure patients with AF. It has been suggested that not only does heart failure lead to a predisposition to AF, but AF may also facilitate and worsen the development of heart failure. The coexistence of these cardiac disorders produces a vicious cycle, which leads to advanced pump failure in heart failure patients [42].

6.5 Assessment of Functionally Different Firing of Single-Unit MSNA

6.5.1 Simultaneous Recording of Three Single-Unit Spike Action Potentials

In an animal study, DiBona [6] showed that specific nerve firing frequencies and discrete afferent inputs exert functionally distinct and important electrolyte, humoral, and vascular regulatory responses. In this study, plural spikes were examined simultaneously in the same electrode position. In recording of single-unit MSNA in human, generally, only a single-unit action potential was isolated during recording session. However, recently, we examined up to three single-unit responses to physiological stress simultaneously (Fig. 6.4) and found that single-unit MSNA not necessarily increase and decrease in the same response of multiunit MSNA [8]. These results indicate that single-unit MSNA could reveal functionally different firing frequencies that cannot be detected by multiunit MSNA.

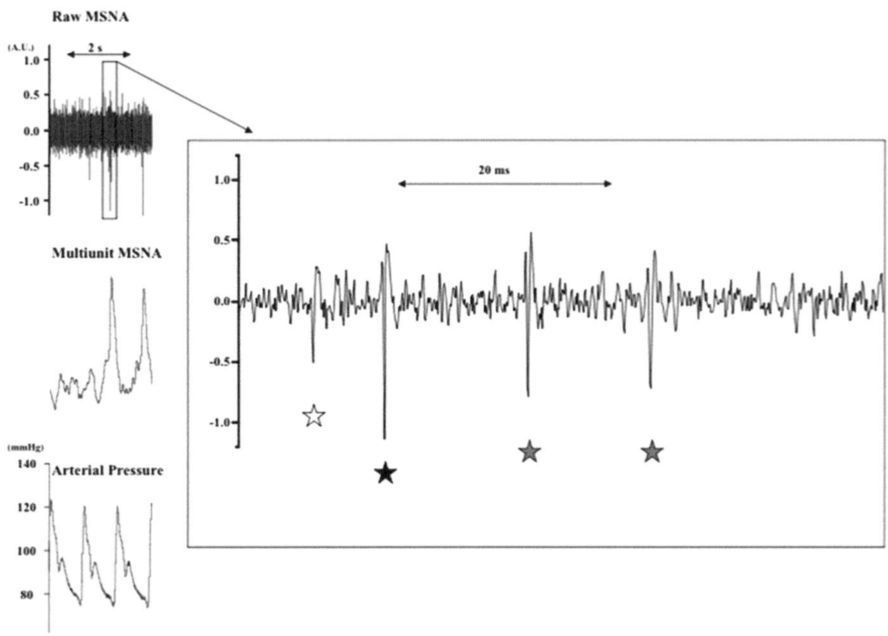

Fig. 6.4 Simultaneous recording of three single units. The identification of three single units (marked as *stars*; 1 discharging twice) based on spike morphology (amplitude and shape) was obtained in a healthy subjects. *AU* arbitrary units

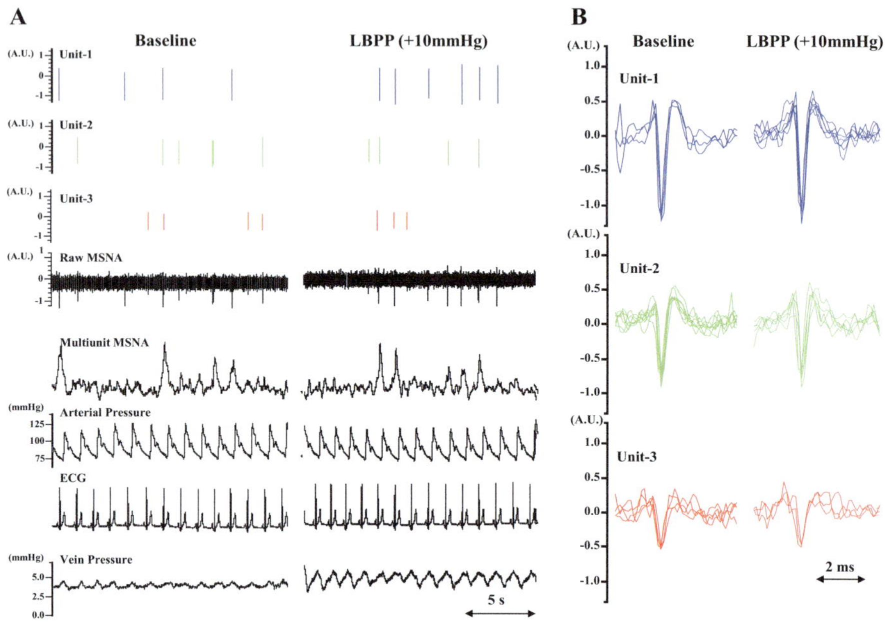

Fig. 6.5 Representative tracing in one acquired before and during nonhypertensive lower body positive pressure (LBPP; +10 mmHg). (**a**) typical recording of single- and multiunit MSNA, arterial pressure and estimated central vein pressure (CVP), and electrocardiography. Examination of the single units demonstrates paradoxical firing in unit 1 and anticipated firing of units 2 and 3. (**b**) identified single units superimposed

6.5.2 Single-Unit MSNA Responses to Cardiac Filling Pressure

In humans, cardiopulmonary baroreceptors play an important role in response to the decrease and increase in pressure and/or volume of lower pressure components. When the lower pressure increases, the cardiopulmonary baroreflex is thought to be inhibitory to sympathetic nerve activity resulting from the increase in afferent vagal tone. However, in subjects with normal cardiac function, the reduction of atrial pressure by nonhypotensive lower body negative pressure (LBNP) elicits discordant peripheral and cardiac sympathetic responses: total body norepinephrine (NE) spillover and MSNA increase reflexively, as anticipated, but cardiac NE spillover and power spectral estimates of sympathetic heart rate modulation do not. In heart failure patients, nonhypotensive LBNP caused an increase in multiunit MSNA but a paradoxical reduction in cardiac NE spillover. The anticipated increase in total body NE spillover was blunted, relative to control subjects as if nonhypotensive LBNP had unloaded two populations of mechanoreceptor reflexes exerting directionally opposite peripheral efferent sympathetic responses [43]. These results suggested that, in response to a selective increase in atrial pressure, single-unit MSNA would reveal a subpopulation of efferent sympathetic neurons with firing patterns opposite to the integrated multiunit MSNA envelope. With single-unit MSNA analysis, we recently found that two subpopulations of single unit within the multiunit MSNA recording exhibited opposite firing characteristics (Fig. 6.5) during lower body positive pressure (LBPP) as well as LBNP and demonstrated that different responses of single-unit muscle sympathetic reflex activated by increasing cardiac filling pressure in healthy subjects [8] as well as in heart failure patients [9]. However, the population of single-unit MSNA that increased during LBPP was significantly greater in heart failure patients than that in healthy subject, which may play an important role in augmentation of sympathetic nerve activity in heart failure.

6.6 Conclusions

The analysis of sympathetic nerve activity is important in humans, as increased sympathetic nerve activity has a deleterious effect on mortality and mobility. However, modalities for the assessment of sympathetic nerve activity in humans are limited. Moreover, the underlying significance of differences in central sympathetic neural firing between diseases remains unclear. Compared to the traditional recording of multiunit MSNA, the assessment of central sympathetic nerve activity to the periphery with single-unit MSNA provides additional information regarding the underlying mechanisms of sympathetic firing. In particular, assessment of the firing spike frequency, functionally different firing, and multiple spike firing within one cardiac interval are thought to contribute to finding underlying mechanisms and disease progression. Therefore, these parameters may shed a light on advantageous therapeutic targets in a sympathetic excitation state, such as seen in heart failure patients.

References

1. Vallbo AB, Hagbarth KE, Torebjork HE, Wallin BG. Somatosensory, proprioceptive, and sympathetic activity in human peripheral nerves. Physiol Rev. 1979;59:919–57.
2. Sundlof G, Wallin BG. The variability of muscle nerve sympathetic activity in resting recumbent man. J Physiol. 1977;272:383–97.
3. Macefield VG, Wallin BG, Vallbo AB. The discharge behaviour of single vasoconstrictor motoneurones in human muscle nerves. J Physiol. 1994;481(Pt 3):799–809.
4. Macefield VG, Wallin BG. Respiratory and cardiac modulation of single sympathetic vasoconstrictor and sudomotor neurones to human skin. J Physiol. 1999;516(Pt 1):303–14.
5. Kimmerly DS, O'Leary DD, Shoemaker JK. Test-retest repeatability of muscle sympathetic nerve activity: influence of data analysis and head-up tilt. Auton Neurosci. 2004;114:61–71.

Burke SL, Lambert E, Head GA. New approaches to quantifying sympathetic nerve activity. Curr Hypertens Rep. 2011;13:249–57.

6. DiBona GF. Neural control of the kidney: functionally specific renal sympathetic nerve fibers. Am J Physiol Regul Integr Comp Physiol. 2000;279:R1517–24.
7. Macefield VG, Elam M, Wallin BG. Firing properties of single postganglionic sympathetic neurones recorded in awake human subjects. Auton Neurosci. 2002;95:146–59.
8. Millar PJ, Murai H, Morris BL, Floras JS. Microneurographic evidence in healthy middle-aged humans for a sympathoexcitatory reflex activated by atrial pressure. Am J Physiol Heart Circ Physiol. 2013;305:H931–8.
9. Millar PJ, Murai HEART FAILUREloras JS. Paradoxical muscle sympathetic reflex activation in human heart failure. Circulation. 2015;131:459–68.
10. Murai H, Takata S, Maruyama M, Nakano M, Kobayashi D, Otowa K, et al. The activity of a single muscle sympathetic vasoconstrictor nerve unit is affected by physiological stress in humans. Am J Physiol Heart Circ Physiol. 2006;290:H853–60.
11. Macefield VG, Rundqvist B, Sverrisdottir YB, Wallin BG, Elam M. Firing properties of single muscle vasoconstrictor neurons in the sympathoexcitation associated with congestive heart failure. Circulation. 1999;100:1708–13.
12. Dibona GF, Sawin LL. Functional significance of the pattern of renal sympathetic nerve activation. Am J Physiol. 1999;277:R346–53.
13. Lambert EA, Schlaich MP, Dawood T, Sari C, Chopra R, Barton DA, et al. Single-unit muscle sympathetic nervous activity and its relation to cardiac noradrenaline spillover. J Physiol. 2011;589:2597–605.
14. Lambert E, Dawood T, Schlaich M, Straznicky N, Esler M, Lambert G. Single-unit sympathetic discharge pattern in pathological conditions associated with elevated cardiovascular risk. Clin Exp Pharmacol Physiol. 2008;35:503–7.
15. Huggett RJ, Scott EM, Gilbey SG, Bannister J, Mackintosh AF, Mary DA. Disparity of autonomic control in type 2 diabetes mellitus. Diabetologia. 2005;48:172–9.
16. Elam M, McKenzie D, Macefield V. Mechanisms of sympathoexcitation: single-unit analysis of muscle vasoconstrictor neurons in awake OSAS subjects. J Appl Physiol. 2002;93:297–303.
17. Lambert E, Hotchkin E, Alvarenga M, Pier C, Richards J, Barton D, et al. Single-unit analysis of sympathetic nervous discharges in patients with panic disorder. J Physiol. 2006;570:637–43.
18. Graham LN, Smith PA, Huggett RJ, Stoker JB, Mackintosh AF, Mary DA. Sympathetic drive in anterior and inferior uncomplicated acute myocardial infarction. Circulation. 2004;109:2285–9.
19. Murai H, Takamura M, Maruyama M, Nakano M, Ikeda T, Kobayashi D, et al. Altered firing pattern of single-unit muscle sympathetic nerve activity during handgrip exercise in chronic heart failure. J Physiol. 2009;587:2613–22.
20. Cohn JN, Levine TB, Olivari MT, Garberg V, Lura D, Francis GS, et al. Plasma norepinephrine as a guide to prognosis in patients with chronic congestive heart failure. N Engl J Med. 1984;311:819–23.
21. Francis GS, Benedict C, Johnstone DE, Kirlin PC, Nicklas J, Liang CS, et al. Comparison of neuroendocrine activation in patients with left ventricular dysfunction with and without congestive heart failure. A substudy of the Studies of Left Ventricular Dysfunction (SOLVD). Circulation. 1990;82:1724–9.
22. Grassi G, Seravalle G, Bertinieri G, Stella ML, Turri C, Mancia G. Sympathetic response to ventricular extrasystolic beats in hypertension and heart failure. Hypertension. 2002;39:886–91.
23. Barretto AC, Santos AC, Munhoz R, Rondon MU, Franco FG, Trombetta IC, et al. Increased muscle sympathetic nerve activity predicts mortality in heart failure patients. Int J Cardiol. 2009;135:302–7.
24. Bohm M, Flesch M, Schnabel P. Beta-adrenergic signal transduction in the failing and hypertrophied myocardium. J Mol Med (Berl). 1997;75:842–8.
25. Communal C, Singh K, Pimentel DR, Colucci WS. Norepinephrine stimulates apoptosis in adult rat ventricular myocytes by activation of the beta-adrenergic pathway. Circulation. 1998;98:1329–34.

26. Chaudhri B, Del Monte F, Hajjar RJ, Harding SE. Interaction between increased SERCA2a activity and beta -adrenoceptor stimulation in adult rabbit myocytes. Am J Physiol Heart Circ Physiol. 2002;283:H2450–7.
27. Mark AL, Victor RG, Nerhed C, Wallin BG. Microneurographic studies of the mechanisms of sympathetic nerve responses to static exercise in humans. Circ Res. 1985;57:461–9.
28. Notarius CF, Atchison DJ, Floras JS. Impact of heart failure and exercise capacity on sympathetic response to handgrip exercise. Am J Physiol Heart Circ Physiol. 2001;280: H969–76.
29. Sterns DA, Ettinger SM, Gray KS, Whisler SK, Mosher TJ, Smith MB, et al. Skeletal muscle metaboreceptor exercise responses are attenuated in heart failure. Circulation. 1991;84:2034–9.
30. Negrao CE, Rondon MU, Tinucci T, Alves MJ, Roveda F, Braga AM, et al. Abnormal neurovascular control during exercise is linked to heart failure severity. Am J Physiol Heart Circ Physiol. 2001;280:H1286–92.
31. Braunwald E. Shattuck lecture-cardiovascular medicine at the turn of the millennium: triumphs, concerns, and opportunities. N Engl J Med. 1997;337:1360–9.
32. Ando S, Dajani HR, Senn BL, Newton GE, Floras JS. Sympathetic alternans. Evidence for arterial baroreflex control of muscle sympathetic nerve activity in congestive heart failure. Circulation. 1997;95:316–9.
33. Grassi G, Seravalle G, Bertinieri G, Mancia G. Behaviour of the adrenergic cardiovascular drive in atrial fibrillation and cardiac arrhythmias. Acta Physiol Scand. 2003;177:399–404.
34. Wasmund SL, Li JM, Page RL, Joglar JA, Kowal RC, Smith ML, et al. Effect of atrial fibrillation and an irregular ventricular response on sympathetic nerve activity in human subjects. Circulation. 2003;107:2011–5.
35. Welch WJ, Smith ML, Rea RF, Bauernfeind RA, Eckberg DL. Enhancement of sympathetic nerve activity by single premature ventricular beats in humans. J Am Coll Cardiol. 1989;13:69–75.
36. Smith ML, Ellenbogen KA, Eckberg DL. Baseline arterial pressure affects sympathoexcitatory responses to ventricular premature beats. Am J Physiol. 1995;269:H153–9.
37. Floras JS. Sympathetic nervous system activation in human heart failure: clinical implications of an updated model. J Am Coll Cardiol. 2009;54:375–85.
38. Elam M, Macefield V. Multiple firing of single muscle vasoconstrictor neurons during cardiac dysrhythmias in human heart failure. J Appl Physiol. 2001;91:717–24.
39. Baman TS, Lange DC, Ilg KJ, Gupta SK, Liu TY, Alguire C, et al. Relationship between burden of premature ventricular complexes and left ventricular function. Heart Rhythm. 2010;7:865–9.
40. Taieb JM, Maury P, Shah D, Duparc A, Galinier M, Delay M, et al. Reversal of dilated cardiomyopathy by the elimination of frequent left or right premature ventricular contractions. J Interv Card Electrophysiol. 2007;20:9–13.
41. Ikeda T, Murai H, Kaneko S, Usui S, Kobayashi D, Nakano M, et al. Augmented single-unit muscle sympathetic nerve activity in heart failure with chronic atrial fibrillation. J Physiol. 2012;590:509–18.
42. Ehrlich JR, Nattel S, Hohnloser SH. Atrial fibrillation and congestive heart failure: specific considerations at the intersection of two common and important cardiac disease sets. J Cardiovasc Electrophysiol. 2002;13:399–405.
43. Azevedo ER, Newton GE, Floras JS, Parker JD. Reducing cardiac filling pressure lowers norepinephrine spillover in patients with chronic heart failure. Circulation. 2000;101:2053–9.

Part II
Heart Rate Variability (HRV)

Chapter 7
Introduction to Heart Rate Variability

Junichiro Hayano

Abstract Beat-to-beat intervals of cardiac sinus rhythm are not constant but show complex and continuous fluctuations called heart rate variability (HRV). Because HRV disappears with cardiac denervation by complete autonomic blockades or cardiac transplantation, HRV is thought to originate from the brain and to transfer to the heart through the autonomic nervous system. HRV includes a plenty of information not only about autonomic neural cardiac regulations but also about health state and hazard that are captured by the brain. To extract information that meets with particular purposes, various methods have been developed for the analysis of HRV. This chapter explains the basic mechanisms generating HRV and introduces the purposes and corresponding methods for HRV analyses.

Keywords Time series analysis • Power spectral analysis • Complex demodulation • Heart rate dynamics

7.1 What Is Heart Rate Variability?

7.1.1 Definitions

The rhythm of heart beat at rest is essentially regular, and irregularity of the rhythm is called arrhythmia. In fact, however, cardiac cycle length even under sinus rhythm shows continuous fluctuations, as indicated by the presence of term "physiological arrhythmia." Respiratory sinus arrhythmia (RSA) is the representative among such physiological arrhythmias, and it has first been described by Ludwig [1] in 1847. Cardiac cycle length shows various physiological fluctuations ranging from circadian rhythm of heart rate to such changes in beat-to-beat intervals as RSA.

Cardiac cycle length and its physiological fluctuations are mediated by the neurohumoral factors, particularly autonomic nervous system, that regulates the rate of discharge of the sinus node. Under complete pharmacological autonomic

J. Hayano, M.D. (✉)
Department of Medical Education, Nagoya City University Graduate School of Medical Sciences, 1 Kawasumi, Mizuho-cho, Mizuho-ku, Nagoya, Aichi 467-8602, Japan
e-mail: Hayano@med.nagoya-cu.ac.jp

S. Iwase et al. (eds.), *Clinical Assessment of the Autonomic Nervous System*,
DOI 10.1007/978-4-431-56012-8_7

blocking with sufficient doses of sympathetic and vagal blockades, the beat-to-beat fluctuations of cardiac cycle length disappear, and the heart ticks a regular rhythm like a metronome [2, 3]. This is also the case for transplanted heart that has no neural connections with the recipient [4]. These indicate that physiological fluctuations in beat-to-beat cardiac cycle length are mediated by the fluctuations in autonomic neural outflow to the sinus node.

In a narrow sense, the term heart rate variability (HRV) is defined as the amount of variability in beat-to-beat fluctuations of cardiac cycle length under normal sinus rhythm. The term is also used as almost the same meaning as physiological arrhythmia. In contrast, fluctuations in cardiac cycle length that are caused by ectopic rhythms and beats, conduction blocks, and sinus dysfunction are not included in HRV and should be excluded from the analysis of HRV. In a broad sense, however, cardiac cycle length fluctuations caused by a part of pathological arrhythmias have also been categorized as studies of HRV. For example, heart rate turbulence (HRT) that is known as a powerful predictor of mortality risk after acute myocardial infarction analyzes the cardiac cycle length fluctuations flowing ectopic beats observed during Holter ECG monitoring [5, 6]. The methods developed for HRV analysis have been used for ventricular response cycle length fluctuations during atrial fibrillation and have produced useful indices for both pathophysiologic understandings and clinical risk stratifications of this arrhythmia [7–11].

7.1.2 HRV and Autonomic Nervous Functions

Heart rate is regulated antagonistically by the sympathetic and vagal divisions of autonomic nervous system. Heart rate reflects the static balance between their activities to the sinus node, and HRV reflects the fluctuations in the balance. HRV, however, includes information that cannot be obtained from heart rate itself, i.e., pure vagal modulation of heart rate separated from sympathetic influence.

The principle of autonomic functional assessment by HRV may be compared to the estimation of driver's manipulations of accelerator and brake pedals from the time series recording of changes in vehicle's speed, where heart rate corresponds to vehicle's speed, sympathetic activity to accelerator, and vagal activity to brakes. Different from automatic transmission vehicles that creep when the driver takes off both accelerator and brake pedals, heart rate increases to a value called "intrinsic heart rate" when both sympathetic and vagal activities are blocked. The intrinsic heart rate is higher in young people and declines with aging; the means are 107 bpm for 20-year, 101 bpm for 30-year, 90 bpm for 50-year, and 78 bpm for 70-year olds. This indicates that young people maintain their cardiac vagal activity at a certain level continuously during rest to keep their heart rate below the intrinsic heart rate.

During vehicles are moving, if you tap the accelerator rapidly, vehicle's speed shows almost no changes. In contrast, if you tap the brake pedals, the speed changes faithfully with the manipulations. This is explained as the difference in frequency characteristic of transfer functions between two systems. Similar difference in frequency characteristic exists between sympathetic and vagal modulations of heart rate. Changes in sympathetic activity cause the effects through several phosphorylating enzymatic processes in the intracellular signal transduction mechanisms existing downstream of beta-adrenergic receptors. Consequently, the sympathetic modulation of heart rate cannot transfer fluctuations at frequencies >0.15 Hz (cycle length <6.7 s) [12]. While on the other hand, changes in vagal activity cause the effects simply by the conformation change of the membrane potassium channels with Ach. Consequently, the vagal modulation of heart rate can transfer fluctuations at frequency up to 1 Hz [12].

When a decrease was observed in the speed of a vehicle, there may be three possibilities that the driver (1) took his/her foot off accelerator, (2) pressed the brake pedal, or (3) both. If the deceleration was faster than a certain level, however, it is attributable only to the effect of (2). Similarly, when heart rate fluctuations at frequencies >0.15 Hz were observed, it is not explained by sympathetic modulation and it is attributable only to the effect of vagal modulation. These indicate that we can separate frequency components of heart rate fluctuations purely mediated by the vagal activities from those affected by sympathetic activities.

Methods for analyzing the fluctuations of time series data by resolving them into components by the difference in frequency are called power spectral analyses. Although there are several methods for this analysis, including fast Fourier transformation (FFT) [13], autoregressive (AR) model [14], and maximum entropy methods (MEM), the principle is the same as spectroscopic analysis by prisms that separate light into components with their colors (wave frequencies). Power spectral analyses resolve fluctuation into frequency components and quantify the power or amplitude of each component.

7.1.3 Purposes of HRV Analysis

The analysis of HRV is currently used for two major purposes: (1) assessment of cardiac autonomic functions and (2) risk stratification of patients with cardiac and other diseases.

The former includes the assessment of stress or arousal level, relaxation depth, and sleep quality as well as clinical evaluation for autonomic dysfunctions. For these purposes, short-term (5–10 min) ECG data that are recorded under standardized conditions (room temperature, body posture, time after food, alcohol and caffeine intake, exercise and smoking, and breathing frequency according to the circumstances) are used. HRV analyzed from short-term ECG data thus obtained is referred to as short-term HRV.

The risk stratification includes the prediction of mortality and adverse prognosis. For these purposes, long-term (24 h) ambulatory ECG data recorded under daily activities are used. HRV analyzed from long-term ECG data thus obtained is referred to as long-term HRV.

7.2 Mechanisms Generating HRV

7.2.1 Short-Term HRV

Power spectrum of short-term HRV at rest includes two major frequency components (Fig. 7.1): low-frequency (LF) component (0.04–0.15 Hz) and high-frequency (HF) component (>0.15 Hz) [2, 15–17]. This is a characteristic of HRV that is unlike vehicle's speed indicating that HRV does not comprise only random fluctuations but includes frequency components with specific frequencies. The cutoff frequency (0.15 Hz) separating between two components has been defined from the frequency characteristic of transfer function for sympathetic heart rate modulation discussed above [12]. Although the frequency of LF component has been defined as 0.04–0.15 Hz, it usually appears between 0.06 and 0.11 Hz.

7.2.1.1 HF Component of HRV

The HF component of HRV usually corresponds to RSA, and thus, its frequency is identical to breathing frequency (e.g., when breathing frequency is 15 cycle/min, the frequency of HF component is 15 cycle/60 s = 0.25 Hz). This means that when breathing frequency decreases below 9 cycle/min (0.15 Hz), HRV caused by RSA is detected as a part of LF component and physiologic HF component disappears; HF component under such conditions, if observed, should be interpreted as a different physiologic entity from RSA or an artifact.

The mechanisms generating RSA is discussed in Chap. 8 in detail. Briefly, RSA is mediated by the vagus originated from the nucleus ambiguus in the medulla oblongata that is modulated by the input from the respiratory center and generates

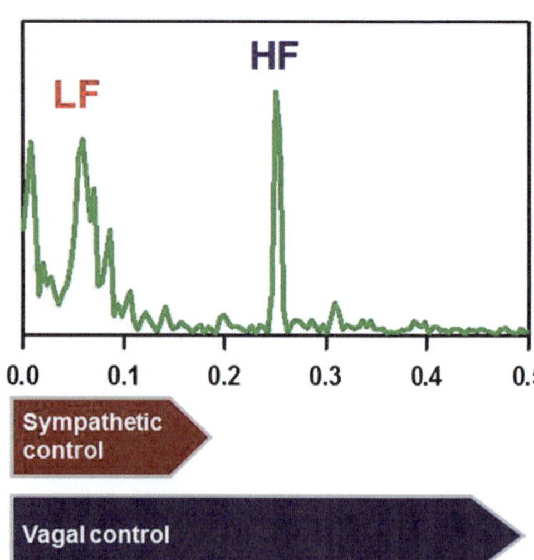

Fig. 7.1 Power spectrum of short-term HRV and frequency ranges modulated by sympathetic and vagal heart rate controls. While cardiac vagal control system can modulate heart rate in the entire frequency band, cardiac sympathetic control system can modulate heart rate at frequencies <0.15 Hz. Consequently, while the LF component of HRV is mediated by both sympathetic and vagal nerves, the HF component is mediated purely by the vagus

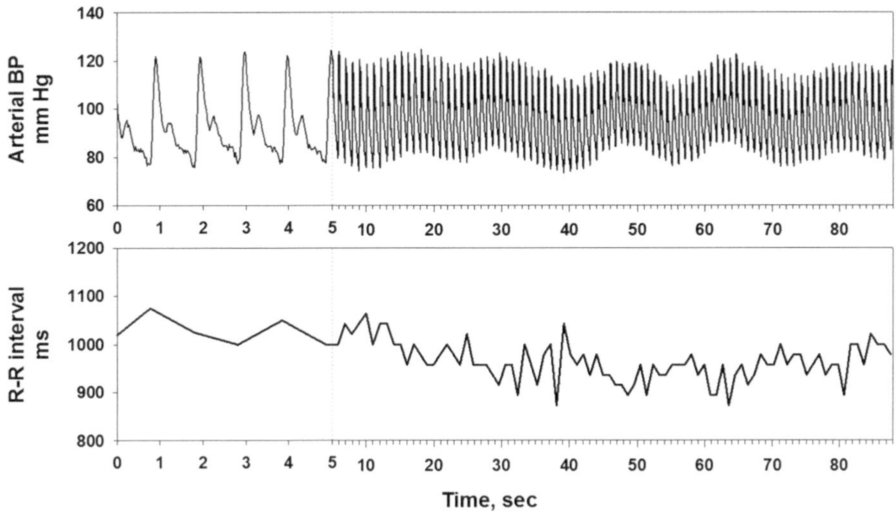

Fig. 7.2 Arterial blood pressure and R-R interval of ECG in a healthy subject. Mayer wave is observed in the compressed strip of arterial blood pressure (*upper panel* after 5 s) at a period around 13 s. In the trend of R-R intervals (*lower panel*), fluctuations corresponding HF (period around 3 s) and LF (period around 13 s) components are observed. The LF fluctuation of R-R interval shows several-second delay behind Mayer wave of arterial blood pressure

vagal outflow fluctuating with respiration [18]. By this mechanism, vagal flow increases during expiration and decreases during inspiration, generating RSA. Although respiratory fluctuation exists also in sympathetic outflow, it does not transfer to HRV when the breathing frequency is above 0.15 Hz.

7.2.1.2 LF Component of HRV

The LF component of HRV is thought to be heart rate variation caused by Mayer wave [19] in blood pressure fluctuation through the arterial baroreceptor reflex mechanism (Fig. 7.2) [20, 21]. Mayer wave is a component of physiological arterial blood pressure fluctuation with a period around 10 s. It is also called the third-order variation of blood pressure and has been found by Cion in 1874 and described by Mayer in 1876 [19].

The mechanisms generating Mayer wave are still controversial, and there are theories proposing peripheral, central, and resonance origins. Blood vessel contraction by sympathetic vasomotor function is known to occur with 5-s delay after sympathetic neural activation. Simulation studies have reported that such delay systems cause spontaneous fluctuation in the baroreceptor reflex feedback loop at a period of 10 s [20]. The LF component of HRV is decreased in patients with low baroreceptor reflex sensitivity independently of the presence of sympathetic innervation to the heart [21].

7.2.2 Long-Term HRV

Although the LF and HF components are the major constituents of short-term HRV, long-term HRV comprises nonperiodic components with a broad range of spectrum as the major constituents [22]. For descriptive purposes, fluctuations at frequencies lower than LF component are divided into two components: very-low-frequency (VLF) component (0.003–0.04 Hz) and ultralow frequency (ULF) component ($\leq$0.003 Hz). Unlike LF and HF components, VLF and ULF are nonperiodic components that form no distinct peaks in power spectrum. The nonperiodic component of long-term HRV is also called 1/*f* fluctuation or fractal component, because it has power negatively correlated with frequency in the power spectrum and it furnishes the properties of fractal dynamics including long-term negative correlation and scale-independent self-affine structure [23].

Long-term HRV from ambulatory ECG recordings under daily activities includes the influences of circadian and ultradian variations in physiological functions, physical and mental activities, and environmental factors. Accordingly, analysis of long-term HRV does not suit for the evaluations of specific autonomic functions, but it may be useful for the evaluations of overall performance of autonomic regulations. In fact, long-term HRV provides prognostic information, particularly all-cause mortality risk in patients with cardiac and other diseases, which have stronger predictive power than short-term HRV indices [24–26].

7.3 Data Collection for HRV

7.3.1 Data Sources of HRV

HRV can be analyzed when at least one lead of continuous ECG recording is available. The ECG needs to be stored as digital data at a sampling frequency of no less than 125 Hz (desirably, 500–1000 Hz) to avoid artificial cycle length fluctuations caused by under sampling. The ECG data are converted into time series of beat-to-beat cycle length by measuring all R-R intervals (Fig. 7.3). For this purpose, accurate automated classifications (normal sinus rhythm, atrial or ventricular ectopic beat, and artifact) are requited for all beats, the results of classification should be reviewed, and all errors in the annotation need to be edited completely.

Although pulse wave signal may be used as a surrogate of ECG, several limitations should be recognized, which include difficulty in beat classifications, limited accuracy in measurement of beat-to-beat cycle length, and modifications by the frequency characteristic of pulse wave conduction [27].

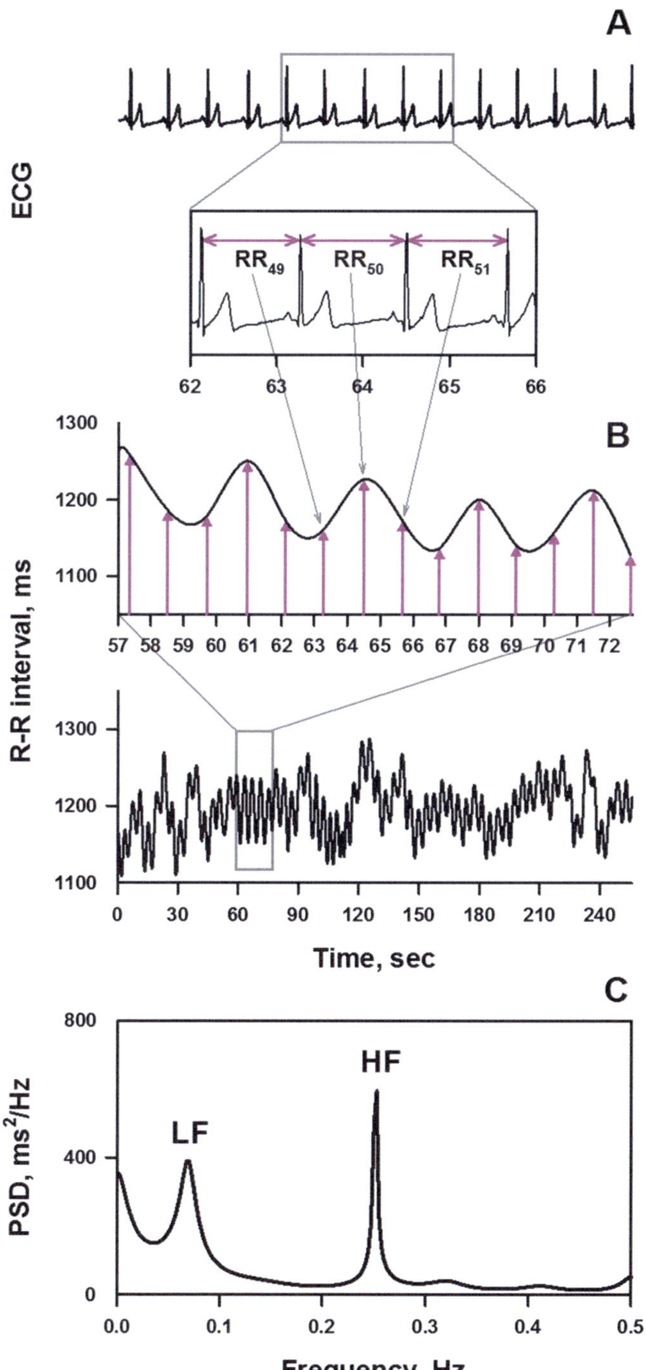

Fig. 7.3 Measurement of HRV from ECG. For the analysis of HRV, time series of beat-to-beat cycle lengths are measured as R-R intervals under sinus rhythm. The temporal position of each R-R interval is usually defined as the position of subsequent R wave. For the purpose of frequency

7.3.1.1 Standardization for Short-Term HRV Measurement

HRV is affected by various intrinsic and extrinsic factors such as environmental temperature, physical activities [28, 29], mental activities [30, 31], food intake [32], smoking [33, 34], and sleep/awake rhythm. For the assessment of autonomic function, subjects need to avoid strenuous exercise, smoking, alcohol and caffeine intake from the previous night, and food intake from 3 h before the study, and the measurement should be performed at constant time of day in an air-conditioned and calm experimental room after >15 min supine rest for equilibrium. Although the length of recording is determined by the purposes, continuous 5-min recording after the stabilization of heart rate is the standard.

For short-term HRV, controlled breathing with metronome signal may be used so that breathing frequency of subjects is kept at >0.15 Hz and heart rate to respiration ratio at >2. There are three reasons for this:

1. To evaluate cardiac vagal function separately from sympathetic influences, breathing frequency needs to be kept at >0.15 Hz (the upper frequency limit of sympathetic heart rate control).
2. The magnitude of HF component decreases with increasing breathing frequency independently of cardiac vagal activity.
3. If heart rate to respiration ratio decreases to <2, respiratory fluctuation in cardiac vagal activity, if exists, is not reflected by HRV.

Thus, when breathing frequency is not controlled, the frequency of HF component and heart rate should be checked if these conditions are satisfied.

7.3.1.2 ECG Recordings for Long-Term HRV

For the assessment of long-term HRV, 24-h Holter ECG recordings are useful. ECG signals are digitized at 125–500 Hz in the recorders, and all R waves are classified automatically. Recent Holter ECG scanners have software for HRV analysis as the standard or optional function. For 24-h long-term HRV, there may be defect of data due to insufficient recording length or to temporary electrode troubles. In such case, care should be needed for the effects of day/night difference in HRV. Because the long-term HRV indices have been standardized as 24-h data, when the substantial data defects occurred disproportionally during daytime or nighttime, it causes bias of sampling.

Fig. 7.3 (continued) domain analyses, R-R interval time series are interpolated and resampled at equidistantly devided time points

7.4 Methods for HRV Analysis

A variety of measures have been used for quantifying the characteristics of HRV. They are classified by the method of analysis as shown in Table 7.1. From time domain analysis, statistical and geometric measures are calculated. These measures mainly applied to 24-h long-term recordings and used for risk stratification for mortality and adverse prognosis among patients with cardiac and other diseases. Frequency domain analysis is used for both short-term and long-term recordings. Short-term measures of HRV are used for the assessment of autonomic function, and long-term measures are used for risk stratification among patients with cardiac and other diseases. Nonlinear and fractal dynamics analyses are used mainly for long-term recordings and provide prognostic indices such as α_1 of detrended fluctuation analysis (DFA) [36, 37], which is a powerful predictor of mortality risk in patients after myocardial infarction [39] and those with end-stage renal disease on chronic hemodialysis therapy [40].

Study of HRV has started in 1970s, but new methods of analysis and novel measures have been still proposed almost every year. Non-Gaussianity index of λ reflects an increase in large abrupt changes in heart rate, and an increase in this measure predicts increased mortality risk among patients with chronic heart failure [38, 41]. This measure is discussed in Chap. 9 in detail. Among most of other measures of HRV whose decrease is associated with increased health risk, λ is the only measure whose increase predicts increased risk. Because λ decreases with β-blockers, this measure is thought to be a unique index reflecting sympathetic over activities. Deceleration capacity (DC) [26] and heart rate turbulence (HRT) [5, 6] are currently the most powerful predictors of mortality risk after acute myocardial infarction. Figure 7.4 shows examples of analysis of long-term HRV in male patients after acute myocardial infarction; one survived >5 years, and the other died suddenly 25 months after the Holter monitoring.

7.4.1 Time Domain Analysis

For time domain analysis, N-N interval data were used without interpolation. Because time domain analysis depends only on the order of data but not on the temporal positions, the statistical and geometric measures are less affected by the defect of data caused by the removal of abnormal data caused by ectopic beats, etc. Conversely, such abnormal data need to be removed because the contaminations of such data could affect substantially to statistical measures. Geometric measures are robust to the contaminations of abnormal data.

To select and interpret the time domain measures appropriately, the knowledge of crude correspondences between time domain and frequency domain measures is useful, although the relationships are affected by heart rate. SDNN, HRV triangular index, and TINN correspond to total power and SDNN index to the mean of the

Table 7.1 Measures of HRV

Variable	Unit	Definition
1. Time domain analysis		
1.1 Statistical measures		
Mean NN	ms	Mean of all N-N intervals[a] during 24 h
SDNN	ms	Standard deviation of all N-N intervals during 24 h
SDANN	ms	Standard deviation of the averages of N-N intervals in all 5 min segments during 24 h
RMSSD	ms	The square root of the mean of squared differences between adjacent N-N intervals during 24 h
SDNN index	ms	Mean of the standard deviations of all N-N intervals for all 5 min segments during 24 h
SDSD	ms	Standard deviation of differences between adjacent N-N intervals during 24 h
NN50 count		Number of pairs of adjacent N-N intervals differing by more than 50 ms during 24 h
pNN50	%	NN50 count divided by the total number of all N-N intervals during 24 h
1.2 Geometric measures		
HRV triangular index	ms	Total number of all N-N intervals during 24 h divided by the height of the histogram of all N-N intervals measured on a discrete scale with bin width of 7.8125 ms (1/128 s)
TINN	ms	Baseline width of the minimum square difference triangular interpolation of the highest peak of the histogram of all N-N intervals during 24 h
2. Frequency domain analysis		
2.1 Measures for analysis of short-term recordings (5 min)		
Total power	ms^2	The variance of N-N intervals during 5 min
VLF	ms^2	Power in very-low-frequency range (≤0.04 Hz)
LF	ms^2	Power in low-frequency range (0.04–0.15 Hz)
LF amp	ms	Mean amplitude of LF: sqrt (2*LF)
LF norm	n.u.	LF power in normalized unit: LF/(Total power – VLF)*100
HF	ms^2	Power in high-frequency range (0.15–0.4 Hz)
HF amp	ms	Mean amplitude of HF: sqrt (2*HF)
HF norm	n.u.	HF power in normalized unit: HF/(Total power – VLF)*100
LF/HF		Power ratio between LF and HF: LF(ms^2)/HF(ms^2)
2.2 Measures for long-term recordings (24 h)		
Total power	ms^2	The variance of N-N intervals during 24 h[b]
VLF	ms^2	Power in very-low-frequency range (≤0.04 Hz)[b]
LF	ms^2	Power in low frequency range (0.04–0.15 Hz)[b]
HF	ms^2	Power in high-frequency range (0.15–0.4 Hz)[b]

(continued)

Table 7.1 (continued)

Variable	Unit	Definition
β		Power-law scaling exponent: slope of the linear regression of the spectrum in a log-log scale below 0.04 Hz
3. Nonlinear and fractal dynamics		
ApEn		Complexity of fluctuation measured by approximate entropy [35]
DFA index		Measures of short-term (4–11 beats, α_1) and long-term (>11 beats, α_2) fractal correlations calculated by detrended fluctuation analysis (DFA) [36, 37]
λ		Non-Gaussianity index of probability density function for abrupt changes in heart rate [38]
4. Others		
DC	ms	Deceleration capacity measured by phase rectified signal averaging [26]
HRT		Heart rate turbulence measured by the decrease in R-R interval immediately after ectopic beats (turbulence onset, TO) and the following recovery slop (turbulence slope, TS) [5, 6]

[a]N-N interval indicates R-R interval between two consecutive sinus QRS waves
[b]These measures are often presented in natural logarithmic values

total powers for all 5 min segments during 24 h. RMSSD, SDSD, NN50 count, and pNN50 reflect N-N interval variations in the frequency band analyzed as HF component.

7.4.2 Frequency Domain Analysis

Frequency domain analysis in general indicates spectral analysis. In spectral analysis, fluctuations observed in time series data are recognized as a set of elementary waves (Fig. 7.5). Spectral analysis provides the mean frequency and amplitude (≈ sqrt [2*power]) of each elementary wave averaged over the entire length of data segment that was analyzed. In case of HRV, VLF, LF, and HF components are elementary waves, and the waveform of original time series data is the ensemble of these elementary waves.

Periodic wave elements form peaks in spectrum; the position of peak represents frequency, and the height (in FFT spectrum) or area (in AR model and MEM spectra) represents amplitude/power of the elementary wave. The term "power"

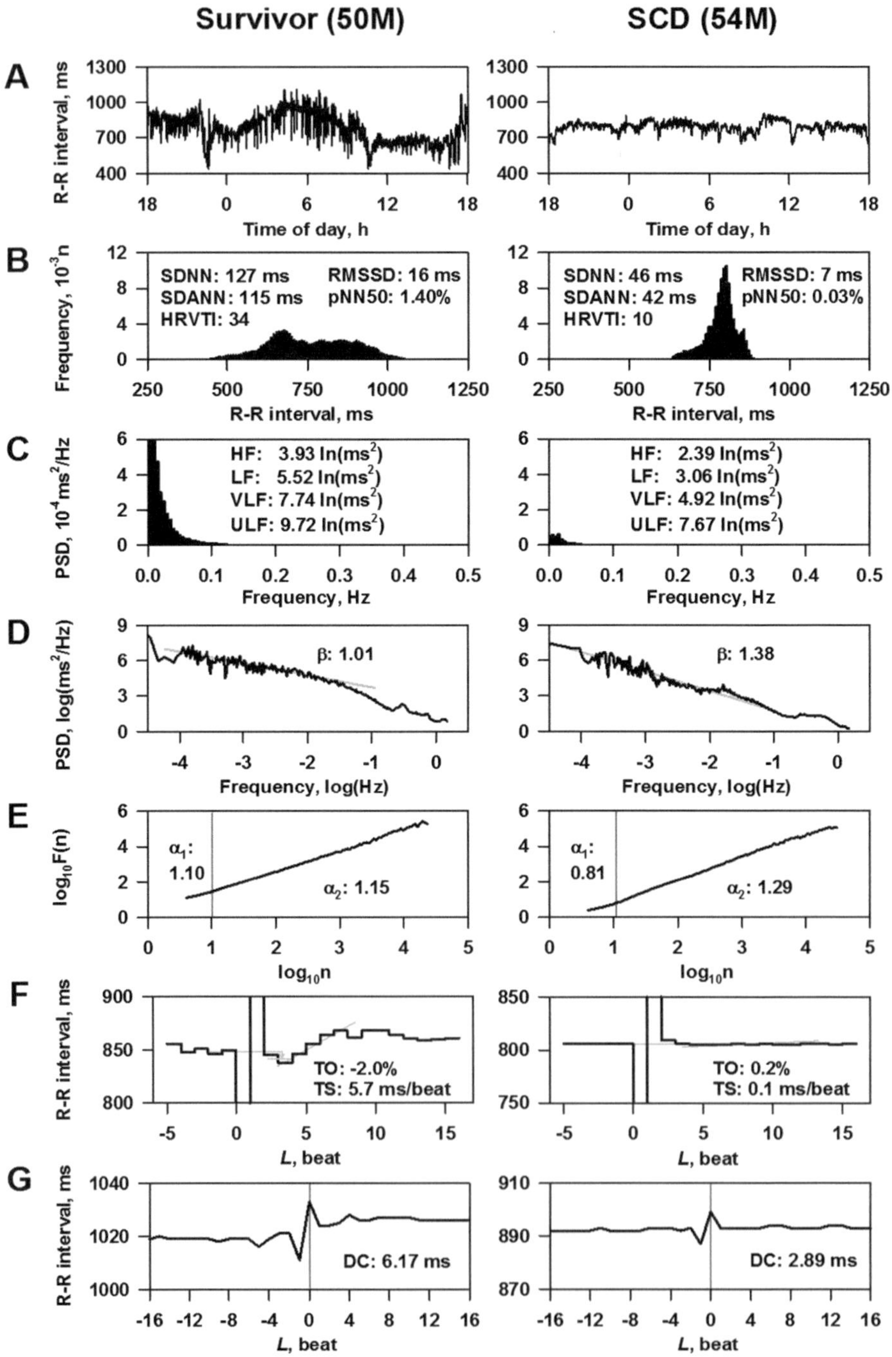

Fig. 7.4 Long-term 24-h HRV in patients after acute myocardial infarction. The patient on the left side survived for >5 years, while the patients on the right side died suddenly 25 months after Holter ECG monitoring. (**a**) 24-h trend graphs of N-N intervals, (**b**) histograms of N-N intervals and time domain measures, (**c**) power spectra of N-N intervals and frequency domain measures,

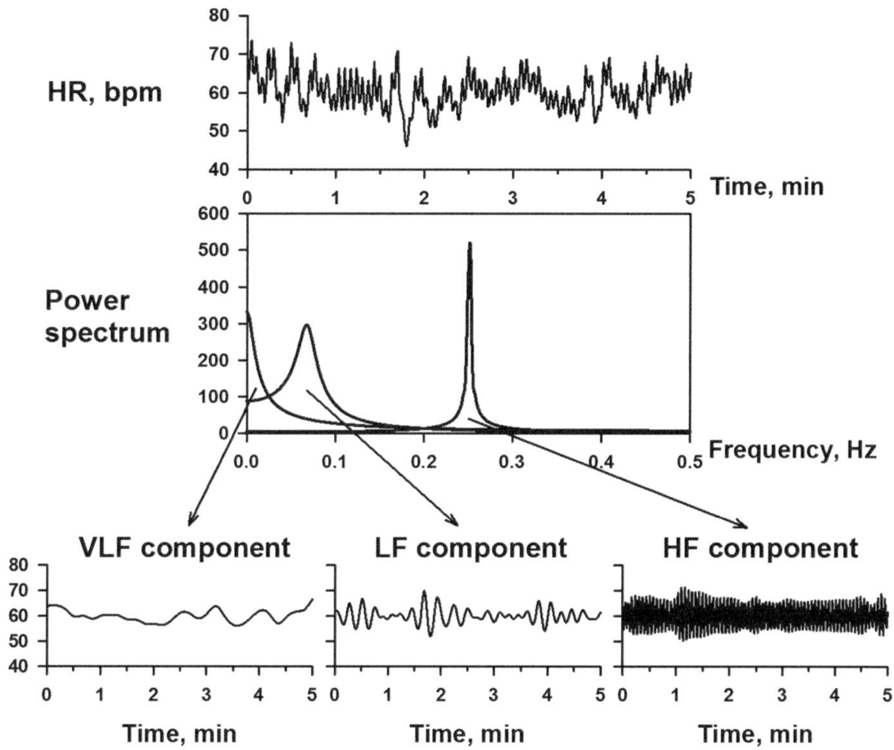

Fig. 7.5 Power spectral analysis of HRV. In spectral analysis, fluctuation observed in time series data is recognized as a set of elementary waves. As the results of analysis, the mean frequency and amplitude (≈ sqrt [2*power]) of each elementary wave averaged over the entire length of segment analyzed. In case of HRV, VLF, LF, and HF components are elementary waves, and the waveform of original time series data is the ensemble of these elementary waves

means variance caused by the wave, and when the wave can be assumed as a sinusoid, power and amplitude can be transformed to each other by the equation

$$\text{Amplitude} = \text{sqrt}(2^{*}\text{power}).$$

Waves and their spectra are summable. Thus, even if two waves with different frequencies are added, the characteristics (such as frequency and amplitude) of individual wave are preserved. Also, the variance or total power of combined wave is the sum of the power of two original waves.

Fig. 7.4 (continued) (**d**) power spectra in log-log scale and spectral exponent β, (**e**) detrended fluctuation analysis (DFA) and short-term and long-term fractal scaling exponents, α_1 and α_2, (**f**) heart rate turbulence and turbulence onset (TO) and turbulence slope (TS), and (**g**) deceleration capacity (DC). See Table 7.1 for the explanation of measures

7.4.3 Interpolation and Resampling

Frequency domain analyses of HRV require interpolation and resampling of time series data before analysis. Including N-N interval and arterial blood pressure, time series data of the circulatory system arise one datum per heartbeat. Because the intervals of heartbeat are fluctuating as HRV, the sampling intervals of data points become unequal, which causes problem in spectral analysis that requires equidistantly sampled time series data.

Data interpolation is the method for estimate value at time points between two consecutive observations. Many methods have been proposed as the methods for interpolating N-N intervals, which include step, linear, and spline interpolations. Although there is no ideal method for interpolation, the difference in the interpolation method is known to cause no substantial difference in the final results of spectral analysis. Interpolation is also used for the portions of data defect caused by removal of ectopic beats and noise. The interpolated N-N interval time series data are resampled at appropriate frequency (such as 2 Hz), yielding equidistantly sampled N-N interval time series for which spectral analysis is performed.

7.4.4 Analysis of Dynamic Autonomic Functions

7.4.4.1 Limitation of Spectral Analysis

Autonomic nervous system shows dynamic responses to various stimuli. Although such responses are reflected in HRV, they are not detected by the conventional methods of spectral analysis. This is because spectral analysis assumes that data are in a steady state throughout the analyzed period, and it provides the characteristics of waves representing their averages over the period. To overcome this limitation, methods such as spectral analysis for overlapping shifting windows have been devised, but there are limitations for detecting rapid autonomic responses with high temporal resolution.

7.4.4.2 Analysis of Time-Dependent Changes in Frequency Component of HRV

The methods for analyzing continuous changes in frequency and amplitude of HRV are divided into two categories:

1. Analysis of time/frequency distributions
2. Complex demodulation (CDM)

The former includes short-term FFT, wavelet transformation, and instantaneous AR spectral analysis. These methods estimate information in three dimensions

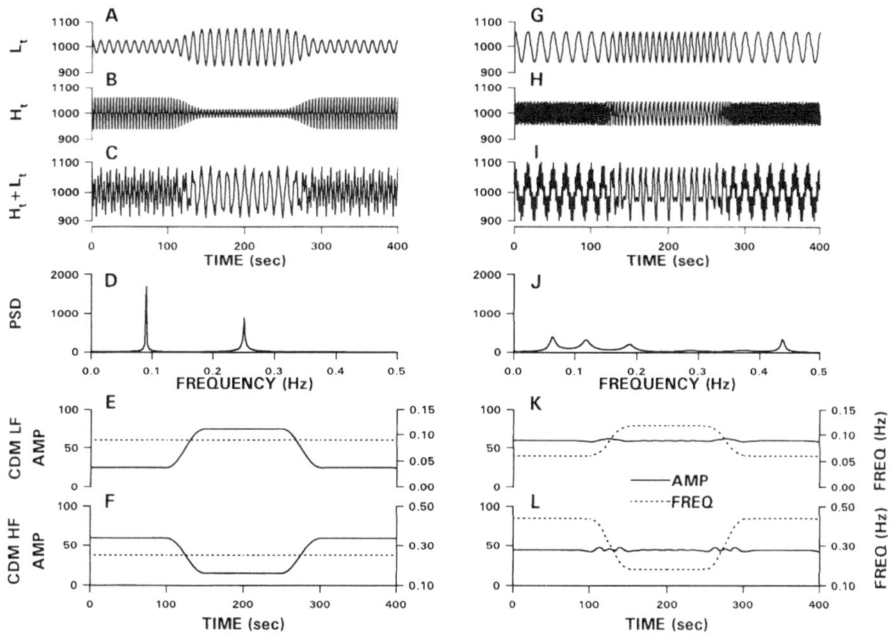

Fig. 7.6 Complex demodulation (CDM) of simulated data with time-dependent changes in amplitude (*left-side* panels **a**–**f**) and in frequency (*right-side* panels **g**–**l**). Lt, simulated low-frequency (LF) components with a fluctuating amplitude and a fixed frequency of 0.09 Hz and (**a**) with a constant amplitude and a fluctuating frequency between 0.06 and 0.12 Hz; (**g**) Ht, simulated high-frequency (HF) components with a fluctuating amplitude and a fixed frequency of 0.25 Hz and (**b**) with a constant amplitude and a fluctuating frequency between 0.16 and 0.44 Hz; (**h**) Ht + Lt: time series data generated by adding two component signals above (C = A + B, I = G + H); PSD, autoregressive power spectral density of the generated time series (panels **d** and **j**); and CDM LF and CDM HF, instantaneous amplitude (AMP, *solid line*) and frequency (FREQ, *dashed line*) of the LF (panels **e** and **k**); and the HF (panels **f** and **l**) components obtained by CDM (Modified from reference [43])

(time, frequency, and power) from N-N interval time series, that is, information in two dimensions (time and N-N interval). Thus, they need to assume the conditions of analysis. Also, even if time/frequency distribution is determined by these methods, time-dependent changes in frequency and amplitude of LF and HF components need to be extracted from that after all. In this view point, CDM that is a simpler and directly extracts necessary results may be often advantageous.

7.4.4.3 Complex Demodulation of HRV

The principle of CDM analysis of HRV is compared with the demodulation of electric wave by radios in amplitude modulation (AM) broadcasting system [42, 43]. In AM broadcasting, audio signal is converted into the changes in amplitude of carrier electric wave whose frequency is unique to each broadcast

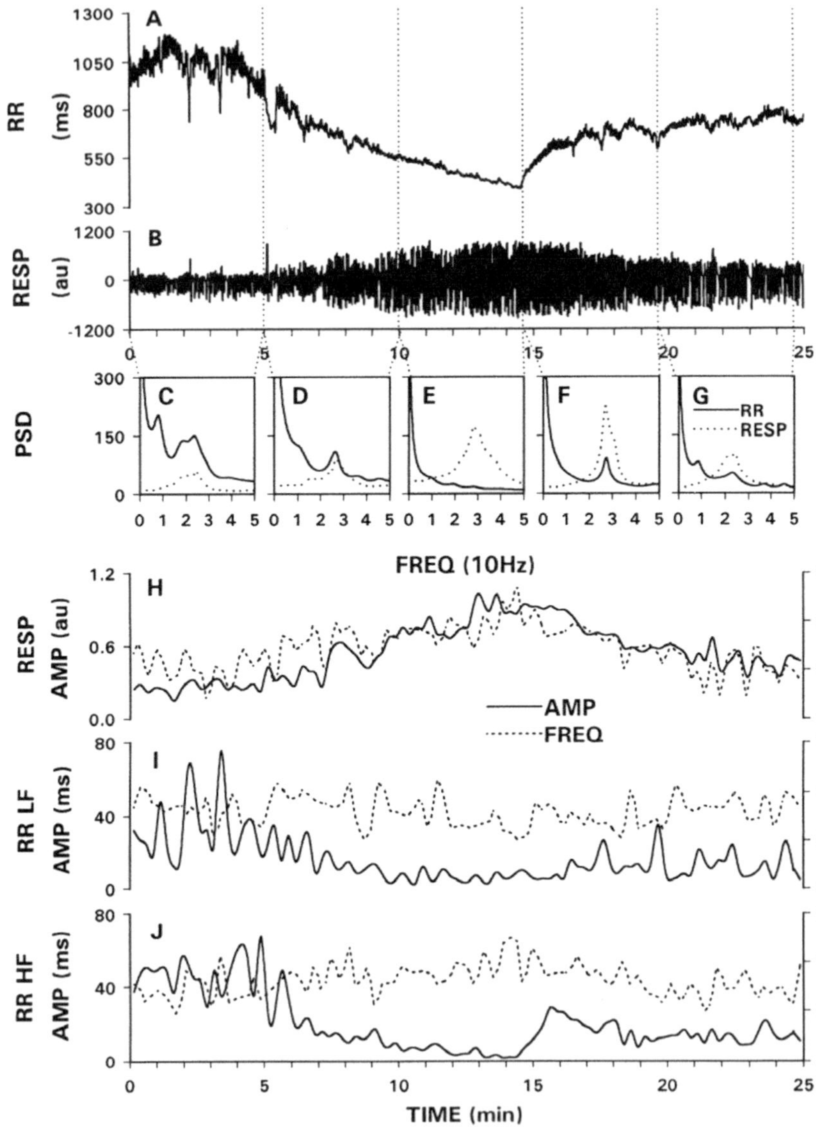

Fig. 7.7 CDM of R-R interval variability and spirogram (RESP) signal obtained during a supine ergometer exercise test with increasing workload (20 W/min) in a healthy young male subject. Panels **a** and **b**: R-R interval (RR) and spirogram (RESP). Panels **c–g**: autoregressive power spectral density (PSD) of R-R interval (*solid line*, RR) and spirogram (*dashed line*, RESP). Panels **h**, **i**, and **j**: instantaneous amplitude (*solid line*) and frequency (*dashed line*) of spirogram (RESP), and the LF and HF components of R-R interval variability (RR LF and RR HF) (Modified from reference [43])

station. This relationship is similar to vagal modulation of the amplitude of HF component of HRV. AM radio obtains audio signals of selected broadcast station by extracting time-dependent changes in amplitude of the electric wave of the station.

If similar procedure is done for HRV in HF frequency band, time-dependent changes in vagal modulation can be obtained as continuous data.

Figure 7.6 shows simulation of CDM analysis. On both sides, simulated time series data of N-N intervals were generated by combining two time series data simulating LF and HF components, on which spectral analysis and CDM were performed. On the left side, the amplitude of LF and HF components shows time-dependent changes, while on the right side, the frequency of both components shows time-dependent changes. CDM detects the changes in amplitude and frequency in each component faithfully.

Figure 7.7 shows CDM analysis of R-R interval and respiration signals during symptom-limited rump exercise in a healthy young male subject. Progressive decreases in the amplitude of HF and LF components with increasing workload are observed.

CDM detects time-dependent changes in amplitude and frequency of LF component of HRV at a time resolution of 15 s and those of HF component at a higher resolution [42]. This method is useful for the analysis of autonomic responses to stress and those accompanying various pathologic episodes [44].

7.5 Conclusion

Although analysis of HRV provide powerful, useful, and unique tools for the assessment of autonomic functions, understandings of its physiology and methodology are important for the selection of suitable methods and measures and for the appropriate interpretation of results.

References

1. Ludwig C. Beiträge zur Kenntniss des Einflusses der Respirationsbewegungen auf den Blutlauf im Aortensysteme. Arch Anat Physiol Leipzig. 1847;13:242–302.
2. Pomeranz B, Macaulay RJ, Caudill MA, Kutz I, Adam D, Gordon D, et al. Assessment of autonomic function in humans by heart rate spectral analysis. Am J Physiol. 1985;248(1 Pt 2): H151–3.
3. Hayano J, Sakakibara Y, Yamada A, Yamada M, Mukai S, Fujinami T, et al. Accuracy of assessment of cardiac vagal tone by heart rate variability in normal subjects. Am J Cardiol. 1991;67(2):199–204.
4. Sands KEF, Appel ML, Lilly LS, Schoen FJ, Mudge Jr GH, Cohen RJ. Power spectrum analysis of heart rate variability in human cardiac transplant recipients. Circulation. 1989;79:76–82.
5. Schmidt G, Malik M, Barthel P, Schneider R, Ulm K, Rolnitzky L, et al. Heart-rate turbulence after ventricular premature beats as a predictor of mortality after acute myocardial infarction. Lancet. 1999;353:1390–6.
6. Bauer A, Malik M, Schmidt G, Barthel P, Bonnemeier H, Cygankiewicz I, et al. Heart rate turbulence: standards of measurement, physiological interpretation, and clinical use: international society for holter and noninvasive electrophysiology consensus. J Am Coll Cardiol. 2008;52(17):1353–65.

7. Hayano J, Yamasaki F, Sakata S, Okada A, Mukai S, Fujinami T. Spectral characteristics of ventricular response to atrial fibrillation. Am J Physiol. 1997;273:H2811–6.
8. Yamada A, Hayano J, Sakata S, Okada A, Mukai S, Ohte N, et al. Reduced ventricular response irregularity is associated with increased mortality in patients with chronic atrial fibrillation. Circulation. 2000;102:300–6.
9. Hayano J, Ishihara S, Fukuta H, Sakata S, Mukai S, Ohte N, et al. Circadian rhythm of atrioventricular conduction predicts long-term survival in patients with chronic atrial fibrillation. Chronobiol Int. 2002;19(3):633–48.
10. Sato K, Yamasaki F, Furuno T, Hamada T, Mukai S, Hayano J, et al. Rhythm-independent feature of heart rate dynamics common to atrial fibrillation and sinus rhythm in patients with paroxysmal atrial fibrillation. J Cardiol. 2003;42(6):269–76.
11. Watanabe E, Kiyono K, Hayano J, Yamamoto Y, Inamasu J, Yamamoto M, et al. Multiscale entropy of the heart rate variability for the prediction of an ischemic stroke in patients with permanent atrial fibrillation. PLoS One. 2015;10(9):e0137144.
12. Berger RD, Saul JP, Cohen RJ. Transfer function analysis of autonomic regulation. I: canine atrial rate response. Am J Physiol. 1989;256:H142–52.
13. Cooley JW, Turkey JW. An algorithm for the machine calculation of complex Fourier series. Math Comput. 1965;19:297–301.
14. Akaike H. Power spectrum estimation through autoregressive model fitting. Ann Inst Stat Math. 1969;21:407–19.
15. Sayers BM. Analysis of heart rate variability. Ergonomics. 1973;16:17–32.
16. Akselrod S, Gordon D, Ubel FA, Shannon DC, Barger AC, Cohen RJ. Power spectrum analysis of heart rate fluctuation: a quantitative probe of beat-to-beat cardiovascular control. Science. 1981;213:220–2.
17. Pagani M, Lombardi F, Guzzetti S, Rimoldi O, Furlan R, Pizzinelli P, et al. Power spectral analysis of heart rate and arterial pressure variabilities as a marker of sympatho-vagal interaction in man and conscious dog. Circ Res. 1986;59:178–93.
18. Hayano J, Yasuma F. Hypothesis: respiratory sinus arrhythmia is an intrinsic resting function of cardiopulmonary system. Cardiovasc Res. 2003;58(1):1–9.
19. Penaz J. Mayer waves: history and methodology. Automedica. 1978;2:135–41.
20. Madwed JB, Albrecht P, Mark RG, Cohen RJ. Low-frequency oscillation in arterial pressure and heart rate: a simple computer model. Am J Physiol. 1991;256:H1573–9.
21. Moak JP, Goldstein DS, Eldadah BA, Saleem A, Holmes C, Pechnik S, et al. Supine low-frequency power of heart rate variability reflects baroreflex function, not cardiac sympathetic innervation. Heart Rhythm: Off J Heart Rhythm Soc. 2007;4(12):1523–9.
22. Yamamoto Y, Nakamura Y, Sato H, Yamamoto M, Kato K, Hughson RL. On the fractal nature of heart rate variability in humans: effects of vagal blockade. Am J Physiol. 1995;269:R830–7.
23. Saul JP, Albrecht P, Berger RJ. Analysis of long term heart rate variability: methods, 1/f scaling and implications. Comput Cardiol. 1987;14:419–22.
24. Camm AJ, Malik M, Bigger Jr JT, Breithardt G, Cerutti S, Cohen RJ, et al. Task force of the European society of cardiology and the north American society of pacing and electrophysiology. Heart rate variability: standards of measurement, physiological interpretation and clinical use. Circulation. 1996;93(5):1043–65.
25. Huikuri HV, Mäkikallio TH, Peng CK, Goldberger AL, Hintze U, Moller M, et al. Fractal correlation properties of R-R interval dynamics and mortality in patients with depressed left ventricular function after an acute myocardial infarction. Circulation. 2000;101:47–53.
26. Bauer A, Kantelhardt JW, Barthel P, Schneider R, Makikallio T, Ulm K, et al. Deceleration capacity of heart rate as a predictor of mortality after myocardial infarction: cohort study. Lancet. 2006;367(9523):1674–81.
27. Hayano J, Barros AK, Kamiya A, Ohte N, Yasuma F. Assessment of pulse rate variability by the method of pulse frequency demodulation. Biomed Eng OnLine. 2005;4:62.
28. Arai Y, Saul JP, Albrecht P, Hartley LH, Lilly LS, Cohen RJ, et al. Modulation of cardiac autonomic activity during and immediately after exercise. Am J Physiol. 1989;256:H132–41.

29. Yamamoto Y, Hughson RL, Peterson JC. Autonomic control of heart rate during exercise studied by heart rate variability spectral analysis. J Appl Physiol. 1991;71:1136–42.
30. Sakakibara M, Takeuchi S, Hayano J. Effect of relaxation training on cardiac parasympathetic tone. Psychophysiology. 1994;31:223–8.
31. Sakakibara M, Kanematsu T, Yasuma F, Hayano J. Impact of real-world stress on cardiorespiratory resting function during sleep in daily life. Psychophysiology. 2008;45(4):667–70.
32. Hayano J, Sakakibara Y, Yamada M, Kamiya T, Fujinami T, Yokoyama K, et al. Diurnal variations in vagal and sympathetic cardiac control. Am J Physiol. 1990;258:H642–6.
33. Hayano J, Yamada M, Sakakibara Y, Fujinami T, Yokoyama K, Watanabe Y, et al. Short- and long-term effects of cigarette smoking on heart rate variability. Am J Cardiol. 1990;65:84–8.
34. Niedermaier ON, Smith ML, Beightol LA, Zukowska-Grojec Z, Goldstein DS, Eckberg DL. Influence of cigarette smoking on human autonomic function. Circulation. 1993;88:562–71.
35. Pincus SM, Goldberger AL. Physiological time-series analysis: what does regularity quantify. Am J Physiol. 1994;266:H1643–56.
36. Peng CK, Havlin S, Stanley HE, Goldberger AL. Quantification of scaling exponents and crossover phenomena in nonstationary heartbeat time series. Chaos. 1995;5(1):82–7. doi:10.1063/1.166141.
37. Iyengar N, Peng CK, Morin R, Goldberger AL, Lipsitz LA. Age-related alterations in the fractal scaling of cardiac interbeat interval dynamics. Am J Physiol. 1996;271:R1078–84.
38. Kiyono K, Hayano J, Watanabe E, Struzik ZR, Yamamoto Y. Non-Gaussian heart rate as an independent predictor of mortality in patients with chronic heart failure. Heart Rhythm. 2008;5(2):261–8.
39. Huikuri HV, Mäkikallio TH, Airaksinen KEJ, Seppänen T, Puukka P, Räihä IJ, et al. Power-law relationship of heart rate variability as a predictor of mortality in the elderly. Circulation. 1998;97:2031–6.
40. Suzuki M, Hiroshi T, Aoyama T, Tanaka M, Ishii H, Kisohara M, et al. Nonlinear measures of heart rate variability and mortality risk in hemodialysis patients. Clin J Am Soc Nephrol. 2012;7(9):1454–60. doi:10.2215/CJN.09430911.
41. Hayano J, Kiyono K, Struzik ZR, Yamamoto Y, Watanabe E, Stein PK, et al. Increased non-gaussianity of heart rate variability predicts cardiac mortality after an acute myocardial infarction. Front Physiol. 2011;2:65.
42. Hayano J, Taylor JA, Yamada A, Mukai S, Hori R, Asakawa T, et al. Continuous assessment of hemodynamic control by complex demodulation of cardiovascular variability. Am J Physiol. 1993;264:H1229–38.
43. Hayano J, Taylor JA, Mukai S, Okada A, Watanabe Y, Takata K, et al. Assessment of frequency shifts in R-R interval variability and respiration with complex demodulation. J Appl Physiol. 1994;77:2879–88.
44. Lipsitz LA, Hayano J, Sakata S, Okada A, Morin RJ. Complex demodulation of cardiorespiratory dynamics preceding vasovagal syncope. Circulation. 1998;98:977–83.

Chapter 8
Respiratory Sinus Arrhythmia and Entraining Heartbeats with Cheyne-Stokes Respiration: Cardiopulmonary Works to Be Minimal by Synchronizing Heartbeats with Breathing

Fumihiko Yasuma and Junichiro Hayano

Abstract There are two types of interaction between heartbeats and respiration with different time scales. One is the well-known respiratory sinus arrhythmia (RSA) and another is entraining heartbeats with Cheyne-Stokes respiration (CSR). The latter is the cyclical fluctuation with the clustering heartbeats during the hyperventilation phase of CSR and their scattering during the apnea/hypopnea phase, which we call the **Entrainment** with CSR. The wisdom of the body for maintaining the homeostasis efficiently could be achieved by synchronizing heartbeats with breathing and consequently by saving the cardiopulmonary works. Analogous to that the RSA benefits the pulmonary gas exchange, we have hypothesized that the **Entrainment** with CSR might function to supply "effective" heartbeats during the hyperventilation phase and save "ineffective" heartbeats during the apneic/hypopneic phase to benefit the pulmonary gas exchange according to our previous investigations on the pulmonary gas exchange and RSA, the control of RSA and respiratory muscles during hypercapnia and hypoxia, the role of vagal nerve on RSA and respiratory muscles, and our clinical experiences in heart failure and atrial fibrillation. The concept that the RSA and **Entrainment** with CSR would serve for the cardiopulmonary works to be minimal by synchronizing heartbeats with breathing would be tested in the future.

Keywords Cheyne-Stokes respiration • Entraining heartbeats with Cheyne-Stokes respiration • Pulmonary gas exchange • Respiratory sinus arrhythmia • Vagal nerve

---

F. Yasuma, M.D., Ph.D., FCCP (✉)
Department of Cardiology, Suzuka National Hospital, 3-2-1 Kasado, Suzuka, Mie 513-8501, Japan
e-mail: yasuma@suzulan.hosp.go.jp

J. Hayano, M.D., FCCP
Department of Medical Education, Nagoya City University Graduate School of Medicine, Aichi 513-8501, Japan

S. Iwase et al. (eds.), *Clinical Assessment of the Autonomic Nervous System*,
DOI 10.1007/978-4-431-56012-8_8

8.1 Introduction

There are two distinct types of interaction between heartbeats and respiration with different time scales. One is the well-known respiratory sinus arrhythmia (RSA), which was first described by Ludwig (1816–1895) [1]. He recorded periodic oscillations in the amplitude and timing of arterial pressure with the smoked drum kymograph, in which the pulse rate increased during inspiration and decreased during expiration. Another is entraining heartbeats (= the **Entrainment**) with Cheyne-Stokes respiration (CSR) in patients with advanced congestive heart failure [2, 3], which has been generally overlooked, to our knowledge, among most physiologists.

Let us show the characteristic CSR and entraining heartbeats with CSR in a heart failure patient. Focusing upon the phasic changes in respiration, a cyclical fluctuation with episodes of central apneas/hypopneas alternating with episodes of hyperventilation is found in a gradual waxing and waning fashion in Fig. 8.1 in a time scale of 180 s (3 min) on the abscissa. Focusing upon the phasic changes in circulation during CSR, a cyclical fluctuation with the clustering heartbeats during hyperventilation phase and their scattering during apnea/hypopnea phase (= the **Entrainment** with CSR) on electrocardiogram (ECG) is found in Fig. 8.2 in 60 s (1 min).

The CSR of an oscillatory ventilation has been the focus of investigation among physicians for more than a century [4–7], since its first description by Cheyne (1777–1836) and Stokes (1804–1878) in a heart failure patient [8], and it has been assumed that the underlying mechanism of CSR might be an unstable feedback control of the respiratory system [9–11]. However, the roles of the interacting

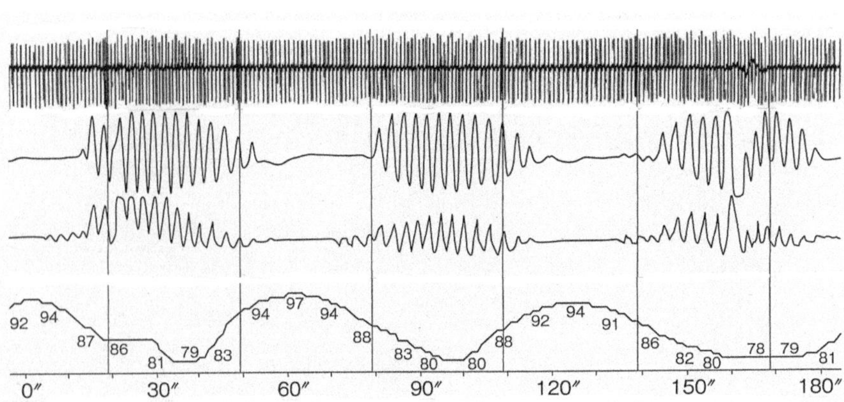

Fig. 8.1 Polygraphic recordings of Cheyne-Stokes respiration in a patient with congestive heart failure in 180 s. From *top* to *bottom*, electrocardiogram, oronasal flow, ribcage motion, and arterial oxygen saturation on pulse oximeter are displayed. A cyclical fluctuation with episodes of central apneas/hypopneas alternating with episodes of hyperventilation in a gradual waxing and waning fashion is seen

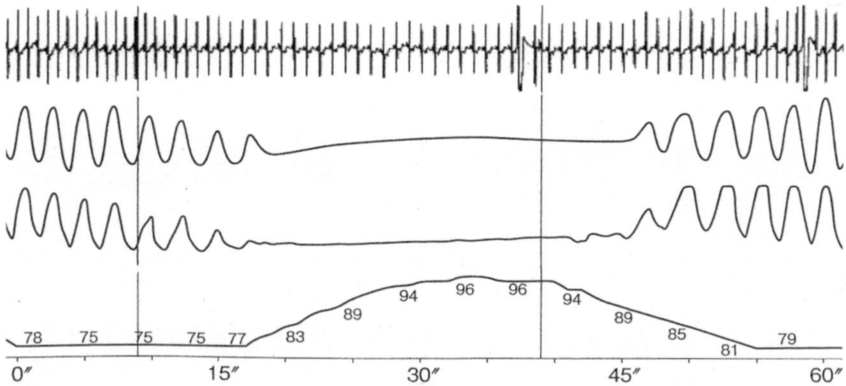

Fig. 8.2 Polygraphic recordings of Cheyne-Stokes respiration in a patient with congestive heart failure in 60 s. Displayed parameters are seen in Fig. 8.1. A clustering of heartbeats during hyperventilation phase and a scattering during apnea/hypopnea phase are seen

respiratory/circulatory centers and, possibly, the vagal nerve contributing to the genesis of the CSR and, thereby, the **Entrainment** with CSR are still unknown.

The authors previously discussed the fundamental question of "Why does the heartbeat synchronize with respiratory rhythm?" to conclude that "RSA or heart rate variability in synchrony with respiration is a biological phenomenon, which may have a positive influence of gas exchange at the level of the lung via efficient ventilation/perfusion matching" [12]. This concept could be applicable to the **Entrainment** with CSR, and hence, the author has recently provided the concept that the efficacy of pulmonary gas exchange might be improved by the **Entrainment** with CSR, the phenomenon of which seems analogous to the RSA [3].

In this article, the authors are proposing the further concept that "RSA and the **Entrainment** with CSR are not simply secondary products of other known reflexes but have physiological roles with the cardiopulmonary works to be minimal by synchronizing heartbeats with breathing," and we are reviewing how this teleological concept has been developed.

For this purpose, after showing the "Comparative Biology of Synchronizing Heartbeats with Breathing (8.2)," we are addressing the "Clinical/Physiological Significances of RSA, CSR, and **Entrainment** with CSR (8.3, 8.4, 8.5, 8.6)." In the "Physiological Significance of RSA (8.4)," we are summarizing our previous five separate experiments in canines (#1–#5 Experiments) along with their implications, respectively. Since the "Physiological Significances of CSR and **Entrainment** with CSR (8.6)" have been seldom investigated, our descriptions are hypothetical.

Herein, the authors are showing the perspectives of the RSA and the **Entrainment** with CSR according to our laboratory experiments and clinical experiences. This review article is written from standpoint of the wisdom of the body in maintaining the homeostasis efficiently [13], which can be achieved by synchronizing heartbeats with breathing [3, 12, 14], and consequently by saving the working loads of breathing [15] and circulation [16].

8.2 Comparative Biology of Synchronizing Heartbeats with Breathing

In the comparative biology, a cardiorespiratory interaction analogous to the RSA and the **Entrainment** with CSR is widely observed among many species of vertebrates from mammals, birds, and reptiles to water-breathing fish [17].

In mammals, the dog has a prominent RSA [18–20], and the human fetus obtains the RSA activity after a gestational age of ~33 weeks, and the healthy-term newborn infant has a strong RSA, which becomes a major component of heart rate variability in humans [21].

Regarding the **Entrainment** with CSR, the details of this phenomenon have been described, to our knowledge, only in humans with congestive heart failure [2, 3, 22]. However, a respiration-related oscillation in heartbeats, analogous to the RSA and the **Entrainment** with CSR, is observed in spontaneous breathing ducks [23], reptiles [24, 25], and fish. For example, in resting fish, gills are ventilated with pulsatile water flow throughout the respiratory cycle and heartbeat occurs in 1:1 synchrony with the respiration so that it results in coincidence of the periods of maximal flow rate of blood and water at the gills [26, 27]. This cardiorespiratory synchronization in fish is mediated by the vagal nerve and abolished by atropine [28].

The above observations among species suggest that the RSA and the **Entrainment** with CSR may bear a primitive role to maintain the homeostasis effectively by synchronizing heartbeats with breathing, although the role of the vagal nerve, especially, in the genesis of the **Entrainment** with CSR has been unknown.

8.3 Clinical Significance of RSA

A number of studies have been published concerning the mechanisms of RSA, and it has been known that RSA occurs mainly as the result of fluctuation of vagal outflow to the heart, despite the fact that the sympathetic outflow also influences the variability [29]. The RSA might result from modulation of vagal efferent activity to the sinus node through gating of the excitatory input to the vagal motor neurons from the lung inflation afferents [30]. During inspiration, impulses arising from the stretch receptors of the lungs are transmitted through the vagal nerve to inhibit the cardioinhibitory area in the medulla. As the tonic vagal discharge to keep the heart rate slow decreases, the heart rate increases. The degree to which this modulation occurs is a function of the level of tonic vagal discharge; thus, RSA measures would offer an index of cardiac vagal outflow [31, 32], under the most physiological circumstances [14].

The magnitude of RSA is clinically assessed as the amplitude of the high frequency band (HF band) of fluctuation of the R-R intervals (0.15 > Hz), utilizing the frequency analysis of heart rate variability. For accurate assessment, an examinee's tidal breathing should be standardized due to the frequency-dependent characteristics of RSA [33]. Therefore, the amplitude of the HF band, i.e., the magnitude of RSA, has been considered to reflect the cardiac vagal activity, and it has been used as the index of the activity of the parasympathetic nervous system [34, 35]. Previous studies have shown that the RSA activity of "cardiac age" is most exaggerated in young and healthy humans, and the activity generally weakens in the geriatric population [36] and among patients with cardiovascular/pulmonary diseases, diabetes mellitus, and other systemic diseases [12]. The authors have already revealed the possible dissociation between the respiratory cardiac modulation (magnitude of RSA) and the cardiac vagal tone (vagal control of heart rate) in twofolds [14]. One is the bradycardia caused by the pharmacological baroreceptor stimulation [37], and the other is the magnified RSA without changes in heart rate and blood pressure caused by the chemoreceptor stimulation with hypercapnia [38, 39], the latter of which is discussed in Sect. 4.4.

8.4 Physiological Significance of RSA

8.4.1 Pulmonary Gas Exchange and RSA (#1 Experiment)

The physiological significance of RSA had been generally unknown until the late twentieth century, when the authors demonstrated experimentally that the efficacy of pulmonary gas exchange is improved by RSA in an animal model study [40]. Herein, we summarize the authors' experiment on the RSA and pulmonary gas exchange in anesthetized canines [40], the species of which has the strong RSA [18–20].

8.4.1.1 Summary

In anesthetized canines with tracheal intubation, the sympathetic activity was eliminated with the premedication of reserpine, and the endogenous vagal activity was blocked with the left-side cervical vagotomy and right-side local anesthesia at the skull base. The right vagal nerve, innervating the sinus node, was electrically stimulated to control the frequency and timing of heartbeats. The artificial negative pressure ventilation was maintained with the diaphragm pacing (electrophrenic respiration) [41], since the mechanical positive pressure ventilation should reverse the intrathoracic pressure from negative during tidal breathing to positive, reducing the phasic increase of venous return to the thoracic cage during inspiration [42, 43] and affecting the features of RSA [44, 45].

The RSA model with heartbeats clustered during inspiration and scattered during expiration, the inverse-RSA model with heartbeats scattered during inspiration and clustered during expiration, and the control model with regular heartbeats with the identical R-R intervals were generated with the electrical stimulation to the right vagal nerve. Arterial/mixed venous blood and inspiratory/expiratory airflow were sampled for gas analyses. The fraction of alveolar gas volume unable to interface with sufficient blood flow during inspiration (wasted ventilation or the ratio of physiological dead space to tidal volume) [46] and the fraction of capillary blood volume unable to interface with sufficient fresh gas during expiration (wasted blood flow or the fraction of intrapulmonary shunt) [47] were compared among the three models.

As the result, these two parameters reflecting the lack of efficacy in the pulmonary gas exchange were minimal in the RSA model, maximal in the inverse-RSA model, and intermediate in the control model.

8.4.1.2 Implications

The results have suggested that the RSA of a respiratory-circulatory interaction benefits the pulmonary gas exchange and enhances the energy efficacy of circulation by supplying "necessary" heartbeats during inspiration and saving "unnecessary" heartbeats during expiration.

The concept that the efficacy of pulmonary gas exchange is improved by RSA, which could be applicable to the healthy humans [39–49], seems to have gained favor among pulmonary/cardiovascular physicians and anesthesiologists [50, 51]. However, controversy still remains on the effects of RSA on the pulmonary gas exchange [52–54], and the central or peripheral origins of RSA have been under much debate among physiologists [55, 56].

8.4.2 Respiratory Muscles and RSA (#2 Experiment)

We summarize the author's experiment on the respiratory muscles and RSA in the conscious canines [57], which are trained to lie down quietly and sleep in the laboratory [58, 59].

8.4.2.1 Summary

Figure 8.3 shows a tracing of the electroencephalogram (EEG), ECG, respiratory variables, and electromyograms of the diaphragm (EMG_{di}), the transversus abdominis (EMG_{ta}) of the major expiratory muscle, and the external oblique

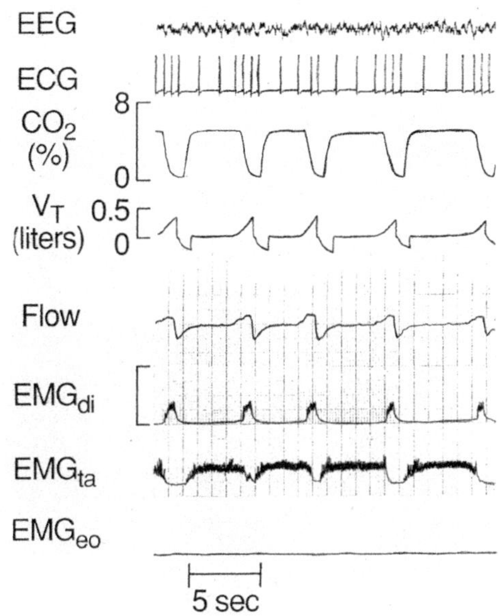

Fig. 8.3 Polygraphic recording during spontaneous breathing in a trained canine. From *top* to *bottom*, electroencephalogram (EEG), electrocardiogram (ECG), expired CO_2 concentration (%), tidal volume (liters), flow, electromyogram (EMG) of the diaphragm (EMG_{di}), EMG of the transversus abdominis (EMG_{ta}), and EMG of the external oblique (EMG_{eo}) are displayed with respiratory frequency of 12 (/min) under spontaneous breathing. Respiratory sinus arrhythmia of a clustering of heartbeats during inspiration and a scattering during expiration is clearly seen with frequency of 63 (/min). Note that the dog is sleeping according to the EEG and behavioral criteria, the EMG_{ta} of the major expiratory muscle is activated and the EMG_{eo} of the minor one is not recruited during tidal breathing

(EMG_{eo}) of the minor expiratory muscle during tidal breathing. A clustering of heartbeats during inspiration with the diaphragmatic contraction and a scattering during expiration with the contraction of transversus abdominis are observed.

The canine sleeps spontaneously (non-REM sleep) according to the EEG and behavioral criteria [58, 60]. From the standpoint of respiratory control, the state of non-REM sleep is important, since the ventilation critically depends on metabolic control, excluding the non-metabolic (behavioral) influences on the respiratory control system [61]. Moreover, in sleeping canines, the effects of anesthetics or drug-induced levels of consciousness do not have any influence on the neural control of respiration, circulation, and their interaction [58].

8.4.2.2 Implications

In spontaneous tidal breathing, the central respiratory control modulates to "breathe in" the air by activating the EMG_{di} during inspiration and to "breathe out" the air by activating only the EMG_{ta}, not the EMG_{eo}, during expiration. It is noteworthy that, only in stimulated breathing, the EMG_{eo} of the minor expiratory muscle is recruited according to the quality and intensity of the stimulus [57]. The interacting respiratory/circulatory centers seem to generate the differential heartbeats between their clustering with the diaphragmatic contraction during the inspiratory phase and their scattering with the contraction of the transversus abdominis during the expiratory phase.

The results of this experiment have led us from the legendary concept that "the respiratory muscles drive the chest wall along the trajectory for which the work of breathing is minimal [15]" to our unique concept that "RSA is not simply the secondary product of other known reflexes but has its own physiological role to minimize the working loads of not only breathing but also circulation" [3, 16].

8.4.3 Respiratory Muscles during Hypercapnia and Hypoxia (#3 Experiment)

We summarize the author's experiment on the respiratory muscles during hypercapnia and hypoxia in trained canines [57].

8.4.3.1 Summary

The canine's respiration and circulation were stimulated by hyperoxic hypercapnia or normocapnic hypoxia with the rebreathing method. As the result, both hypercapnia and hypoxia substantially recruited the diaphragmatic and expiratory muscles. The expiratory muscle activities were greater during hypercapnia than during hypoxia at any comparable levels of minute volume of ventilation (expiratory shift) [62].

However, the two chemical stimuli also resulted in different tidal volume and respiratory frequency at any given minute volume of ventilation. When the EMG activities were reanalyzed as a function of tidal volume, the expiratory muscle activities were the same for a given tidal volume whether induced by hypercapnia or hypoxia, but the diaphragmatic inspiratory activity was consistently greater during hypoxia than during hypercapnia (inspiratory shift) [62].

8.4.3.2 Implications

These findings might support the concept that the stimulations by hypercapnia and hypoxia result in asymmetric activation of the inspiratory and expiratory muscles to achieve the kinematics for minimal work of breathing [15]. From the teleological viewpoint, the central regulation modulates to "breathe in" the air by recruiting the diaphragm substantially during hypoxia, and it modulates to "breathe out" the air by recruiting the expiratory muscles substantially during hypercapnia.

8.4.4 RSA during Hypercapnia and Hypoxia (#4 Experiment)

We summarize the authors' experiments on the RSA during hypercapnia and hypoxia in the trained canines [38, 63, 64].

8.4.4.1 Summary

The canine was stimulated by hypercapnia or hypoxia as in Sect. 4.3 (#3 Experiment). The effects of both stimuli on RSA were analyzed with complex demodulation, a technique with which the frequency shifts and time-dependent changes in amplitude in the rhythmic component of the biological phenomena are assessed continuously [65].

As the result, during hypoxia, the RSA magnitude decreased (after adjusting the effects of respiratory rate and tidal volume), when heart rate, mean arterial blood pressure, respiratory rate, and tidal volume increased [63]. In contrast, during hypercapnia, the RSA magnitude increased (after adjusting the effects of respiratory rate and tidal volume) along with the increased respiratory rate and tidal volume, when the heart rate and mean arterial blood pressure did not change [38].

8.4.4.2 Implications

The above findings provide support for the concept that hypercapnia and hypoxia result in asymmetric activation not only of the respiratory muscles but also the hemodynamics and, further, the RSA [64].

From the teleological viewpoint, during hypoxia, the central regulation via the vagal nerve modulates to "breathe in" the air and subsequently to distribute the O_2 from the pulmonary to systemic circulation with the maximally functioning cardiac pump with sinus tachycardia, whereas the RSA is attenuated. During hypercapnia, in contrast, it modulates to "breathe out" the air and subsequently to eliminate the CO_2 from the pulmonary circulation to the open air with the modestly functioning cardiac pump with the intensified RSA, which suppresses the increase in heart rate during expiration to save the "unnecessary" heartbeats.

8.4.5 Role of Vagal Nerve on Respiratory Muscles and RSA (#5 Experiment)

We summarize the author's experiment on the role of vagal nerve on respiratory muscles and RSA in the trained canines [57].

8.4.5.1 Summary

The effects of reversible vagal blockade on respiratory muscle activation and RSA were investigated by temporarily cooling the surfaces of bilateral exteriorized cervical vagal loops [66].

Figure 8.4 is a tracing after the transient cervical vagal blockade during spontaneous sleep in the same canine of Fig. 8.3. The pattern of EMGs is obviously changed from the spontaneous tidal breathing with a frequency of 12 (/min) to the very slow and deep breathing with a frequency of 3 (/min). Although the EMG_{di} during inspiration is magnified to the maximal tidal volume above 2 l, the EMG_{ta} during expiration is abolished.

Simultaneously, it is noteworthy that the pattern of ECG is also obviously changed from the stable sinus rhythm of 63 (/min) (Fig. 8.3) to the sinus tachycardia of 168 (/min) (Fig. 8.4), while the RSA is eliminated.

8.4.5.2 Implications

When the vagal nerve is blocked temporarily at the level of the cervical vagal trunk, both of the afferent vagal input from the lungs and the efferent vagal output to the sinus node are abolished. The classical Hering-Breuer inflation reflex mediated with the vagal nerve is eliminated [66, 67], and the disappearance in the "inspiratory-off-switch" reflex generates very slow and deep breathing. The findings of the EMGs before and after the vagal blockade suggest that the increases in the expiratory muscle activity are generated by lung inflation, which is known to augment the abdominal motor activity through an afferent vagal mechanism [57, 68].

The findings indicate that afferent vagal stimuli play an important facilitating role in the expiratory muscle activation, and hence in the chest wall movement, to achieve the minimal work of breathing by suppressing the excessiveness of lung inflation.

The findings of the ECG imply that the activity of RSA is to depend on the afferent vagal inflow from the lungs and the respiratory (phasic) modulation of

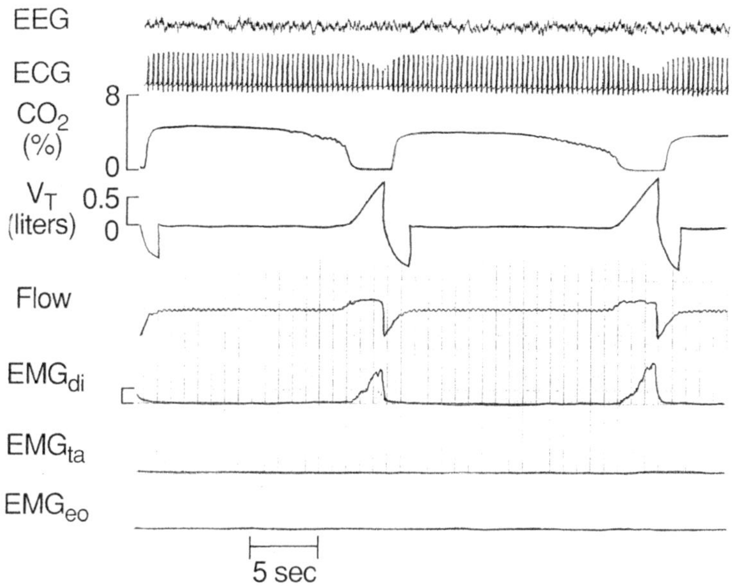

Fig. 8.4 Polygraphic recording during vagal blockade in a trained canine. Displayed parameters are as in Fig. 8.3. During vagal blockade, the maximal contraction of the diaphragm with frequency of 3 (/min) and the sinus tachycardia of continuously accelerated heartbeats with frequency of 168 (/min) are seen. Note that the dog is sleeping, the EMG_{di} is magnified during inspiration, the contraction of the EMG_{ta} during expiration is abolished, and the RSA seen in Fig. 8.3 has disappeared

vagal outflow to the sinus node. The findings also imply that the vagal nerve is to play the important roles not only in generating the RSA to achieve the minimal work of circulation, saving "unnecessary" heartbeats during expiration, but also in maintaining the homeostasis with the cardiopulmonary works to be minimal, which might serve for the wisdom of the body [13].

8.5 Clinical Significances of CSR and Entrainment with CSR

8.5.1 Clinical Significance of CSR

In an approximately 30 % of congestive heart failure patients, CSR of a cyclical fluctuation with episodes of central apneas/hypopneas alternating with episodes of hyperventilation is found [69–71], and it is more often found during nighttime sleep than daytime wakefulness [72, 73].

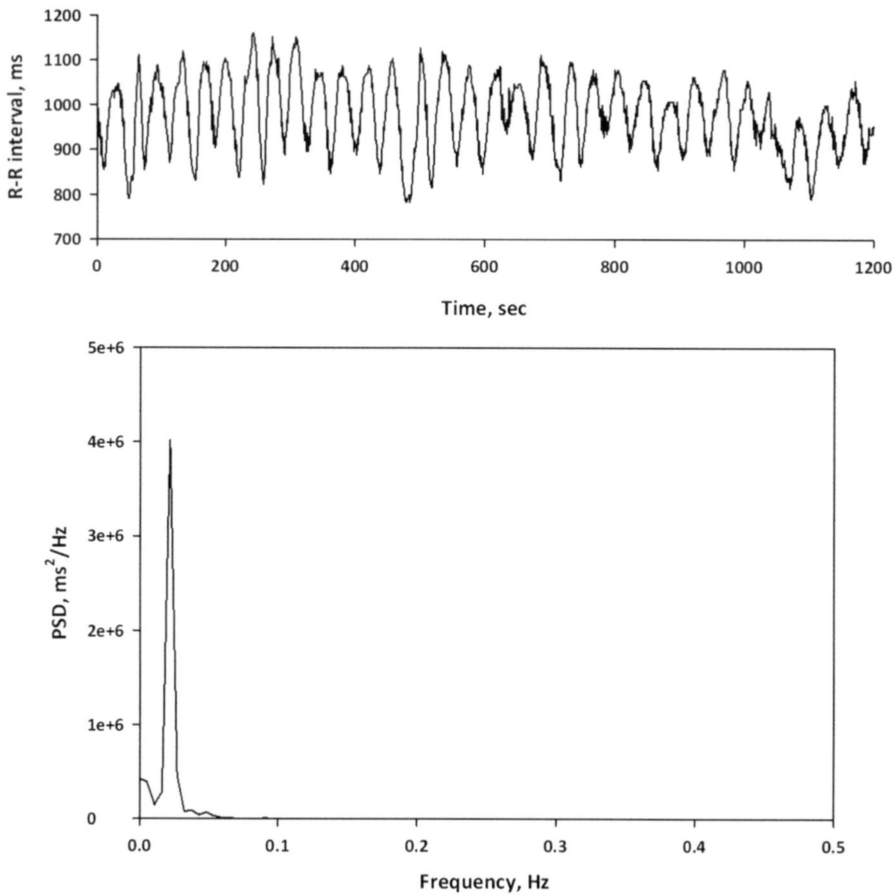

Fig. 8.5 Power spectral densities of heart rate (and respiratory variables) in a representative heart failure patient with and without Cheyne-Stokes respiration (CSR). A peak oscillatory density in heartbeats (and respiratory variables) is occurring at the very low frequency band during CSR, which disappears when the CSR does not occur

8.5.2 Clinical Significance of Entrainment with CSR

Figure 8.5 shows a representative spectral analysis of heart rate variability in a heart failure patient with CSR, which exhibits the peak frequency at the very low frequency (VLF) band of <0.04 Hz.

In heart failure patients, the RSA with its peak at the HF band disappears [14] and, instead, the heart rate variability with its peak at the VLF band appears [22]. The power spectral peak in R-R interval and blood pressure dominates the VLF band, and it occurs at precisely the same frequency as the peak in respiration, which would be a reflection of entraining heartbeats with CSR (**Entrainment** with CSR). However, there has not been definitive evidence that the heart rate and blood pressure oscillations at the VLF band are caused by CSR [74], which could be triggered by obstructive sleep apnea [75], or other physiological circumstances [76].

In a simulated CSR in healthy volunteers, the instructed fluctuated ventilation amplifies oscillations in heart rate in the absence of hypoxia or arousal from sleep. In the heart failure patients without CSR, the peak at the VLF band is not observed [2, 74].

8.5.3 Implications

These findings imply that the **Entrainment** with CSR in congestive heart failure might be a physiological synchronization of circulation with respiration within each cycle length of respiratory periodicity, which could be comparable to the RSA within each cycle length of tidal breathing.

8.6 Physiological Significances of CSR and Entrainment with CSR

8.6.1 Physiological Significance of CSR

The animal model of CSR, to our knowledge, has been rarely prepared except for those in anesthetized dogs quite long time ago [4, 7], and the computer simulation models of CSR have been studied lately [9–11].

From the standpoint of cardiopulmonary efficacy, the mathematical model of the CSR indicates that an intermittent working of the respiratory organs followed by their resting has more advantages than continuous work of tidal breathing [77]. In patients with advanced heart failure, the maximal inspiratory and expiratory pressures are reduced by approximately 25–50 % [78], and the RSA of respiratory fluctuation in R-R interval on ECG has almost disappeared [14].

Therefore, it could be assumed that the cyclic working and resting (hyperventilation and central apnea/hypopnea) by the CSR would offset the risks of developing respiratory muscle fatigue [79] and disappearing RSA with the mechanism of an improved ventilation-perfusion efficiency [40].

8.6.2 Physiological Significance of Entrainment with CSR

The associate phenomenon of the **Entrainment** with CSR during CSR seemingly has been out of focus of the investigation among most physiologists.

In patients with congestive heart failure, stroke volume has been reported to increase by 25 % during the hyperventilation phase compared with the apneic/hypopneic phase [80]. Accordingly, the cardiac output increases, at least 25 % during CSR, which could be generated by the **Entrainment** with CSR. Moreover, in the majority of heart failure patients, the CSR is more enhanced during sleep than wakefulness [72, 73], as is the RSA more enhanced during sleep than wakefulness [12, 14, 81]. It has been known that the CSR is often exaggerated during exercise rather than resting [82, 83]. Additionally, patients with atrial fibrillation are more likely to have CSR than those with regular sinus rhythm [84, 85].

8.6.3 Implications

These findings imply that not only the alveolar gas volume interfacing with sufficient blood flow but also the capillary blood volume interfacing with sufficient fresh gas should be increased during the CSR hyperventilation phase. Hence, the frequent coexistence of CSR in congestive heart failure is likely to improve the efficacy of pulmonary gas exchange by entraining heartbeats with CSR.

These findings also imply that the CSR could offset the lack of RSA by entraining heartbeats with CSR in congestive heart failure and atrial fibrillation, in both of which prevalence of CSR is greater than the non-affected humans [69–71, 84, 85]. It is assumed that the cardiopulmonary efficacy could be improved to compensate for the inability to adopt the physiological changes induced by sleep and by exercise in heart failure or in atrial fibrillation.

8.7 Conclusions

The authors addressed the concept previously that "RSA benefits the pulmonary gas exchange to enhance the energy efficacy of circulation by supplying "necessary" heartbeats during inspiration and by saving "unnecessary" heartbeats during expiration." Analogously, we have proposed the further concept herein that "the **Entrainment** with CSR might function to supply "effective" heartbeats during the hyperventilation phase and save "ineffective" heartbeats during the apneic/hypopneic phase with the cardio-pulmonary works to be minimal" by synchronizing heartbeats with breathing, which would be tested in the future in integrative physiology [86].

References

1. Ludwig C. Beiträge zur Kenntniss des Einflusses der Respirationsbewegungen auf den Blutlauf im Aortensysteme. Arch Anat Physiol Leipzig. 1847;13:242–302.
2. Lorenzi-Filho G, Dajani HR, Leung RS, Floras JS, Bradley TD. Entrainment of blood pressure and heart rate oscillations by periodic breathing. Am J Respir Crit Care Med. 1999;159:1147–54.
3. Yasuma F. Is Cheyne-Stokes respiration friend or foe of heart failure? Thorax. 2013;68:106–7.
4. Eyster JAE. Clinical and experimental observations upon Cheyne-Stokes respiration. J Exp Med. 1906;8:565–613.
5. Douglas CG, Haldane JS. The causes of periodic or Cheyne-Stokes breathing. J Physiol. 1909;38:401–19.
6. Pryor WW. Cheyne-Stokes respiration in patients with cardiac enlargement and prolonged circulation time. Circulation. 1951;4:233–8.
7. Guyton AC, Crowell JW, Moore JW. Basic oscillating mechanism of Cheyne-Stokes breathing. Am J Physiol. 1956;187:395–8.
8. Cheyne J. A case of apoplexy in which the fleshy part of the heart was converted into fat. Dublin Hosp Rep. 1818;2:216–22.
9. Cherniack NS, Longobardo GS. Abnormalities in respiratory rhythm. In: Cherniack NS, Widdicombe JG, editors. Handbook of physiology, the respiratory system, Control of breathing part 2, vol. 2. Washington, DC: American Physiological Society; 1986. p. 729–49.
10. Khoo MC, Gottschalk A, Pack AI. Sleep-induced periodic breathing and apnea: a theoretical study. J Appl Physiol. 1991;70:2014–24.
11. Cherniack NS, Longobardo G, Evangelista CJ. Causes of Cheyne-Stokes respiration. Neurocrit Care. 2005;3:271–9.
12. Yasuma F, Hayano J. Respiratory sinus arrhythmia: why does the heart beat synchronize with respiratory rhythm? Chest. 2004;125:683–90.
13. West BJ. The wisdom of the body; a contemporary review. Front Physiol. 2010;1:1. doi:10.3389/fphys.2010.00001.
14. Hayano J, Yasuma F. Hypothesis: respiratory sinus arrhythmia is an intrinsic resting function of cardiovascular system. Cardiovasc Res. 2003;58:1–9.
15. Wilson TA, Angelillo M, Legrand A, De Troyer A. Muscle kinematics for minimal work of breathing. J Appl Physiol. 1999;87:554–60.
16. Ben-Tal A, Shamailov SS, Paton JF. Evaluating the physiological significance of respiratory sinus arrhythmia: looking beyond ventilation-perfusion efficiency. J Physiol. 2012;590 (Pt 8):1989–2008.
17. Taylor EW, Jordan D, Coote JH. Central control of the cardiovascular and respiratory systems and their interactions in vertebrates. Physiol Rev. 1999;79:855–916.
18. Hamlin RL, Smith CR, Smetzer DL. Sinus arrhythmia in the dog. Am J Physiol. 1966;210:321–8.
19. Scher AM, Young AC. Reflex control of heart rate in the unanesthetized dog. Am J Physiol. 1970;218:780–9.
20. Pagani M, Lombardi F, Guzzetti O, Rimoldi O, Furlan R, Pizzinelli P, et al. Power spectral analysis of heart rate and arterial blood pressure variabilities as a marker of sympatho-vagal interaction in man and conscious dog. Circ Res. 1986;59:178–93.
21. Thompson CR, Brown GS, Gee H, Taylor EW. Heart rate variability in healthy term newborns: the contribution of respiratory sinus arrhythmia. Early Hum Dev. 1993;31:217–28.
22. Trinder J, Merson R, Rosenberg JI, Fitzgerald F, Kleiman J, Bradley TD. Pathophysiological interactions of ventilation, arousals, and blood pressure oscillations during Cheyne-Stokes respiration in patients with heart failure. Am J Respir Crit Care Med. 2000;162:808–13.
23. Butler PJ, Taylor EW. Factors affecting the respiratory and cardiovascular responses to hypercapnic hypoxia in mallard ducks. Respir Physiol. 1983;53:109–27.

24. Porges SW, Riniolo TC, McBride T, Campbell B. Heart rate and respiration in reptiles: contrasts between a sit-and-wait predator and an intensive forager. Brain Cogn. 2003;52:88–96.
25. Campbell HA, Leite CA, Wang T, Skals M, Abe AS, Egginton S, et al. Evidence for a respiratory component, similar to mammalian respiratory sinus arrhythmia, in the heart rate variability signal from the rattlesnake, Crotalus durissus terrificus. J Exp Biol. 2006;209 (Pt 14):2628–36.
26. Satchell GH. The reflex co-ordination of the heart beat with respiration in the dogfish. J Exp Biol. 1960;37:719–31.
27. Campbell HA, Taylor EW, Egginton S. Does respiratory sinus arrhythmia occur in fishes? Biol Lett. 2005;1:484–7.
28. Taylor EW. Nervous control of the heart and cardio-respiratory interactions. In: Hoar WS, Randall DJ, Farrell AP, editors. Fish physiology, vol. XIIV. New York: Academic; 1992. p. 343–p387.
29. Taylor JA, Myers CW, Halliwill JR, Seidel H, Eckberg DL. Sympathetic restraint of respiratory sinus arrhythmia: implications for vagal-cardiac tone assessment in humans. Am J Physiol Heart Circ Physiol. 2001;280:H2804–14.
30. Eckberg DL. The human respiratory gate. J Physiol. 2003;548:339–52.
31. Chess GF, Tam RM, Calaresu FR. Influence of cardiac neural inputs on rhythmic variations of heart period in the cat. Am J Physiol. 1975;228:775–80.
32. Eckberg DL. Human sinus arrhythmia as an index of vagal cardiac outflow. J Appl Physiol Respir Environ Exerc Physiol. 1983;54:961–6.
33. Hayano J, Mukai S, Sakakibara M, Okada A, Takata K, Fujinami T. Effects of respiratory interval on vagal modulation of heart rate. Am J Phyisiol. 1994;267:H33–40.
34. Malik M, Camm AJ. Heart rate variability and clinical cardiology. Br Heart J. 1994;71:3–6.
35. Katona PG, Jih F. Respiratory sinus arrhythmia: noninvasive measure of parasympathetic cardiac control. J Appl Physiol. 1975;39:801–5.
36. Hrushesky WJ, Fader D, Schmitt O, Gilbertsen V. The respiratory sinus arrhythmia: a measure of cardiac age. Science. 1984;224:1001–4.
37. Goldberger JJ, Ahmed MW, Parker MA, Kadish AH. Dissociation of heart rate variability from parasympathetic tone. Am J Physiol. 1994;266:H2152–7.
38. Yasuma F, Hayano J. Augmentation of respiratory sinus arrhythmia in response to progressive hypercapnia in conscious dogs. Am J Physiol Heart Circ Physiol. 2001;280:H2336–41.
39. Sasano N, Vesely AE, Hayano J, Sasano H, Somogyi R, Preiss D, et al. Direct effect of $PaCO_2$ on respiratory sinus arrhythmia in conscious humans. Am J Physiol Heart Circ Physiol. 2002;282:H973–6.
40. Hayano J, Yasuma F, Okada A, Mukai S, Fujinami T. Respiratory sinus arrhythmia. A phenomenon improving pulmonary gas exchange and circulatory efficiency. Circulation. 1996;94:842–7.
41. Yasuma F, Nomura H, Ogawa S, Miyaguchi K, Narita G, Hama Y, et al. Hemodynamic study of negative pressure ventilation using diaphragm pacing. Kokyu To Junkan. 1989;37:977–81 (in Japanese with English abstract).
42. Fewell JE, Abendschein DR, Carlson J, Murray JF, Rapaport E. Continuous positive-pressure ventilation decreases right and left ventricular end-diastolic volumes in the dog. Circ Res. 1980;46:125–32.
43. Daly MDB. Interactions between respiration and circulation. In: Cherniack NS, Widdicombe JG, editors. Handbook of physiology, the respiratory system, Control of Breathing part 2, vol. 2. Washington, DC: American Physiological Society; 1986. p. 529–94.
44. Van de Louw A, Médigue C, Papelier Y, Cottin F. Breathing cardiovascular variability and baroreflex in mechanically ventilated patients. Am J Physiol Regul Integr Comp Physiol. 2008;295:R1934–40.

45. Cooper HE, Clutton-Brock TH, Parkes MJ. Contribution of the respiratory rhythm to sinus arrhythmia in normal unanesthetized subjects during positive-pressure mechanical hyperventilation. Am J Physiol Heart Circ Physiol. 2004;286:H402–11.
46. Severinghaus JW, Stupfel M. Alveolar dead space as an index of distribution of blood flow in pulmonary capillaries. J Appl Physiol. 1957;10:335–48.
47. Hass F, Bergofsky EH. Effect of pulmonary vasoconstriction on balance between alveolar ventilation and perfusion. J Appl Physiol. 1968;24:491–7.
48. Giardino ND, Glenny RW, Borson S, Chan L. Respiratory sinus arrhythmia is associated with efficiency of pulmonary gas exchange in healthy humans. Am J Physiol Heart Circ Physiol. 2003;284:H1585–91.
49. Ito S, Sasano H, Sasano N, Hayano J, Fisher JA, Katsuya H. Vagal nerve activity contributes to improve the efficiency of pulmonary gas exchange in hypoxic humans. Exp Physiol. 2006;91:935–41.
50. Morgan BJ, Croix CMS, Skatrud JB. Influence of respiration on autonomic control of heart rate and blood pressure. In: Bradley TD, Floras JS, editors. Sleep Apnea – implications in cardiovasucular and cerebrovascular disease. New York: Marcel Dekker; 2000. p. 1–p31.
51. Galletly DC, Larsen PD. Relationship between cardioventilatory coupling and respiratory sinus arrhythmia. Brit J Anaesth. 1998;80:164–8.
52. Beda A, Jandre FC, Giannella-Neto A. Changes in dead space can explain part of the reduction in gas exchange efficiency found, not necessarily linked to respiratory sinus arrhythmia. Exp Physiol. 2008;93:513–4.
53. Tzeng YC, Sin PY, Galletly DC. Human sinus arrhythmia: inconsistencies of a teleological hypothesis. Am J Physiol Heart Circ Physiol. 2009;296:H65–70.
54. Sin PY, Webber MR, Galletly DC, Ainslie PN, Brown SJ, Willie CK, et al. Interactions between heart rate variability and pulmonary gas exchange efficacy in humans. Exp Physiol. 2010;95:788–97.
55. Eckberg DL, Karemaker JM. Point: counterpoint: respiratory sinus arrhythmia is due to a central mechanism vs. respiratory sinus arrhythmia is due to the baroreflex mechanism. J Appl Physiol. 2009;106:1740–4.
56. Julien C, Parkes MJ, Tzeng SY, Sin PY, Ainslie PN, van de Borne P, et al. Comments on point: counterpoint: respiratory sinus arrhythmia is due to a central mechanism vs. respiratory sinus arrhythmia is due to the baroreflex mechanism. J Appl Physiol. 2009;106:1745–9.
57. Yasuma F, Kimoff RJ, Kozar LF, England SJ, Bradley TD, Phillipson EA. Abdominal muscle activation by respiratory stimuli in conscious dogs. J Appl Physiol. 1993;74:16–23.
58. Yasuma F, Hayashi H, Shimokata K, Yokota M, Okada T, Kitoh J. Recording of electroencephalograms and electrocardiograms during daytime sleep in trained canines: preparation of the sleeping dogs. Psychiatry Clin Neurosci. 1997;51:237–9.
59. Yasuma F, Kozar LF, Kimoff RJ, Bradley TD, Phillipson EA. Interaction of chemical and mechanical respiratory stimuli in the arousal response to hypoxia in sleeping dogs. Am Rev Respir Dis. 1991;143:1274–7.
60. Phillipson EA, Murphy E, Kozar LF. Regulation of respiration in sleeping dogs. J Appl Physiol. 1976;40:688–93.
61. Phillipson EA. Control of breathing during sleep. Am Rev Respir Dis. 1978;118:909–39.
62. Sears TA, Berger AJ, Phillipson EA. Reciprocal tonic activation of inspiratory and expiratory motoneurones by chemical drives. Nature. 1982;299:728–30.
63. Yasuma F, Hayano J. Impact of acute hypoxia on heart rate and blood pressure variability in conscious dogs. Am J Physiol Heart Circ Physiol. 2000;279:H2344–9.
64. Yasuma F, Hirai M, Hayano J. Differential effects of hypoxia and hypercapnia on respiratory sinus arrhythmia in conscious dogs. Jpn Circ J. 2001;65:738–42.
65. Hayano J, Taylor JA, Mukai S, Okada A, Watanabe Y, Takata K, et al. Assessment of frequency shifts in R-R interval variability and respiration with complex demodulation. J Appl Physiol. 1994;77:2879–88.

66. Phillipson EA, Hickey RF, Bainton CR, Nadel JA. Effect of vagal blockade on regulation of breathing in conscious dogs. J Appl Physiol. 1970;29:475–9.
67. Phillipson EA, Hickey RF, Graf PD, Nadel JA. Hering-Breuer inflation reflex and regulation of breathing in conscious dogs. J Appl Physiol. 1971;31:746–50.
68. Kelsen SG, Altose MD, Cherniack NS. Interaction of lung volume and chemical drive on respiratory muscle EMG and respiratory timing. J Appl Physiol Respir Environ Exerc Physiol. 1977;42:287–94.
69. Yasuma F, Nomura H, Hayashi H, Okada T, Tsuzuki M. Breathing abnormalities during sleep in patients with chronic heart failure. Jpn Circ J. 1989;53:1506–10.
70. Javaheri S, Parker TJ, Liming JD, Corbett WS, Nishiyama H, Wexler L, et al. Sleep apnea in 81 ambulatory male patients with stable heart failure. Types and their prevalences, consequences, and presentations. Circulation. 1998;97:2154–9.
71. Sin DD, Fitzgerald F, Parker JD, Newton G, Floras JS, Bradley TD. Risk factors for central and obstructive sleep apnea in 450 men and women with congestive heart failure. Am J Respir Crit Care Med. 1999;160:1101–6.
72. Hanly P, Zuberi-Khokhar NS. Increased mortality associated with Cheyne-Stokes respiration in patients with congestive heart failure. Am J Respir Crit Care Med. 1996;153:272–6.
73. Ancoli-Israel S, Engler RL, Friedman PJ, Klauber MR, Ross PA, Kripke DF. Comparison of patients with central sleep apnea. With and without Cheyne-Stokes respiration. Chest. 1994;106:780–6.
74. Leung RS, Floras JS, Lorenzi-Filho G, Rankin F, Picton P, Bradley TD. Influence of Cheyne-Stokes respiration on cardiovascular oscillations in heart failure. Am J Respir Crit Care Med. 2003;167:1534–9.
75. Shiomi T, Guilleminault C, Sasanabe R, Hirota I, Maekawa M, Kobayashi T. Augmented very low frequency component of heart rate variability during obstructive sleep apnea. Sleep. 1996;19:370–7.
76. Taylor JA, Carr DL, Myers CW, Eckberg DL. Mechanisms underlying very-low-frequency RR-interval oscillations in humans. Circulation. 1998;98:547–55.
77. Levine M, Cleave JP, Dodds C. Can periodic breathing have advantages for oxygenation? J Theor Biol. 1995;172:355–68.
78. Hughes PD, Polkey MI, Harrus ML, Coats AJ, Moxham J, Green M. Diaphragm strength in chronic heart failure. Am J Respir Crit Care Med. 1999;160:529–34.
79. Naughton MT. Cheyne-Stokes respiration: friend or foe? Thorax. 2012;67:357–60.
80. Maze SS, Kotler MN, Parry WR. Doppler evaluation of changing cardiac dynamics during Cheyne-Stokes respiration. Chest. 1989;95:525–9.
81. Carrol MS, Kenny AS, Patwari PP, Ramirez JM, Weese-Mayer DE. Respiratory and cardiovascular indicators of autonomic nervous system dysregulation in familial dysautonomia. Pediatr Pulmonol. 2012;47:682–91.
82. Olson LJ, Arruda-Olson AM, Somers VK, Scott CG, Johnson BD. Exercise oscillatory ventilation: instability of breathing control associated with advanced heart failure. Chest. 2008;133:474–81.
83. Corrà U, Pistono M, Mezzani A, Braghiroli A, Giordano A, Lanfranchi P, et al. Sleep and exertional periodic breathing in chronic heart failure: prognostic importance and interdependence. Circulation. 2006;113:44–50.
84. Larsen PD, Booth P, Galletly DC. Cardioventilatory coupling in atrial fibrillation. Br J Anaesth. 1999;82:685–90.
85. Leung RS, Bowman ME, Diep TM, Lorenzi-Filho G, Floras JS, Bradley TD. Influence of Cheyne-Stokes respiration on ventricular response to atrial fibrillation in heart failure. J Appl Physiol. 2005;99:1689–96.
86. Lenfant C. Integrative physiology: remember the big picture. Circulation. 1995;91:1901.

Chapter 9
Heart Rate Variability (HRV) and Sympathetic Nerve Activity

Ken Kiyono, Junichiro Hayano, Eiichi Watanabe, and Yoshiharu Yamamoto

Abstract The available epidemiological and clinical data implicate increased sympathetic nervous system activity in increased cardiovascular morbidity and mortality and show that it has strong predictive power for mortality and cardiovascular events. Analysis of heart rate variability (HRV) has been widely used as a noninvasive assessment tool for autonomic nervous system function, and results show that reduced and/or abnormal HRV is associated with an increased risk of mortality in cardiac patients such as patients after acute myocardial infarction and patients with congestive heart failure. However, most indices derived from HRV primarily reflect vagal function. In contrast, few indices have been suggested as markers of sympathetic nervous system activity. This chapter reviews characteristics of HRV that have been proposed as potential markers of cardiac sympathetic activity, such as (in the frequency domain) low-frequency (LF) power, short-term scaling exponent, and non-Gaussianity index. While there is no widely accepted and well-tested HRV-based index of cardiac sympathetic activity, we discuss the key issues for the assessment of cardiac sympathetic activity based on HRV analysis.

Keywords Heart rate variability • Autonomic nervous system • Sympathetic nervous system • Nonlinear index

---

K. Kiyono (✉)
Division of Bioengineering, Graduate School of Engineering Science, Osaka University, 1-3 Machikaneyama, Toyonaka, Osaka 560-8531, Japan
e-mail: kiyono@bpe.es.osaka-u.ac.jp

J. Hayano
Department of Medical Education, Graduate School of Medical Sciences, Nagoya City University, Nagoya, Aichi, Japan

E. Watanabe
Division of Cardiology, Department of Internal Medicine, School of Medicine, Fujita Health University, Toyoake, Aichi, Japan

Y. Yamamoto
Educational Physiology Laboratory, Graduate School of Education, University of Tokyo, Tokyo, Japan

S. Iwase et al. (eds.), *Clinical Assessment of the Autonomic Nervous System*,
DOI 10.1007/978-4-431-56012-8_9

9.1 Introduction

In the complex network of the autonomic nervous system, the sympathetic nervous system plays an important role in the regulation of cardiovascular function and in the pathophysiology of a variety of disease states. To date, many epidemiological and clinical studies have shown that increased activity of the sympathetic nervous system leads to an increase in cardiovascular morbidity and mortality [1, 2]. For instance, persistently elevated heart rate due to an autonomic imbalance in the direction of sympathetic overactivity is a known cardiovascular risk factor and predictor of cardiovascular mortality as well as of all-causes mortality [3–7]. This association has been observed in general populations, in patients with hypertension or diabetes, and in patients with cardiac diseases [3–5, 7–11]. As shown by measurements of muscle sympathetic nerve activity [12, 13] and of circulating plasma catecholamines [14, 15], the available experimental evidence indicates that sympathetic nervous system activity increases with age [16]. This age-related change has been considered to be causally involved in the increasing incidence with age of a variety of cardiovascular disorders, including heart failure, essential hypertension, and ventricular arrhythmias [17, 18].

Moreover, the pathophysiological significance of sympathetic overactivity is highlighted by the observation that, in patients with coronary disease, the use of the short-acting calcium antagonist nifedipine in moderate to high doses causes an increase in total mortality [19]. This could be due to a reflex increase in sympathetic activity that would be expected from the pharmacodynamics of this agent [20]. On the other hand, it has been shown that the β-adrenoceptor blocking agents (β-blockers) used to antagonize the effects of sympathetic overactivity are particularly useful for the treatment of cardiovascular disease [21–23]. In patients with severe heart failure, sympathetic activity is markedly elevated, and high noradrenaline levels are predictors of mortality [24–26]. Several large, randomized, placebo-controlled clinical trials have shown that β-blockers, when combined with angiotensin-converting-enzyme (ACE) inhibitors, substantially reduce mortality, decrease sudden death, and improve symptoms in heart failure patients with impaired systolic function [27, 28]. The beneficial effect of β-blockers in chronic heart failure could be achieved by a number of possible mechanisms, such as upregulation of myocardial beta 1 receptor density and/or reductions in heart rate, blood pressure, myocardial oxygen consumption, and circulating levels of vasoconstrictors [29].

In addition, studies using animal models have shown that sympathetic nervous activity plays an important pathogenic role as a trigger of ventricular arrhythmias and sudden cardiac death. Myocardial ischemia demonstrably increases the cardiac sympathetic activity that contributes to malignant ventricular arrhythmias [30–32]. For example, Schwartz et al. demonstrated that transient coronary occlusion provokes a reflex increase in cardiac sympathetic activity that directly promotes ventricular fibrillation [33, 34]. Moreover, in the setting of a healed myocardial infarction, a combination of acute myocardial ischemia and physiologically elicited sympathetic overactivity can consistently induce malignant ventricular arrhythmias [30, 31].

Accumulating evidence suggests that sympathetic overactivity is a potential cause of fatal cardiovascular events and is presumably the major contributor to arrhythmic events. Cardiac sympathetic activity in particular seems to be important because sympathetic outflow is not necessarily uniformly distributed across all organs, and the cardiac sympathetic nervous system is associated with a prognosis of heart failure [2]. Therefore, noninvasively assessing sympathetic input specifically to the heart is of great importance.

A widely used noninvasive tool for the assessment of autonomic nervous system activity is analysis of heart rate variability (HRV) [35]. Even under resting conditions, heart rate is modulated almost continuously by the activity of the sympathetic and/or parasympathetic (vagal) nervous systems. Therefore it is possible to evaluate autonomic nervous system activity based on characteristics of the HRV time series. Moreover, abnormality in the HRV characteristics, such as reduced variability, is associated with higher mortality in cardiac patients [36–38]. Hence, the HRV characteristics are expected to serve as diagnostic and prognostic markers of various cardiovascular disorders and to have widespread application. However, it is important to note that most HRV indices, including commonly used time-domain and frequency-domain indices, primarily reflect vagal function. On the other hand, there is no widely accepted and well-tested index of sympathetic nervous system activity based on measurements of heart rate. Establishment of a powerful HRV index probing sympathetic activity still remains an important and difficult task.

Among commonly used HRV indices, the low-frequency (LF: 0.04–0.15 Hz) power of the HRV Fourier spectrum has been presumed to reflect some aspects of cardiac sympathetic modulation, and the ratio of LF power to high-frequency (HF: 0.15–0.40 Hz) power (LF/HF ratio) indicates the sympathovagal balance [35, 39]. In short-term HRV recordings (~ minutes), the relative power contribution of the LF component is considered to reflect sympathetic modulation [40]. However, many studies using pharmacological, physiological, and psychological manipulations affecting sympathetic activity, and hence, presumably, HRV, have challenged the putative association of cardiac sympathetic activity with LF power [41–44]. The interpretation of LF power and the LF/HF ratio in long-term HRV recordings in cardiac patients is also controversial. For instance, in patients after acute myocardial infarction (AMI), a decrease, not an increase, in LF power and LF/HF ratio is commonly observed and is associated with an increased risk of all-cause mortality [45–48]. The interpretation and clinical significance of the LF component have aroused intense interest and persistent controversy.

To characterize the nonlinear and complex behavior of HRV, a variety of nonlinear indices has been proposed [49, 50]. Nonlinear indices are expected to provide information on HRV characteristics complementary to that from conventional time- and frequency-domain indices as well as better diagnostic and prognostic information. Among such nonlinear indices, a few, such as the short-term scaling exponent α_1 [51–53] and the non-Gaussianity index [54–56], are suggested to be associated with sympathetic activity. However, a mathematical link between

α_1 and the LF/HF ratio can be shown [57, 58]. Thus, the interpretation of α_1 encounters problems similar to the case of the LF/HF ratio. On the other hand, an association of increases in the non-Gaussianity index $\lambda_{25\text{sec}}$ with cardiac sympathetic overactivity has been suggested based on the following observations:

1. Increased $\lambda_{25\text{sec}}$ was consistently associated with increased mortality risk in patients with congestive heart failure (CHF) and with increased cardiac mortality risk after acute myocardial infarction (AMI) [54, 55].
2. The $\lambda_{25\text{sec}}$ showed no substantial correlation with the HRV indices reflecting vagal heart rate regulation in patients with CHF and following AMI [54, 55].
3. The $\lambda_{25\text{sec}}$ was decreased post-AMI in patients taking β-blockers [55].
4. A marked increase in $\lambda_{25\text{sec}}$ observed in patients with CHF was not observed in patients with multiple system atrophy or Parkinson disease who had sympathetic dysfunction [56].

In this chapter, we review HRV indices proposed as potential markers of cardiac sympathetic activity: frequency-domain indices (Sect. 9.2), short-term scaling exponent α_1 (Sect. 9.3.1), and the non-Gaussianity index (Sect. 9.3.2). While there is no widely accepted and well-tested HRV index of sympathetic activity, we discuss the key issues and ideas for evaluating cardiac sympathetic activity based on HRV analysis.

9.2 Low-Frequency (LF) Power of Heart Rate Variability

Frequency-domain HRV analysis based on power spectral estimation has been widely employed to assess sympathetic activity [35]. In healthy subjects under controlled conditions, a typical HRV power spectrum displays two well-defined peaks corresponding to oscillating components in high-frequency (HF; 0.15–0.4 Hz) and low-frequency (LF; 0.04–0.15 Hz) bands (Fig. 9.1). The HF component reflects synchronization between respiration cycles and HRV, also referred to as respiratory sinus arrhythmia (RSA); and the LF component is associated with Mayer waves (at approximately 0.1 Hz), which are oscillations related to regulation of blood pressure and vasomotor tone. Spectral powers in the HF and LF bands have most commonly been used in practical methods of noninvasive assessment of autonomic nervous system activity.

HF power is generally accepted to reflect RSA, which is mediated by parasympathetic activity. Therefore, HF power is taken as an index of cardiac parasympathetic tone. However, it has been shown that sympathetic cardiac blockade augments vagally mediated heart rate oscillations, including the HF component [59, 60]. Therefore, the HF power is, in some degree, affected by sympathetic tone and may not be a pure parasympathetic index [61].

In contrast, the interpretation of LF-power-related indices is controversial. Many studies have presumed that LF power reflects both sympathetic and vagal influences or that LF power in normalized units provides an index of cardiac sympathetic tone,

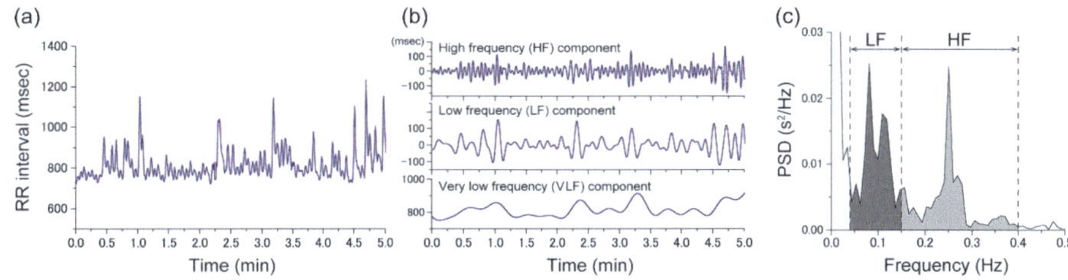

Fig. 9.1 Frequency-domain HRV analysis. (**a**) Time series of R–R intervals in a 31-year-old male subject. The time series was resampled at evenly spaced intervals (2 Hz). (**b**) The decomposition of the same time series into high-frequency (HF: 0.15–0.40 Hz), low-frequency (LF: 0.04–0.15 Hz), and very-low-frequency (0–0.04 Hz) components. These three traces were obtained by inverse Fourier transforming the corresponding frequency bands. (**c**) Power spectral density (power spectrum) of the same HRV time series. HF and LF areas under the power spectrum are equivalent to the variance of the HF- and LF-component time series, respectively, shown in (**b**)

and the ratio of LF to HF power (LF/HF ratio) indicates sympathovagal balance [35, 39]. In this view, an increase in LF power is assumed to indicate an increase in sympathetic activity, and a higher value of the LF/HF ratio indicates a shift of sympathovagal balance toward a sympathetic predominance. In short-term HRV recordings under experimental conditions affecting autonomic response, the relative power contribution of the LF component is considered a marker of sympathetic activity [39, 40]. However, many studies using pharmacological, physiological, or psychological manipulations affecting sympathetic activity and HRV have challenged the putative association of cardiac sympathetic activity with LF power [41, 42]. One important fact is that vagal blockade dramatically reduces LF power [62], whereas sympathetic blockade has no significant effect [42]. The available data suggest that over a wide frequency range including the LF and HF bands, the HRV power spectrum is mainly determined by the parasympathetic system [44]. It has also been suggested that LF power is an index of baroreceptor reflex gain (or sensitivity) because of the contributions from oscillations in blood pressure and vasomotor activity to this band [43, 63, 64].

More controversial is the interpretation of LF power and the LF/HF ratio in patients with a marked reduction of ventricular function. In such patients, whereas sympathetic activity is clearly elevated, a decrease in LF power and the LF/HF ratio is commonly observed and is associated with increased risk of mortality [47, 52]. Therefore, the clinical significance of the association of cardiac sympathetic overactivity with LF power-related indices is questionable.

9.3 Nonlinear HRV Indices and Sympathetic Activity

In the framework of linear stochastic processes as described by autoregressive models, the power spectrum provides a complete characterization of the process if the stationarity condition is fulfilled [65]. However, many real-world signals including HRV time series cannot be fully characterized based on the assumption of stationary linear processes and display much more complex behavior. To quantify such nonlinear properties, concepts and analysis methods developed in nonlinear dynamics and statistical physics have been applied to HRV analysis. As a result, various types of nonlinear HRV indices have been proposed: e.g., scaling exponents characterizing fractal and long-range correlation characteristics [51, 66], multifractal properties [67], Poincaré plot-based indices [68], entropy measures [69–71], symbolic pattern statistics [70, 72], and non-Gaussian properties [73]. Some nonlinear indices are expected to provide information on HRV characteristics complementary to that from traditional linear methods, as well as better diagnostic and prognostic information. However, their interpretations are not yet fully elucidated. Among nonlinear indices suggesting clinical significance, a few, such as the short-term scaling exponent α_1 [52, 53] and the non-Gaussianity index [54–56], are suggested to be associated with sympathetic activity. In this section, we discuss the association of sympathetic activity with these nonlinear indices.

9.3.1 Short-Term Scaling Exponent

The short-term scaling exponent α_1 is estimated by detrended fluctuation analysis (DFA; see Fig. 9.2 for details) [51]. The DFA is used to quantify long-range temporal correlations and fractal scaling properties of time series, as observed in, for example, fractional Brownian motion and fractional Gaussian noise. This method was developed to accurately quantify long-range temporal correlations embedded in nonstationary time series.

In HRV analysis, short-term (<11 beats) and long-term (>11 beats) scaling exponents, α_1 and α_2, have commonly been calculated [47]. Decreased short-term scaling exponent α_1 has been shown to have a more powerful prognostic power than traditional HRV indices in various post-AMI populations [74–76]. In the Danish Investigations of Arrhythmia and Mortality ON Dofetilide (DIAMOND-MI) trial, patients post-AMI with $\alpha_1 < 0.75$ had a significantly increased risk of mortality (relative risk 3.0), and the new index was more strongly related to outcome than traditional HRV indices [47]. Additionally, in a study of consecutive patients with acute MI, reduced short-term fractal exponent was independently associated with recurrent nonfatal coronary events [77, 78].

Although the above findings suggest clinical significance for the short-term scaling properties of HRV, the physiological interpretation of α_1 remains unclear. In some studies, α_1 has been suggested to reflect the sympathovagal balance, similar to the LF/HF ratio [52, 53]. This is understandable in light of the mathematical relation between α_1 and the LF/HF ratio. Willson et al. showed mathematically that a scaling exponent estimated by DFA corresponds to standard spectral analysis [57, 58]. Based on the mathematical argument, it can be shown that α_1 is approximately related to the LF/HF ratio by the expression:

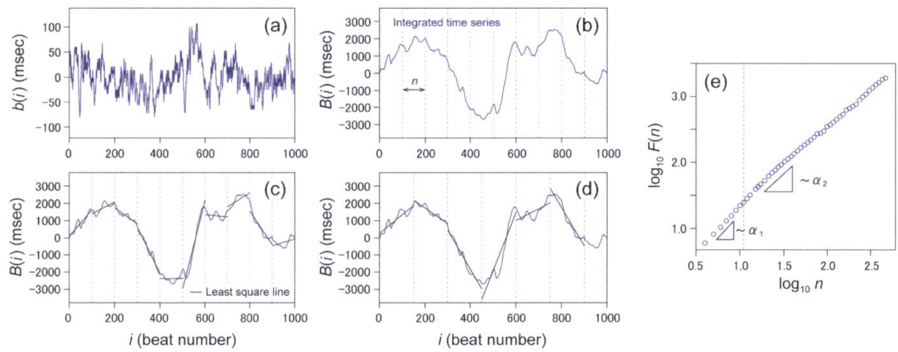

Fig. 9.2 Assessment of scaling exponents by detrended fluctuation analysis. The standard DFA procedure is as follows: (1) after subtracting the mean from each data point, (**a**) R–R interval time series $\{b(i)\}$ is integrated (**b**); (2) the integrated time series $\{B(i)\}$ is divided into equal-sized, nonoverlapping segments of length n (**b**); (3) in each segment, the mean squared deviation from the least-squares polynomial fit of degree k is calculated ($k = 1$ in **c**); the mean squared deviations are then averaged overall segments and the square root $F(n)$ is calculated; this computation (steps [2] and [3]) is repeated over multiple time scales (window sizes) to characterize the relationship between $F(n)$ and n (**d**); a linear relationship on a log–log plot of $F(n)$ as function of n indicates the power–law scaling range; in this scaling range, the fluctuations can be characterized by a scaling exponent α, the slope of the linear relation between $\log F(n)$ and $\log n$. In HRV analysis, short-term (<11 beats) and long-term (>11 beats) scaling exponents, α_1 and α_2, have commonly been calculated (**e**)

$$\alpha_1 \sim \frac{2}{1 + 1/(\mathrm{LF}/\mathrm{HF\,ratio})},$$

where the symbol "~" denotes proportionality. In agreement with this expression, a strong correlation between α_1 and the LF/HF ratio has been observed empirically. Therefore, α_1 reflects aspects similar to those characterized by the LF/HF ratio, although they are not exactly the same. As in the interpretation of the LF/HF ratio, the association of α_1 with sympathetic activity appears to be questionable.

9.3.2 Non-Gaussianity Index

Kiyono et al. proposed the non-Gaussianity index as potentially related to cardiac sympathetic over activity [54–56]. This index is calculated by multiscale probability density function (PDF) analysis to characterize the intermittent and non-Gaussian behavior of time series as shown in Fig. 9.3 [79, 80]. This type of multiscale analysis method was developed to characterize the intermittent fluctuations observed in hydrodynamic turbulence [81]. In this method, after detrending via least-squares polynomial fitting, the intermittent behavior of the HRV, such as the heterogeneity of the variance and burst-like patterns, is analyzed through the process of deformation of a non-Gaussian probability distribution (Fig. 9.3b, c)

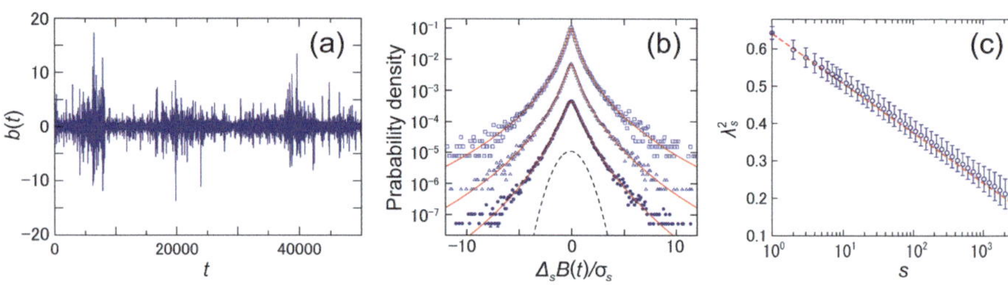

Fig. 9.3 Multiscale probability density function (PDF) analysis for intermittent time series. (**a**) An example of intermittent fluctuations generated numerically by a random cascade model [80]. (**b**) Probability-distribution deformation process across scales s of partial sums ($s = 1$, 4, 16 from *top* to *bottom*). Estimated PDFs (markers) from the time series (**a**) are plotted on log–linear scales. The observed non-Gaussian shape at each scale is quantified by the non-Gaussianity index λ_s, which is defined as a parameter of a non-Gaussian distribution model (*solid lines*) [80]. A large value of λ_s indicates a non-Gaussian shape with markedly fat tails and a peak around the mean value. For comparison, a Gaussian distribution is also shown (*dashed lines*). (**c**) Scale dependence of λ_s^2. In this plot, a gradual decline in λ_s^2 with increasing scale s indicates the process of convergence to a Gaussian distribution

across scales of coarse graining (local averaging), and the observed non-Gaussian shape at each scale is quantified by the non-Gaussianity index λ_s. The latter is defined as a parameter of a non-Gaussian distribution model (see Figs. 9.3 and 9.4 for details). The greater the λ_s, the greater the proportion of large deviations over what is expected from the Gaussian distribution.

In a study using the non-Gaussianity index, Kiyono et al. reported that, in patients with chronic heart failure (CHF), increased non-Gaussianity in the 24-h ambulatory HRV predicted increased mortality risk, while none of the conventional HRV indices, including those reflecting vagal heart rate control, were predictive of death [54]. Moreover, Hayano et al. reported that, following acute myocardial infarction, increased non-Gaussianity index $\lambda_{25\text{sec}}$ at a scale of 25 s was associated with increased cardiac mortality risk, with predictive power independent of clinical risk factors and other HRV predictors [55].

The aspects characterized by $\lambda_{25\text{sec}}$ are related to the amplitude modulation properties of the LF component shown in Fig. 9.1b. The detrending procedure in the multiscale PDF analysis acts as a high-pass filter that removes low-frequency components below a given time scale, and the calculation of partial sums (or increments of integrated series) acts as a low-pass filter. In the calculation of $\lambda_{25\text{sec}}$, the combination of these procedures acts as a band-pass filter that passes frequency components mainly in the LF band. Increased $\lambda_{25\text{sec}}$ indicates an increase in the heterogeneity of the amplitude of the LF oscillation as seen in Figs. 9.3a, and 9.4, middle right. Note that the $\lambda_{25\text{sec}}$ can change independently of the value of LF power, the latter quantifying the variance of the LF component. Thus, the $\lambda_{25\text{sec}}$ can provide information complementary to that provided by conventional frequency-domain indices.

The relationship between the autonomic nervous system and $\lambda_{25\text{sec}}$ can be understood in light of the following facts: (1) heart rate fluctuations at the scales

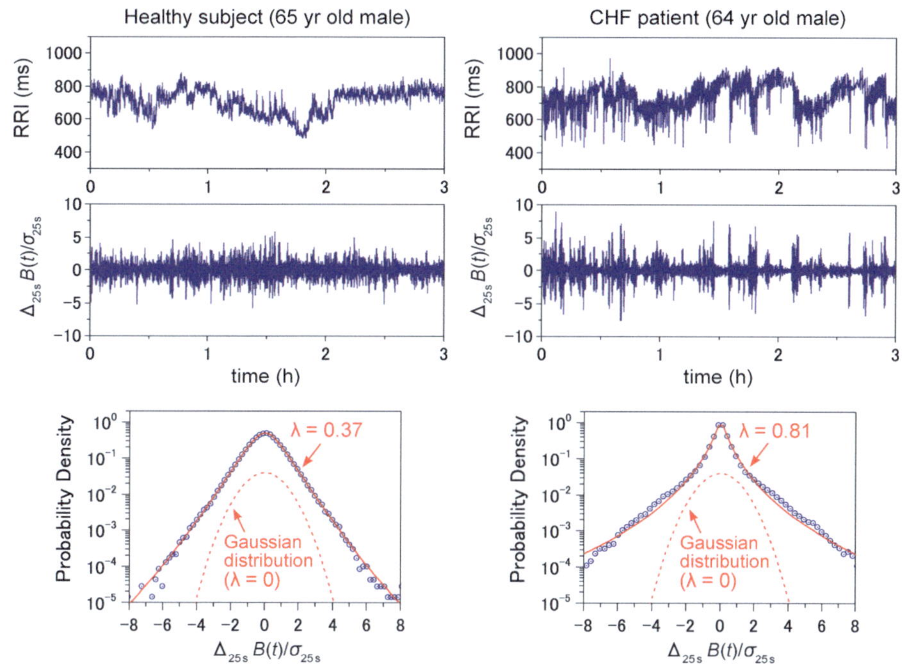

Fig. 9.4 Representative examples of non-Gaussian heart rate fluctuations in a healthy subject and in a patient with congestive heart failure. Shown are time series of normal-to-normal R–R intervals (*top row*), standardized time series of coarse-grained heart rate variations $\{\Delta_{25\text{sec}}B(t)\}$ (*middle row*), and standardized probability density functions (PDFs) of $\Delta_{25\text{sec}}B(t)$ (*bottom row*). Estimated values of the non-Gaussianity index $\lambda_{25\text{sec}}$ are shown in each panel in the *bottom* row. The *solid lines* represent the PDF approximated by a non-Gaussian model [79] with the parameter $\lambda_{25\text{sec}}$. The procedure used for this analysis is as follows: (1) time series of R–R intervals are interpolated and resampled at 4 Hz, yielding interpolated time series $\{b(t)\}$; after subtracting the mean from interpolated time series, integrated time series $\{B(t)\}$ are obtained by integrating $\{b(t)\}$ over the entire length; (2) the integrated time series $\{B(t)\}$ are divided into overlapping segments of length 2 *s* with 50 % overlap, where *s* is the scale of coarse graining ($s = 25$ s in Fig. 9.4); in each segment, the local trend is eliminated by a third-order polynomial fit; (3) coarse-grained variation $\Delta_s B(t)$ is measured as the increment with a time lag *s* of integrated and detrended time series; (4) $\{\Delta_s B(t)\}$ is standardized by its standard deviation to quantify the probability density function; then the non-Gaussianity index λ_s is estimated based on the *q*-th order moment of $\{\Delta_s B(t)\}$ as $\lambda_s^2(q) = \frac{2}{q(q-2)}\left[\ln\left(\frac{\sqrt{\pi}E[|\Delta_s B(t)|^q]}{2^{q/2}}\right) - \ln\Gamma\left(\frac{q+1}{2}\right)\right]$. In the *bottom row*, λ_s is estimated based on the 0.25th order moment ($q = 0.25$) to emphasize the center part of PDF

corresponding to the HF and LF bands are mediated almost exclusively by neural autonomic mechanisms [35, 82]; (2) the $\lambda_{25\text{sec}}$ showed no substantial correlation with HRV indices reflecting vagal heart rate regulation [54, 55]; and (3) the $\lambda_{25\text{sec}}$ was decreased in post-AMI patients taking β-blockers [55]. Based on these facts, Hayano et al. suggested that $\lambda_{25\text{sec}}$ at least partly captures heart rate fluctuations mediated by intermittent activations of cardiac sympathetic input [55].

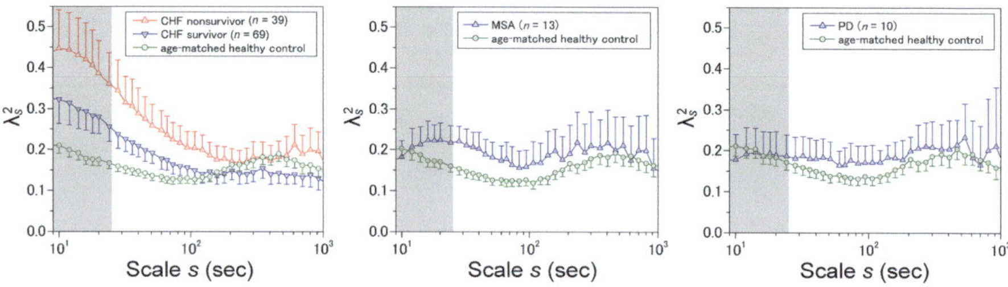

Fig. 9.5 Scale dependence of the non-Gaussianity index λ_s. (**a**) Results for patients with congestive heart failure (CHF), both for survivors ($n = 69$) and nonsurvivors ($n = 39$) within a follow-up period of 33 ± 17 months. (**b**) Results for patients with multiple system atrophy (MSA). (**c**) Results for patients with Parkinson disease (PD). Age-matched controls were selected from a database of healthy subjects. *Error bars* indicate the 95 % confidence intervals of the group means. The *gray, shaded* range corresponds to the low-frequency band ($s \leq 25$ s). Marked increases in λ_s in the *gray shaded scales* are observed only in patients with CHF, particularly nonsurvivors (**a**)

In addition, Kiyono et al. reported that, in a study of ambulatory HRV that included patients with multiple system atrophy (MSA) and patients with Parkinson disease (PD), the marked increase in $\lambda_{25\text{sec}}$ observed in patients with CHF was not observed in the patients with MSA or PD who had sympathetic dysfunction [56]. Both MSA and PD are progressive neurodegenerative disorders, although the autonomic lesion in MSA is preganglionic sympathetic failure [83], and that of PD is ganglionic and postganglionic [84, 85]. As shown in Fig. 9.5a, compared with the age-matched healthy control group, the mean values of λ_s in patients with CHF display marked increases at the relatively small scales covering the LF band. On the other hand, as shown in Fig. 9.5b, c, such marked increases were not observed in patients with MSA or PD. This result also supports the view that an increase of λ_s at scales corresponding to the LF band ($s \leq 25$ s) could be a hallmark of cardiac sympathetic overdrive.

Although the above findings suggest an association of cardiac sympathetic activity with the non-Gaussianity of the HRV, most studies evaluating HRV using the non-Gaussianity index have been limited by relatively small cohorts. Therefore, to confirm the clinical and prognostic significance of the non-Gaussianity of HRV, additional prospective studies with larger cohorts will be required.

9.4 Conclusion

The majority of studies evaluating HRV in cardiac patients have shown that reduced and/or abnormal HRV is associated with an increased risk of mortality. In these studies, most HRV indices that predict mortality risk are considered to primarily reflect reduced or impaired vagal function. In contrast, few HRV indices are related to sympathetic function. In this chapter, we try to give a better understanding of the association between sympathetic nervous system activity and such HRV indices. Among HRV indices that have been suggested as markers of sympathetic activity, increased non-Gaussianity of HRV may reflect the deleterious effects of elevated cardiac sympathetic activity in cardiac patients. In addition, some other nonlinear indices not mentioned in this chapter, such as approximate and sample entropies [85, 86], have been suggested to be associated with sympathetic activation. However, to explore the physiological and clinical significance of such indices, a more extensive program of validation will be necessary in the future.

References

1. Esler M, Kaye D. Sympathetic nervous system activation in essential hypertension, cardiac failure and psychosomatic heart disease. J Cardiovasc Pharmacol. 2000;35:S1–7.
2. Malpas SC. Sympathetic nervous system overactivity and its role in the development of cardiovascular disease. Physiol Rev. 2010;90:513–57. doi:10.1152/physrev.00007.2009.
3. Perret-Guillaume C, Joly L, Benetos A. Heart rate as a risk factor for cardiovascular disease. Prog Cardiovasc Dis. 2009;52:6–10. doi:10.1016/j.pcad.2009.05.003.
4. Palatini P, Benetos A, Grassi G, Julius S, Kjeldsen SE, Mancia G, et al. Identification and management of the hypertensive patient with elevated heart rate: statement of a European Society of Hypertension Consensus Meeting. J Hypertens. 2006;24:603–10.
5. Palatini P. Elevated heart rate in cardiovascular diseases: a target for treatment? Prog Cardiovasc Dis. 2009;52:46–60. doi:10.1016/j.pcad.2009.05.005.
6. Lang CC, Gupta S, Kalra P, Keavney B, Menown I, Morley C, et al. Elevated heart rate and cardiovascular outcomes in patients with coronary artery disease: clinical evidence and pathophysiological mechanisms. Atherosclerosis. 2010;212:1–8. doi:10.1016/j.atherosclero sis.2010.01.029.
7. Custodis F, Reil JC, Laufs U, Bohm M. Heart rate: a global target for cardiovascular disease and therapy along the cardiovascular disease continuum. J Cardiol. 2013;62:183–7. doi:10. 1016/j.jjcc.2013.02.018.
8. Goldberg RJ, Larson M, Levy D. Factors associated with survival to 75 years of age in middle-aged men and women. The Framingham Study. Arch Intern Med. 1996;156:505–9.
9. Kannel WB. Office assessment of coronary candidates and risk factor insights from the Framingham study. J Hypertens Suppl. 1991;9:S13–9.
10. Hozawa A, Ohkubo T, Kikuya M, Ugajin T, Yamaguchi J, Asayama K, et al. Prognostic value of home heart rate for cardiovascular mortality in the general population: the Ohasama study. Am J Hypertens. 2004;17:1005–10.
11. Bohm M, Swedberg K, Komajda M, Borer JS, Ford I, Dubost-Brama A, et al. Heart rate as a risk factor in chronic heart failure (SHIFT): the association between heart rate and outcomes in a randomised placebo-controlled trial. Lancet. 2010;376:886–94. doi:10.1016/S0140-6736 (10)61259-7.
12. Sundlof G, Wallin BG. Human muscle nerve sympathetic activity at rest. Relationship to blood pressure and age. J Physiol. 1978;274:621–37.
13. Ng AV, Callister R, Johnson DG, Seals DR. Age and gender influence muscle sympathetic nerve activity at rest in healthy humans. Hypertension. 1993;21:498–503.
14. Ziegler MG, Lake CR, Kopin IJ. Plasma noradrenaline increases with age. Nature. 1976;261:333–5.
15. Barnes RF, Raskind M, Gumbrecht G, Halter JB. The effects of age on the plasma catecholamine response to mental stress in man. J Clin Endocrinol Metab. 1982;54:64–9.
16. Seals DR, Esler MD. Human ageing and the sympathoadrenal system. J Physiol. 2000;528 (Pt 3):407–17.
17. Esler MD, Thompson JM, Kaye DM, Turner AG, Jennings GL, Cox HS, et al. Effects of aging on the responsiveness of the human cardiac sympathetic nerves to stressors. Circulation. 1995;91:351–8.

18. Kaye D, Esler M. Sympathetic neuronal regulation of the heart in aging and heart failure. Cardiovasc Res. 2005;66:256–64.
19. Furberg CD, Psaty BM, Meyer JV. Nifedipine dose-related increase in mortality in patients with coronary heart disease. Circulation. 1995;92:1326–31.
20. Frohlich ED, McLoughlin MJ, Losem CJ, Ketelhut R, Messerli FH. Hemodynamic comparison of two nifedipine formulations in patients with essential hypertension. Am J Cardiol. 1991;68:1346–50.
21. Thattassery E, Gheorghiade M. Beta blocker therapy after acute myocardial infarction in patients with heart failure and systolic dysfunction. Heart Fail Rev. 2004;9:107–13.
22. Manurung D, Trisnohadi HB. Beta blockers for congestive heart failure. Acta Med Indones. 2007;39(1):44–8.
23. Kezerashvili A, Marzo K, De Leon J. Beta blocker use after acute myocardial infarction in the patient with normal systolic function: when is it "ok" to discontinue? Curr Cardiol Rev. 2012;8:77–84.
24. Cohn JN, Levine TB, Olivari MT, Garberg V, Lura D, Francis GS, et al. Plasma norepinephrine as a guide to prognosis in patients with chronic congestive heart failure. N Engl J Med. 1984;311:819–23.
25. Cohn JN. Plasma norepinephrine and mortality. Clin Cardiol. 1995;18:I9–12.
26. de Lucia C, Femminella GD, Gambino G, Pagano G, Allocca E, Rengo C, et al. Adrenal adrenoceptors in heart failure. Front Physiol. 2014;5:246. doi:10.3389/fphys.2014.00246.
27. Gheorghiade M, Colucci WS, Swedberg K. β-blockers in chronic heart failure. Circulation. 2003;107:1570–5. doi:10.1161/01.CIR.0000065187.80707.18.
28. Kubon C, Mistry NB, Grundvold I, Halvorsen S, Kjeldsen SE, Westheim AS. The role of beta-blockers in the treatment of chronic heart failure. Trends Pharmacol Sci. 2011;32:206–12. doi:10.1016/j.tips.2011.01.006.
29. Gilbert EM, Abraham WT, Olsen S, Hattler B, White M, Mealy P, et al. Comparative hemodynamic, left ventricular functional, and antiadrenergic effects of chronic treatment with metoprolol versus carvedilol in the failing heart. Circulation. 1996;94:2817–25.
30. Schwartz PJ, Billman GE, Stone HL. Autonomic mechanisms in ventricular fibrillation induced by myocardial ischemia during exercise in dogs with healed myocardial infarction. An experimental preparation for sudden cardiac death. Circulation. 1984;69:790–800.
31. Vanoli E, De Ferrari GM, Stramba-Badiale M, Hull Jr SS, Foreman RD, Schwartz PJ. Vagal stimulation and prevention of sudden death in conscious dogs with a healed myocardial infarction. Circ Res. 1991;68:1471–81.
32. Zhou S, Jung BC, Tan AY, Trang VQ, Gholmieh G, Han SW, et al. Spontaneous stellate ganglion nerve activity and ventricular arrhythmia in a canine model of sudden death. Heart Rhythm. 2008;5:131–9.
33. Malliani A, Schwartz PJ, Zanchetti A. A sympathetic reflex elicited by experimental coronary occlusion. Am J Physiol. 1969;217:703–9.
34. Schwartz PJ, Foreman RD, Stone HL, Brown AM. Effect of dorsal root section on the arrhythmias associated with coronary occlusion. Am J Physiol. 1976;231:923–8.
35. Camm J, Malik M, Bigger Jr JT, Breithardt G, Cerutti S, Cohen RJ, et al. Heart rate variability: standards of measurement, physiological interpretation and clinical use. Task Force of the European Society of Cardiology and the North American Society of Pacing and Electrophysiology. Circulation. 1996;93:1043–65.
36. Kleiger RE, Miller JP, Bigger Jr JT, Moss AJ. Decreased heart rate variability and its association with increased mortality after acute myocardial infarction. Am J Cardiol. 1987;59:256–62.
37. La Rovere MT, Pinna GD, Maestri R, Mortara A, Capomolla S, Febo O, et al. Short-term heart rate variability strongly predicts sudden cardiac death in chronic heart failure patients. Circulation. 2003;107:565–70.
38. Buccelletti E, Gilardi E, Scaini E, Galiuto L, Persiani R, Biondi A, et al. Heart rate variability and myocardial infarction: systematic literature review and metanalysis. Eur Rev Med Pharmacol Sci. 2009;13:299–307.

39. Malliani A, Pagani M, Lombardi F, Cerutti S. Cardiovascular neural regulation explored in the frequency domain. Circulation. 1991;84:482–92.
40. Furlan R, Guzzetti S, Crivellaro W, Dassi S, Tinelli M, Baselli G, et al. Continuous 24-hour assessment of the neural regulation of systemic arterial pressure and RR variabilities in ambulant subjects. Circulation. 1990;81:537–47.
41. Hopf HB, Skyschally A, Heusch G, Peters J. Low-frequency spectral power of heart rate variability is not a specific marker of cardiac sympathetic modulation. Anesthesiology. 1995;82:609–19.
42. Taylor JA, Carr DL, Myers CW, Eckberg DL. Mechanisms underlying very-low-frequency RR-interval oscillations in humans. Circulation. 1998;98:547–55.
43. Goldstein DS, Bentho O, Park MY, Sharabi Y. Low-frequency power of heart rate variability is not a measure of cardiac sympathetic tone but may be a measure of modulation of cardiac autonomic outflows by baroreflexes. Exp Physiol. 2011;96:1255–61. doi:10.1113/expphysiol.2010.056259.
44. Reyes del Paso GA, Langewitz W, Mulder LJ, van Roon A, Duschek S. The utility of low frequency heart rate variability as an index of sympathetic cardiac tone: a review with emphasis on a reanalysis of previous studies. Psychophysiology. 2013;50:477–87. doi:10.1111/psyp.12027.
45. Bigger Jr JT, Fleiss JL, Steinman RC, Rolnitzky LM, Kleiger RE, Rottman JN. Frequency domain measures of heart period variability and mortality after myocardial infarction. Circulation. 1992;85:164–71.
46. Bigger Jr JT, Fleiss JL, Rolnitzky LM, Steinman RC. Frequency domain measures of heart period variability to assess risk late after myocardial infarction. J Am Coll Cardiol. 1993;21:729–36. Erratum in: J Am Coll Cardiol 1993;21:1537.
47. Huikuri HV, Makikallio TH, Peng CK, Goldberger AL, Hintze U, Moller M. Fractal correlation properties of R-R interval dynamics and mortality in patients with depressed left ventricular function after an acute myocardial infarction. Circulation. 2000;101:47–53.
48. Stein PK, Domitrovich PP, Huikuri HV, Kleiger RE, Cast Investigators. Traditional and nonlinear heart rate variability are each independently associated with mortality after myocardial infarction. J Cardiovasc Electrophysiol. 2005;16:13–20.
49. Voss A, Schulz S, Schroeder R, Baumert M, Caminal P. Methods derived from nonlinear dynamics for analysing heart rate variability. Philos Trans A Math Phys Eng Sci. 2009;367:277–96.
50. Sassi R, Cerutti S, Lombardi F, Malik M, Huikuri HV, Peng C-K, et al. Advances in heart rate variability signal analysis: joint position statement by the e-Cardiology ESC Working Group and the European Heart Rhythm Association co-endorsed by the Asia Pacific Heart Rhythm Society. Europace. 2015;17:1341–53. doi:10.1093/europace/euv015.
51. Peng CK, Havlin S, Stanley HE, Goldberger AL. Quantification of scaling exponents and crossover phenomena in nonstationary heartbeat time series. Chaos. 1995;5:82–7.
52. Makikallio TH, Huikuri HV, Makikallio A, Sourander LB, Mitrani RD, Castellanos A, et al. Prediction of sudden cardiac death by fractal analysis of heart rate variability in elderly subjects. J Am Coll Cardiol. 2001;37:1395–402.
53. Tapanainen JM, Thomsen PE, Kober L, Torp-Pedersen C, Makikallio TH, Still AM, et al. Fractal analysis of heart rate variability and mortality after an acute myocardial infarction. Am J Cardiol. 2002;90:347–52.
54. Kiyono K, Hayano J, Watanabe E, Struzik ZR, Yamamoto Y. Non-Gaussian heart rate as an independent predictor of mortality in patients with chronic heart failure. Heart Rhythm. 2008;5:261–8. doi:10.1016/j.hrthm.2007.10.030.
55. Hayano J, Kiyono K, Struzik ZR, Yamamoto Y, Watanabe E, Stein PK, et al. Increased non-gaussianity of heart rate variability predicts cardiac mortality after an acute myocardial infarction. Front Physiol. 2011;2:65. doi:10.3389/fphys.2011.00065.

56. Kiyono K, Hayano J, Kwak S, Watanabe E, Yamamoto Y. Non-gaussianity of low frequency heart rate variability and sympathetic activation: lack of increases in multiple system atrophy and Parkinson disease. Front Physiol. 2012;3:34. doi:10.3389/fphys.2012.00034.
57. Willson K, Francis DP, Wensel R, Coats AJ, Parker KH. Relationship between detrended fluctuation analysis and spectral analysis of heart-rate variability. Physiol Meas. 2002;23:385–401.
58. Willson K, Francis DP. A direct analytical demonstration of the essential equivalence of detrended fluctuation analysis and spectral analysis of RR interval variability. Physiol Meas. 2003;24:N1–7.
59. Cacioppo JT, Berntson GG, Binkley PF, Quigley KS, Uchino BN, Fieldstone A. Autonomic cardiac control. II. Noninvasive indices and basal response as revealed by autonomic blockades. Psychophysiology. 1994;31:586–98.
60. Taylor JA, Myers CW, Halliwill JR, Seidel H, Eckberg DL. Sympathetic restraint of respiratory sinus arrhythmia: implications for vagal-cardiac tone assessment in humans. Am J Physiol Heart Circ Physiol. 2001;280:H2804–14.
61. Grossman P, Taylor EW. Toward understanding respiratory sinus arrhythmia: relations to cardiac vagal tone, evolution and biobehavioral functions. Biol Psychol. 2007;74:263–85.
62. Pomeranz B, Macaulay RJ, Caudill MA, Kutz I, Adam D, Gordon D, et al. Assessment of autonomic function in humans by heart rate spectral analysis. Am J Physiol. 1985;248: H151–3.
63. Moak JP, Goldstein DS, Eldadah BA, Saleem A, Holmes C, Pechnik S, et al. Supine low-frequency power of heart rate variability reflects baroreflex function, not cardiac sympathetic innervation. Heart Rhythm. 2007;4:1523–9.
64. Rahman F, Pechnik S, Gross D, Sewell L, Goldstein DS. Low frequency power of heart rate variability reflects baroreflex function, not cardiac sympathetic innervation. Clin Auton Res. 2011;21:133–41. doi:10.1007/s10286-010-0098-y.
65. Hamilton JD. Time series analysis. Princeton: Princeton University Press; 1994.
66. Huikuri HV, Makikallio TH, Airaksinen KE, Seppanen T, Puukka P, Raiha IJ, et al. Power-law relationship of heart rate variability as a predictor of mortality in the elderly. Circulation. 1998;97:2031–6.
67. Ivanov PC, Amaral LA, Goldberger AL, Havlin S, Rosenblum MG, Struzik ZR, et al. Multifractality in human heartbeat dynamics. Nature. 1999;399:461–5.
68. Tulppo MP, Mäkikallio TH, Takala TE, Seppänen T, Huikuri HV. Quantitative beat-to-beat analysis of heart rate dynamics during exercise. Am J Physiol. 1996;271:H244–52.
69. Richman JS, Moorman JR. Physiological time-series analysis using approximate entropy and sample entropy. Am J Physiol Heart Circ Physiol. 2000;278:H2039–49.
70. Porta A, Guzzetti S, Montano N, Furlan R, Pagani M, Malliani A, et al. Entropy, entropy rate, and pattern classification as tools to typify complexity in short heart period variability series. IEEE Trans Biomed Eng. 2001;48:1282–91.
71. Costa M, Goldberger AL, Peng C-K. Multiscale entropy analysis of physiologic time series. Phys Rev Lett. 2002;89:062102.
72. Yang AC, Hseu SS, Yien HW, Goldberger AL, Peng CK. Linguistic analysis of the human heartbeat using frequency and rank order statistics. Phys Rev Lett. 2003;90:108103.
73. Kiyono K, Struzik ZR, Aoyagi N, Sakata S, Hayano J, Yamamoto Y. Critical scale invariance in a healthy human heart rate. Phys Rev Lett. 2004;93:178103.
74. Jokinen V, Tapanainen JM, Seppänen T, Huikuri HV. Temporal changes and prognostic significance of measures of heart rate dynamics after acute myocardial infarction in the beta-blocking era. Am J Cardiol. 2003;92:907–12.
75. Perkiömäki JS, Bloch Thomsen PE, Kiviniemi AM, Messier MD, Huikuri HV, CARISMA Study Investigators. Risk factors of self-terminating and perpetuating ventricular tachyarrhythmias in post-infarction patients with moderately depressed left ventricular function, a CARISMA sub-analysis. Europace. 2011;13:1604–11. doi:10.1093/europace/eur166.

76. Gang UJ, Jøns C, Jørgensen RM, Abildstrøm SZ, Messier MD, Haarbo J, et al. Risk markers of late high-degree atrioventricular block in patients with left ventricular dysfunction after an acute myocardial infarction: a CARISMA substudy. Europace. 2011;13:1471–7. doi:10.1093/europace/eur165.
77. Perkiömäki JS, Jokinen V, Tapanainen J, Airaksinen KE, Huikuri HV. Autonomic markers as predictors of nonfatal acute coronary events after myocardial infarction. Ann Noninvasive Electrocardiol. 2008;13:120–9. doi:10.1111/j.1542-474X.2008.00211.x.
78. Kiyono K, Struzik ZR, Aoyagi N, Yamamoto Y. Multiscale probability density function analysis: non-Gaussian and scale-invariant fluctuations of healthy human heart rate. IEEE Trans Biomed Eng. 2006;53(1):95–102.
79. Kiyono K, Struzik ZR, Yamamoto Y. Estimator of a non-Gaussian parameter in multiplicative log-normal models. Phys Rev E. 2007;76:041113.
80. Frish U. Turbulence. Cambridge: Cambridge University Press; 1995.
81. Altimiras J. Understanding autonomic sympathovagal balance from short-term heart rate variations. Are we analyzing noise? Comp Biochem Physiol A Mol Integr Physiol. 1999;124:447–60.
82. Sone M, Yoshida M, Hashizume Y, Hishikawa N, Sobue G. Alpha-synuclein-immunoreactive structure formation is enhanced in sympathetic ganglia of patients with multiple system atrophy. Acta Neuropathol. 2005;110:19–26.
83. Braune S, Reinhardt M, Bathmann J, Krause T, Lehmann M, Lücking CH. Impaired cardiac uptake of meta-[123I]iodobenzylguanidine in Parkinson's disease with autonomic failure. Acta Neurol Scand. 1998;97:307–14.
84. Braune S, Reinhardt M, Schnitzer R, Riedel A, Lücking CH. Cardiac uptake of [123I]MIBG separates Parkinson's disease from multiple system atrophy. Neurology. 1999;53:1020–5.
85. Vikman S, Mäkikallio TH, Yli-Mäyry S, Pikkujämsä S, Koivisto AM, Reinikainen P, et al. Altered complexity and correlation properties of R-R interval dynamics before the spontaneous onset of paroxysmal atrial fibrillation. Circulation. 1999;100:2079–84.
86. Porta A, Gnecchi-Ruscone T, Tobaldini E, Guzzetti S, Furlan R, Montano N. Progressive decrease of heart period variability entropy-based complexity during graded head-up tilt. J Appl Physiol. 2007;103:1143–9.

Chapter 10
Heart Rate Variability and Cardiac Diseases

Eiichi Watanabe, Ken Kiyono, Yoshiharu Yamamoto, and Junichiro Hayano

Abstract Heart rate variability (HRV) is a noninvasive methodology for evaluating the autonomic nervous system modulation of the sinoatrial node and to identify patients at risk of cardiac disorders. The contemporary therapeutic use for the HRV analysis is for risk stratification in patients with post-myocardial infarction or heart failure who are prone to have arrhythmic death and who would benefit from implantable cardioverter-defibrillators (ICDs). Although multiple HRV measurements have been developed to achieve a better risk stratification, the HRV measurements have been rarely tested to see whether they harbor a significant power to serve as a practical risk predictor. To date, the only reliable metric to predict the benefit from an ICD is a severely reduced ejection fraction; however, the predictive value of the ejection fraction is relatively low. Because of the high negative predictive value of the HRV, a combination with the ejection fraction may be helpful to identify candidates who are unlikely to benefit from ICD therapy. A more sophisticated risk approach that combines the HRV and other known clinical measures should be developed to provide accurate estimates of the risk to allow patients to make informed treatment decisions. Another requirement is to explore the novel HRV measurements for atrial fibrillation. The prevalence of atrial fibrillation rapidly increases among the aging population and is independently associated with a higher risk of ischemic strokes and excess mortality. The HRV measurements specific for atrial fibrillation may facilitate the risk stratification in such a high-risk population.

E. Watanabe (✉)
Department of Cardiology, Fujita Health University School of Medicine, 1-98 Dengakugakubo, Kutsukake-cho, Toyoake, Aichi 470-1192, Japan
e-mail: enwatan@fujita-hu.ac.jp

K. Kiyono
Division of Bioengineering, Graduate School of Engineering Science, Osaka University, Osaka, Japan

Y. Yamamoto
Educational Physiology Laboratory, Graduate School of Education, University of Tokyo, Tokyo, Japan

J. Hayano
Department of Medical Education, Nagoya City University Graduate School of Medical Sciences, Nagoya, Aichi, Japan

S. Iwase et al. (eds.), *Clinical Assessment of the Autonomic Nervous System*, DOI 10.1007/978-4-431-56012-8_10

Keywords Autonomic nervous system • Mortality • Arrhythmia • Sinus rhythm • Atrial fibrillation

10.1 Diagnostic Utilities of Heart Rate Variability

10.1.1 Ischemic Heart Disease

10.1.1.1 Before the Reperfusion Era

Historically, the heart rate variability (HRV) analysis has been used to assess the risk of mortality and arrhythmic outcomes in patients with a post-myocardial infarction (MI). Wolf et al. [1] first observed that a higher heart rate and reduced heart rate variability in patients with an acute MI were related to in-hospital death. They calculated the variance of 30 consecutive RR intervals using a 1 min ECG recording in 176 patients at admission to the coronary care unit. Patients with an RR interval variance of <32 ms had a higher mortality compared to those with that ≥32 ms. However, they did not examine the prognostic value of those measures after discharge. Kleiger et al. subsequently demonstrated the prognostic value of the HRV in post-MI patients enrolled in the Multicenter Post-Infarction Program [2]. The data from 808 participants showed that those who had a reduced standard deviation of all normal to normal RR (NN) intervals (SDNN) of <50 ms had a higher mortality than those with an SDNN of ≥50 ms, with a crude annual mortality of 13.3 % vs. 4.8 % (Fig. 10.1). Bigger et al. [3] evaluated the predictive value of the frequency-domain measures in the Multicenter Post-Infarction Program patients. They studied 24 h Holter ECGs recorded 2 weeks after an acute MI in 715 patients and analyzed six HRV measurements (ULF, VLF, LF, HF, total power, and LF/HF). The 4-year mortality rate was 38.5 % (all-cause death 119 [16.7 %], cardiac death 88 [12.3 %], and arrhythmic death 68 [9.5 %]). They found that all six measurements well predicted the mortality. Of those, the VLF power was more strongly associated with arrhythmic death than with all-cause or cardiac death after adjusting for the covariates. Bigger et al. subsequently demonstrated the prognostic value of the power-law relationship in the Multicenter Post-Infarction Program patients [4]. To derive the power-law relationship, the frequency-domain data are plotted [log (power) vs. log (frequency)], and the inverse slope of this plot helps to define the complexity of heart rate fluctuations. They found that the power-law regression parameters are excellent predictors of all-cause death or arrhythmic death and predict these outcomes better than the traditional power spectral bands.

The heart rate turbulence (HRT) implicates the baroreflex-mediated autonomic influence on the sinoatrial node [5, 6]. In healthy subjects a ventricular premature ectopic beat provokes a biphasic reaction of an early acceleration and late deceleration of the heart rate, while in high-risk subjects such a reaction is diminished or even completely nonexistent. Schmidt et al. [7] showed that the HRT is even a

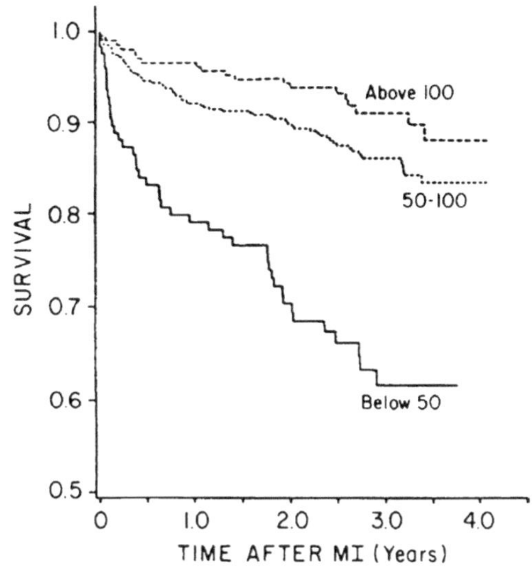

Fig. 10.1 Kaplan-Meier survival curves from the Multicenter Post-Infarction Study demonstrating a decreased survival among patients with an SDNN of <50 ms (Reprinted from Kleiger et al. [2] with permission)

stronger risk predictor than the traditional time- or frequency-domain variables using the Multicenter Post-Infarction Program patients.

10.1.1.2 After the Reperfusion Era

Timely reperfusion after an acute MI facilitates the cardiomyocyte salvage and decreases the cardiac morbidity and mortality. The crude annual mortality from the above studies decreased from approximately 10 % [2, 3] to 2–3 % in the current studies [8–10], probably due to the multidisciplinary approach in acute MIs [11]. Data on the risk stratification in recent MI patients using traditional and new methods have been reported. The GISSI study evaluated the prognostic value of the HRV in 567 patients with an acute MI who underwent thrombolytic therapy [8]. They had 52 deaths (9.2 %) during a 1000-day follow-up. They found that the time-domain analysis (SDNN, NN50+, and rMSSD) identified high-risk groups, but the relative risk of these measures was at most 3. The ATRAMI study, where 63 % of the patients received reperfusion therapy after an acute MI, showed that patients with an SDNN of <70 ms or baroreflex sensitivity of <3.0 ms per mm Hg carried a significant multivariate risk of cardiac mortality (3.2 [95 % CI 1.42–7.36] and 2.8 [1.24–6.16], respectively) after adjusting for the left ventricular ejection fraction (LVEF) and ventricular ectopy [9]. The DIAMOND substudy assessed the use of various fractal analysis methods of HRV from 24-h Holter recordings to predict death in 446 patients with an acute MI and a depressed left ventricular function [12]. They found that the short-term scaling exponent $\alpha 1$ was the most powerful measurement for predicting both arrhythmic and non-arrhythmic death.

Autonomic dysfunction occurring early after an MI has been shown to improve over time [13].

Data regarding the denervation during the early phase of an acute MI and reinnervation during the late phase of an MI have been demonstrated in humans [14, 15]. Furthermore, early revascularization, the use of β-blockers, and angiotensin-converting enzyme inhibitor ameliorate the remodeling and modulate the arrhythmia substrate following an MI [11]. The above three studies [8, 9, 12] confirmed the ability of the HRV analysis for risk stratification in the early phase after revascularization, whereas recent results demonstrated that the HRV measured during the late phase after an acute MI (10–14 weeks) is more predictive of mortality than those measured during the early phase after an MI (2–4 weeks) [16]. Recently, Huikuri et al. examined the relationship between the magnitude of recovery of the HRV or HRT recorded during the early and late phases of an MI. They found that an attenuated recovery of the VLF and LF components of the HRV and HRT was associated with a risk for serious arrhythmias but not with non-arrhythmic death [17]. Early measurement of the HRV to identify those at high risk should likely be repeated later in order to assess the risk of fatal arrhythmic events in the late phase.

Recently, several new HRV measurements have been proposed and their predictive values have been reported. Bauer et al. [18] explored a novel HRV measurement, termed the acceleration capacity (AC) and deceleration capacity (DC) of the heart rate. They developed and validated the predictive accuracy of the AC and DC using several large post-MI cohorts. They noted that an impaired DC of $\leq$2.5 ms was an efficient predictor of mortality in patients with a reduced LVEF ($<$30 %) as well as a preserved LVEF. They subsequently tested the incremental value of the DC adding it to the HRT in other cohorts of post-MI patients [19]. They found that patients with both an abnormal HRT and abnormal DC (called severe autonomic failure [SAF]) had a higher mortality irrespective of the LVEF.

More recently, Kisohara et al. [20] proposed a modified AC and DC, which were calculated with the time scales of T (window size defining heart rate) and s (wavelet scale) from 1 to 500 s and compared their prognostic values with the conventional measures proposed by Bauer (AC (conv) and the DC (conv)) that were calculated with $(T, s) = [1,2$ (beat)] [18] (detailed explanation will be in the Part II, Chap. 7). They used a cohort of 708 post-MI patients, with a crude annual mortality of 2.7 %. They found that a decrease in the DC for the minute-order long-term heart rate dynamics is a strong predictor of mortality, and their predictive power was independent of the AC (conv) and DC (conv). This finding bolsters the role of a new DC in the risk assessment among the post-MI patients and provides impetus for further study of the incremental value of the DC in a reduced and preserved LVEF population, respectively.

Kiyono et al. [21] previously reported a new HRV measure termed the non-Gaussianity index (lambda: λ) that characterizes an increased probability of large heart rate deviations from its trend (detailed explanation will be in the Part II, Chap. 9) [21]. A previous study reported that an increased λ is an independent mortality predictor among patients with chronic heart failure [22]. Hayano et al. [23] recently examined the predictive value of λ in 670 post-MI patients. During a median follow-up period of 25 months, 45 (6.7 %) patients died (32 cardiac

Fig. 10.2 Kaplan-Meier curves for the cardiac death after an AMI. The patients were stratified by a $\lambda 25$ s > 0.6 and by the combination of a $\lambda 25$ s > 0.6 and abnormal HRT. *AMI* acute myocardial infarction, *HRT* heart rate turbulence (Reprinted from Hayano et al. [23] with permission)

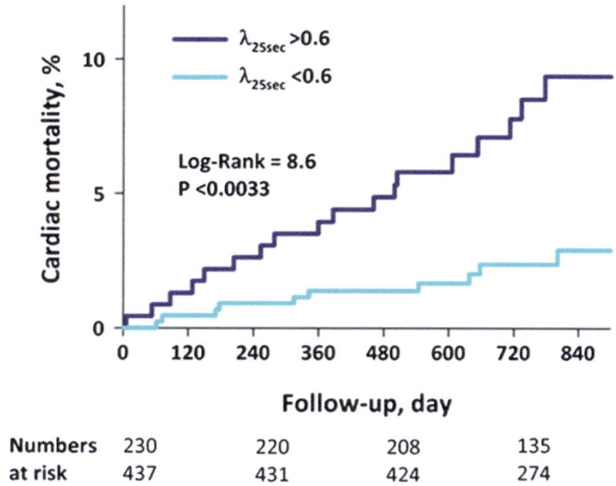

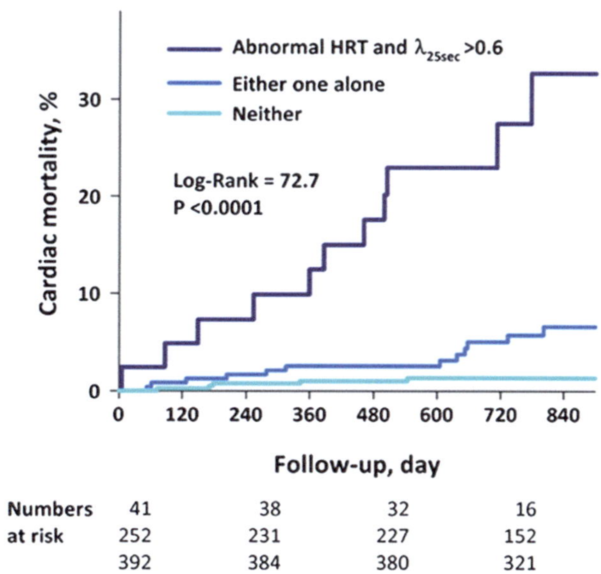

and 13 noncardiac deaths). An increased λ predicted exclusively cardiac death after adjustments for the covariates. The combination of an increased λ and abnormal HRT provided the best predictive model for cardiac death (Fig. 10.2).

10.1.2 Heart Failure

The epidemiology of heart failure (HF) has been changing, with an increasing proportion of patients being diagnosed with HF and a preserved left ventricular ejection fraction (HFpEF), in addition to the classic HF with a reduced left ventricular ejection fraction (HFrEF). Compared to those with HFrEF, the patients with HFpEF are more often female, older, less likely to have coronary artery disease, and more likely to have hypertension [24]. Despite advancement in the heart failure treatment, the mortality rate of a HFpEF is similar to that of an HFrEF [24]. The mode of death depends on the New York Heart Association functional class. Patients with mild to moderate HF more frequently die suddenly, while those with severe HF are more likely to die of pump failure [25].

10.1.2.1 Risk Stratification in Heart Failure with Reduced Left Ventricular Ejection Fraction

Numerous studies have examined the prognostic role of various HRV measurements in those with an HFrEF for years [26]. Although a decreased time-domain HRV analysis, mostly the SDNN, has been considered as an independent risk marker for all-cause death or pump failure death [27–31], data on predicting sudden cardiac death is not consistent [32].

Much greater controversy exists in regard to the spectral analysis for the risk stratification in patients with HFrEF. Galinier et al. [29] showed that ischemic heart disease and a reduced LF power during the daytime were associated with sudden death. La Rovere [33] analyzed the short-term spectral components (8 min) during controlled breathing. They developed and validated the ability of a reduced short-term LF power to predict sudden death. Guzetti et al. [34] showed that a decreased VLF power at night was independently related to death from progressive pump failure, while the reduction in the LF power during the night was linked to sudden death. The role of the HRV in predicting the mortality of HF patients was recently confirmed by the GISSI-HF Holter substudy [35]. They found that the LF, VLF, and an abnormal HRT were associated with arrhythmic events defined as sudden death or appropriate discharges from an implantable cardioverter-defibrillator (ICD). These studies indicated that a spectral analysis for the risk stratification in patients with HFrEF seems to be a better marker of sudden death compared to a time-domain analysis. Finally, Maestri et al. [36] showed that nonlinear HRV indices provide important prognostic information on top of clinical data. They found that two variables (from empirical mode decomposition and symbolic dynamics families) added prognostic information to the clinical model.

10.1.2.2 Risk Stratification in Heart Failure Patients with a Preserved Left Ventricular Ejection Fraction

The HFpEF patients are not covered by current indications for ICDs. In this patient population, the HRT seems to be more useful in the risk stratification of sudden death rather than the HRV. The aforementioned GISSI-HF Holter substudy [35] demonstrated that an abnormal HRT in patients with an LVEF of >30 % was associated with a fourfold increased risk of arrhythmic events. In the HFpEF patients (LVEF >35 %) from the MUSIC study, a combination of an abnormal HRT, decreased HRV, and impaired QT interval dynamics provided the highest accuracy in identifying high-risk groups for the all-cause mortality and sudden death [37]. Further studies are required to test whether nonlinear HRV indices provide prognostic information using a large cohort of patients with HFpEF.

10.1.3 Nonischemic Heart Disease

Patients with nonischemic cardiomyopathy are at an increased risk of sudden death and heart failure death, with a 12–20 % estimated mortality rate at 3 years [38–40]. They have frequent and high-grade ventricular ectopy and a reduced HRV like those with ischemic heart disease. Currently, various risk stratifiers have been reported, but available data, especially in nonischemic patients, are too small to provide clear recommendations. A recent meta-analysis by Goldberger et al. [41] examined the prognostic role in arrhythmic outcomes, in 6088 patients with nonischemic dilated cardiomyopathy from 45 studies. They reviewed and validated the predictive value of various tests including autonomic tests, echocardiographic data, electrophysiologic studies, arrhythmias, surface ECG findings, signal-averaged electrocardiograms, vectrocardiograms, and T-wave alternans. They found that the measurements of depolarization (fragmented QRS) and repolarization (T-wave alternans) had a higher odds ratio, but none of the autonomic tests (HRV, HRT, and baroreflex sensitivity) were significant predictors of arrhythmic outcomes.

10.1.4 Therapeutic Implications of the HRV (ICD Implantations)

The contemporary therapeutic use for HRV analysis in post-MI or HF patients is the selection of patients for prophylactic ICD therapy [42–44]. Presently, no definitive recommendations can be given for the clinical use of noninvasive ECGs or Holter-derived risk markers in selecting candidates for prophylactic ICD therapy.

The contemporary practice is almost solely reliant on the LVEF for the risk stratification; some studies demonstrated the incremental value of the HRV in patients with a decreased LVEF in prospective ICD studies. In DINAMIT [45], 675 recent MI patients (day 6 to day 40) with an LVEF $\leq$35 % and SDNN $\leq$70 ms (or mean 24 h heart rate of $\geq$80 bpm) were randomized to receive or not receive an ICD. There was no significant difference in the survival rate between the groups. The ICD reduced the arrhythmic mortality but increased the non-arrhythmic mortality. In the ALIVE study [46], 3717 post-MI patients with a depressed LVEF were randomized to a placebo and azimilide and classified them into low- and high-risk groups on the basis of the triangular index of the HRV. By a multivariate analysis, a low HRV was associated with the all-cause mortality, but it did not predict the arrhythmic mortality. These data showed that linear HRV measurements in addition to a depressed LVEF may not be a useful marker to isolate patients who will benefit from ICD therapy.

The majority of the studies assessing the prognostic power of the HRV in post-MI or HF patients have used all-cause mortality as the endpoint. Some studies have suggested that a reduced HRV may be specifically related to arrhythmic events and sudden death. The CHARISMA trial assessed the use of a combination of the HRV, HRT, ambient arrhythmias, signal-averaged electrocardiograms, T-wave alternans, and programmed electrical stimulation (PES) to predict fatal arrhythmias when performed 6 weeks after an acute MI [47]. A total of 312 patients with a mean LVEF of 31 % received an implantable ECG loop recorder to detect ventricular fibrillation (VF) or symptomatic sustained ventricular tachycardia (VT). They found that an adjusted hazard ratio of the reduced VLF component and induction of sustained monomorphic VT during the PES was 7.0 and 4.8, respectively. Although multiple HRV measurements have been investigated to achieve a better selection, nonlinear HRV measurements have been neither analyzed in randomized controlled ICD trials nor tested as to whether they harbor a significant power to serve as a practical risk predictor. Recently, Au-yeung et al. [48] performed a retrospective analysis on the HRV measurements and occurrence of VT/VF and sudden death in the heart failure patients enrolled in the SCD-HeFT trial [40]. They found that a combination of the fractal exponent ($\alpha1$ and $\alpha2$) and HRT or LF/HF ratio, number of premature ventricular complexes, or triangular index of the HRV could discriminate survivors and those who had appropriate discharges from their ICDs. Of note, the high negative predictive value from these data identifies patients who are not likely to experience ICD therapies with a 97.3 % certainty.

Sudden death due to VT/VF is a major mechanism of sudden death, and pulseless electrical activity (PEA) and asystole occupy the rest [49]. There has been a progressive decrease in the prevalence of sudden death presenting as VF in recent publications, and a proportional increase in those presenting with non-shockable rhythms [50–52]. It is plausible that, in addition to the differences in the clinical characteristics and arrhythmic substrate, these two distinct mechanisms of sudden death make it difficult to predict the onset of sudden death with a high certainty (Fig. 10.3).

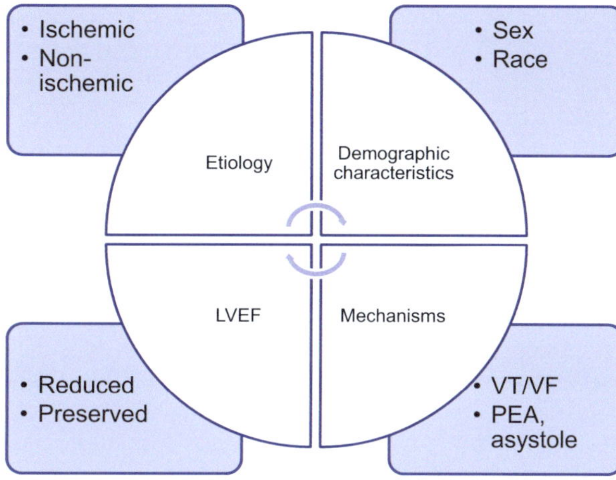

Fig. 10.3 Why is it difficult to predict sudden death? In addition to the differences in the clinical characteristics and arrhythmic substrate, two distinct mechanisms of sudden death (VT/VF and PEA or asystole) make it hard to predict the onset of sudden death with a high certainty. *VT* ventricular tachycardia, *VF* ventricular fibrillation, *PEA* pulseless electrical activity, *LVEF* left ventricular ejection fraction

To date, the only reliable metric to predict the benefit from an ICD is a severely depressed LVEF; however, the predictive value of the LV function is relatively low [53]. HRV has a low positive predictive value, but a high negative predictive value in predicting sudden death. Therefore, we suppose that a multivariable risk modeling should incorporate a Holter-based HRV analysis to refine the ICD patient selection, especially to exclude patients who are *unlikely* to benefit from ICD therapy. On the other hand, a risk model that incorporates a high positive predictive value can identify those that would benefit from ICD therapy. The clinical utility of noninvasive tests will be assessed in an ongoing randomized trial of ICDs (REFINE-ICD; NCT identifier 00673842).

10.2 Other Indications for the HRV

10.2.1 Heart Rate Variability Just Before Sudden Death

If we can recognize specific precursors before the initiation of ventricular tachycardia or ventricular fibrillation (VT/VF) by analyzing the HRV, these findings could be helpful in implementing algorithms for an ICD to improve the diagnosis and therapy of VT/VF. To date, several studies reported characteristic changes in the time- and frequency-domain measurements before the onset of VT/VF. Shusterman et al. [54] analyzed the HRV indices in 53 patients enrolled in the ESVEM trial that included ischemic heart disease patients who had a history of cardiac arrest, documented VF, sustained VT, or syncope. Only 10 % of the patients were on β-blockers. They found that the VLF, LF, and HF declined 30 min before sustained VT compared to that observed in the controls. In addition, the LF and LF/HF ratio declined just before the onset of the VT. Osaka et al. [55] performed frequency-domain analyses in 34 patients experiencing VF, cardiac arrest, or an acute MI and compared that to an age-, sex-, and disease-matched cohort of 191 patients. They found that several hours before the onset of a cardiac event, a transient decrease in the LF/HF was followed by an increase in the LF/HF. However, conflicting results were reported in which neither time- nor frequency-domain analyses failed to find characteristic changes in the RR dynamics before the VT/VF [56]. Recent studies using nonlinear methods found a significant difference in the RR dynamics before the occurrence of VT/VF. Makikallio et al. [57] examined the heart rate dynamics in post-MI patients who had VT/VF during Holter ECGs, where they found that a short-term fractal scaling exponent (α) obtained by a detrended fluctuation analysis and the slope (β) of the power-law regression line of the RR interval dynamics were the most powerful independent predictors differentiating patients with VT/VF from the controls. Because the deviations in the short-term slopes from the 1/f curve occur 15 min to 1 h before the onset of VF, a fractal-like signal behavior may reflect changes in the autonomic modulation of the heart. In another report by Wessel et al. [58], they analyzed the linear and nonlinear parameters using 1,000 beat-to-beat intervals stored in an ICD. They found that the time- and frequency-domain parameters were similar between the patients with and without VT/VF, but the symbolic dynamics and finite-time growth well discriminated those patients.

10.2.2 Atrial Fibrillation

Atrial fibrillation (AF) remains the most common, sustained arrhythmia and is independently associated with a higher risk of an ischemic stroke, as well as a poorer quality of life, higher hospitalization rates, and excess mortality [59]. Most of the HRV studies, however, excluded AF patients from the analysis. In early reports by Stein et al. [60] and Frey et al. [61], they demonstrated that an SD of a 5 min average ventricular response interval during a 24 h recording (SDAVI) was associated with mortality in chronic AF. A previous study by Yamada et al. [62] demonstrated the prognostic values of the linear and nonlinear dynamics of the heart rate in patients with chronic AF. Data on 107 participants showed that reduced nonlinear measurements by means of the Shannon entropy and approximate entropy were independent predictors for death after an adjustment for the clinical variables. The conventional linear measurements, however, were not significant predictors (Fig. 10.4).

Although AF patients are in a prothrombotic state, few studies have examined whether the nonlinear dynamics of the heart rate could predict ischemic strokes in AF. We have recently shown that a novel complexity measurement of the HRV, termed multiscale entropy, is a useful risk stratification measurement of ischemic strokes in patients with permanent AF [63]. Data on 173 participants showed that those who had a larger multiscale entropy in the VLF subrange (90–300 s) were

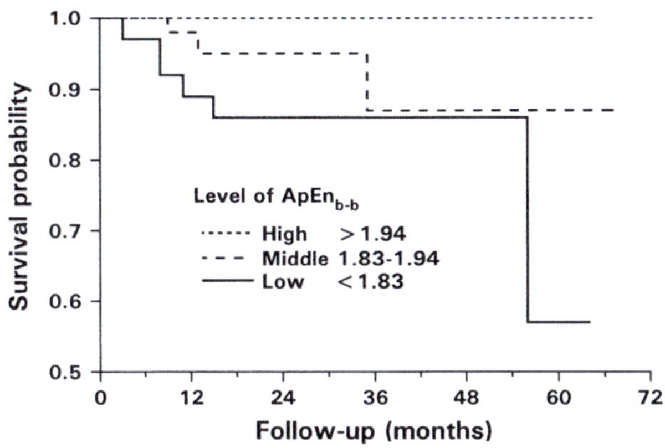

Fig. 10.4 Kaplan-Meier curves for the cardiac mortality in patients stratified by the 33rd and 67th percentile values of the ApEn$_{b\text{-}b}$ (1.83 and 1.94, respectively). The 5-year cardiac mortality rate for the upper, middle, and lower tertiles were 0 %, 13 %, and 43 %, respectively (log-rank test, $P = 0.04$). The ApEn$_{b\text{-}b}$ indicates the average of the approximate entropy of the ventricular response interval for all 1-h segments during 24 h (Reprinted from Yamada et al. [62] with permission)

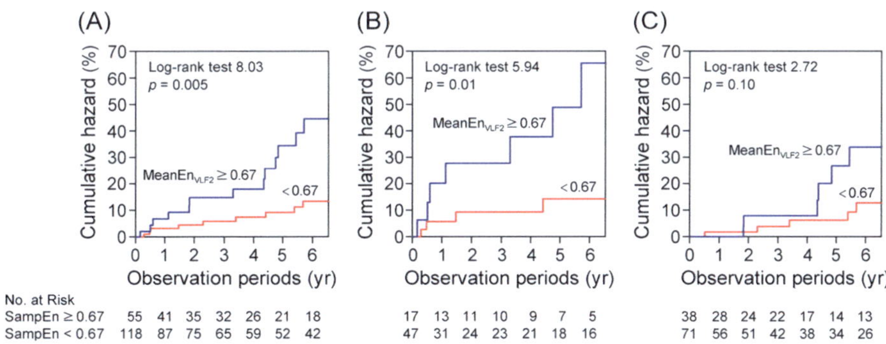

Fig. 10.5 Nelson-Aalen estimate of the ischemic strokes. During the mean follow-up of 47 ± 35 months (median 48 months), ischemic strokes were observed in 22 patients. The patients were dichotomized by the mean value of the sampling entropy of a very low frequency (MeanEn$_{VLF2}$) of 0.67. (**a**) All patients ($n = 173$), (**b**) patients not taking antithrombotic drugs ($n = 64$), and (**c**) patients taking antithrombotic drugs ($n = 109$). There was a significant difference in the ischemic stroke rates in all patients and the patients not receiving antithrombotic drugs (Reprinted from Watanabe et al. [63] with permission)

highly associated with ischemic strokes after an adjustment of the clinical factors and using a wide stroke risk stratification scheme, the CHA$_2$DS$_2$-VASc score [64] (Fig. 10.5).

10.2.3 Long QT Syndrome

Long QT syndrome is an inherited arrhythmic disorder characterized by a prolonged QT interval on the surface ECG and lethal VTs. It is well acknowledged that life-threatening arrhythmias are frequently precipitated by a sympathetic surge such as emotional stress or vigorous exercise. It is not known why symptomatic and asymptomatic patients exist even if they have the same gene mutations. Recently, Bari et al. reported a significant difference in the autonomic tone and autonomic regulation of the QT interval between asymptomatic and symptomatic gene mutation carriers [65, 66].

10.3 Conclusions

Much interest exists in exploring various novel HRV measurements that identify subsets of patients most likely to benefit from medical therapy, but none of these prediction strategies have been proven to be sufficiently discriminative or have received an independent validation for use in clinical practice. A more sophisticated risk approach combining the HRV and other known clinical measurements should be developed to provide accurate estimates of the risk in order to allow patients to make informed treatment decisions.

References

1. Wolf MM, Varigos GA, Hunt D, Sloman JG. Sinus arrhythmia in acute myocardial infarction. Med J Aust. 1978;2(2):52–3.
2. Kleiger RE, Miller JP, Bigger Jr JT, Moss AJ. Decreased heart rate variability and its association with increased mortality after acute myocardial infarction. Am J Cardiol. 1987;59(4):256–62.
3. Bigger Jr JT, Fleiss JL, Steinman RC, Rolnitzky LM, Kleiger RE, Rottman JN. Frequency domain measures of heart period variability and mortality after myocardial infarction. Circulation. 1992;85(1):164–71.
4. Bigger Jr JT, Steinman RC, Rolnitzky LM, Fleiss JL, Albrecht P, Cohen RJ. Power law behavior of RR-interval variability in healthy middle-aged persons, patients with recent acute myocardial infarction, and patients with heart transplants. Circulation. 1996;93(12):2142–51.
5. Cygankiewicz I. Heart rate turbulence. Prog Cardiovasc Dis. 2013;56(2):160–71. doi:10.1016/j.pcad.2013.08.002.
6. Bauer A, Malik M, Schmidt G, Barthel P, Bonnemeier H, Cygankiewicz I, et al. Heart rate turbulence: standards of measurement, physiological interpretation, and clinical use: International Society for Holter and Noninvasive Electrophysiology Consensus. J Am Coll Cardiol. 2008;52(17):1353–65.
7. Schmidt G, Malik M, Barthel P, Schneider R, Ulm K, Rolnitzky L, et al. Heart-rate turbulence after ventricular premature beats as a predictor of mortality after acute myocardial infarction. Lancet. 1999;353(9162):1390–6.

8. Zuanetti G, Neilson JM, Latini R, Santoro E, Maggioni AP, Ewing DJ. Prognostic significance of heart rate variability in post-myocardial infarction patients in the fibrinolytic era. The GISSI-2 results. Gruppo Italiano per lo Studio della Sopravvivenza nell' Infarto Miocardico. Circulation. 1996;94(3):432–6.
9. La Rovere MT, Bigger Jr JT, Marcus FI, Mortara A, Schwartz PJ. Baroreflex sensitivity and heart-rate variability in prediction of total cardiac mortality after myocardial infarction. ATRAMI (Autonomic Tone and Reflexes After Myocardial Infarction) Investigators. Lancet. 1998;351(9101):478–84.
10. Ohno J, Watanabe E, Toyama J, Kawamura T, Ohno M, Kodama I. Risk stratification and survival in post myocardial infarction patients: a large prospective and multicenter study in Japan. Int J Cardiol. 2004;93(2–3):263–8. doi:10.1016/S0167-5273(03)00215-8.
11. Authors/Task Force M, Roffi M, Patrono C, Collet JP, Mueller C, Valgimigli M, et al. ESC guidelines for the management of acute coronary syndromes in patients presenting without persistent ST-segment elevation: task force for the management of acute coronary syndromes in patients presenting without persistent ST-segment elevation of the European Society of Cardiology (ESC). Eur Heart J. 2016;37:267–315. doi:10.1093/eurheartj/ehv320.
12. Huikuri HV, Makikallio TH, Peng CK, Goldberger AL, Hintze U, Moller M. Fractal correlation properties of R-R interval dynamics and mortality in patients with depressed left ventricular function after an acute myocardial infarction. Circulation. 2000;101(1):47–53.
13. Bigger Jr JT, Fleiss JL, Rolnitzky LM, Steinman RC, Schneider WJ. Time course of recovery of heart period variability after myocardial infarction. J Am Coll Cardiol. 1991;18(7):1643–9.
14. Stanton MS, Tuli MM, Radtke NL, Heger JJ, Miles WM, Mock BH, et al. Regional sympathetic denervation after myocardial infarction in humans detected noninvasively using I-123-metaiodobenzylguanidine. J Am Coll Cardiol. 1989;14(6):1519–26.
15. Chen PS, Chen LS, Cao JM, Sharifi B, Karagueuzian HS, Fishbein MC. Sympathetic nerve sprouting, electrical remodeling and the mechanisms of sudden cardiac death. Cardiovasc Res. 2001;50(2):409–16.
16. Jokinen V, Tapanainen JM, Seppanen T, Huikuri HV. Temporal changes and prognostic significance of measures of heart rate dynamics after acute myocardial infarction in the beta-blocking era. Am J Cardiol. 2003;92(8):907–12. doi:10.1016/S0002-9149(03)00968-8.
17. Huikuri HV, Exner DV, Kavanagh KM, Aggarwal SG, Mitchell LB, Messier MD, et al. Attenuated recovery of heart rate turbulence early after myocardial infarction identifies patients at high risk for fatal or near-fatal arrhythmic events. Heart Rhythm. 2010;7(2):229–35. doi:10.1016/j.hrthm.2009.11.004.
18. Bauer A, Kantelhardt JW, Barthel P, Schneider R, Makikallio T, Ulm K, et al. Deceleration capacity of heart rate as a predictor of mortality after myocardial infarction: cohort study. Lancet. 2006;367(9523):1674–81.
19. Bauer A, Barthel P, Schneider R, Ulm K, Muller A, Joeinig A, et al. Improved stratification of autonomic regulation for risk prediction in post-infarction patients with preserved left ventricular function (ISAR-risk). Eur Heart J. 2009;30(5):576–83. doi:10.1093/eurheartj/ehn540.
20. Kisohara M, Stein PK, Yoshida Y, Suzuki M, Iizuka N, Carney RM, et al. Multi-scale heart rate dynamics detected by phase-rectified signal averaging predicts mortality after acute myocardial infarction. Europace. 2013;15(3):437–43. doi:10.1093/europace/eus409.
21. Kiyono K, Struzik ZR, Aoyagi N, Sakata S, Hayano J, Yamamoto Y. Critical scale invariance in a healthy human heart rate. Phys Rev Lett. 2004;93(17):178103.
22. Kiyono K, Hayano J, Watanabe E, Struzik ZR, Yamamoto Y. Non-Gaussian heart rate as an independent predictor of mortality in patients with chronic heart failure. Heart Rhythm. 2008;5(2):261–8. doi:10.1016/j.hrthm.2007.10.030.
23. Hayano J, Kiyono K, Struzik ZR, Yamamoto Y, Watanabe E, Stein PK, et al. Increased non-gaussianity of heart rate variability predicts cardiac mortality after an acute myocardial infarction. Front Physiol. 2011;2:65. doi:10.3389/fphys.2011.00065.

24. Hogg K, Swedberg K, McMurray J. Heart failure with preserved left ventricular systolic function; epidemiology, clinical characteristics, and prognosis. J Am Coll Cardiol. 2004;43 (3):317–27. doi:10.1016/j.jacc.2003.07.046.
25. MERIT-HF Study Group. Effect of metoprolol CR/XL in chronic heart failure: metoprolol CR/XL randomised intervention trial in congestive heart failure (MERIT-HF). Lancet. 1999;353(9169):2001–7.
26. Cygankiewicz I, Zareba W, de Luna AB. Prognostic value of Holter monitoring in congestive heart failure. Cardiol J. 2008;15(4):313–23.
27. Nolan J, Batin PD, Andrews R, Lindsay SJ, Brooksby P, Mullen M, et al. Prospective study of heart rate variability and mortality in chronic heart failure: results of the United Kingdom heart failure evaluation and assessment of risk trial (UK-heart). Circulation. 1998;98(15):1510–6.
28. Ponikowski P, Anker SD, Chua TP, Szelemej R, Piepoli M, Adamopoulos S, et al. Depressed heart rate variability as an independent predictor of death in chronic congestive heart failure secondary to ischemic or idiopathic dilated cardiomyopathy. Am J Cardiol. 1997;79 (12):1645–50.
29. Galinier M, Pathak A, Fourcade J, Androdias C, Curnier D, Varnous S, et al. Depressed low frequency power of heart rate variability as an independent predictor of sudden death in chronic heart failure. Eur Heart J. 2000;21(6):475–82. doi:10.1053/euhj.1999.1875.
30. Boveda S, Galinier M, Pathak A, Fourcade J, Dongay B, Benchendikh D, et al. Prognostic value of heart rate variability in time domain analysis in congestive heart failure. J Interv Card Electrophysiol. 2001;5(2):181–7.
31. Bilchick KC, Fetics B, Djoukeng R, Fisher SG, Fletcher RD, Singh SN, et al. Prognostic value of heart rate variability in chronic congestive heart failure (Veterans Affairs' Survival Trial of Antiarrhythmic Therapy in Congestive Heart Failure). Am J Cardiol. 2002;90(1):24–8.
32. Fauchier L, Babuty D, Cosnay P, Fauchier JP. Prognostic value of heart rate variability for sudden death and major arrhythmic events in patients with idiopathic dilated cardiomyopathy. J Am Coll Cardiol. 1999;33(5):1203–7.
33. La Rovere MT, Pinna GD, Maestri R, Mortara A, Capomolla S, Febo O, et al. Short-term heart rate variability strongly predicts sudden cardiac death in chronic heart failure patients. Circulation. 2003;107(4):565–70. doi:10.1161/01.cir.0000047275.25795.17.
34. Guzzetti S, La Rovere MT, Pinna GD, Maestri R, Borroni E, Porta A, et al. Different spectral components of 24 h heart rate variability are related to different modes of death in chronic heart failure. Eur Heart J. 2005;26(4):357–62. doi:10.1093/eurheartj/ehi067.
35. La Rovere MT, Pinna GD, Maestri R, Barlera S, Bernardinangeli M, Veniani M, et al. Autonomic markers and cardiovascular and arrhythmic events in heart failure patients: still a place in prognostication? Data from the GISSI-HF trial. Eur J Heart Fail. 2012;14 (12):1410–9. doi:10.1093/eurjhf/hfs126.
36. Maestri R, Pinna GD, Accardo A, Allegrini P, Balocchi R, D'Addio G, et al. Nonlinear indices of heart rate variability in chronic heart failure patients: redundancy and comparative clinical value. J Cardiovasc Electrophysiol. 2007;18(4):425–33. doi:10.1111/j.1540-8167.2007.00728.x.
37. Cygankiewicz I, Zareba W, Vazquez R, Bayes-Genis A, Pascual D, Macaya C, et al. Risk stratification of mortality in patients with heart failure and left ventricular ejection fraction >35%. Am J Cardiol. 2009;103(7):1003–10. doi:10.1016/j.amjcard.2008.11.061.
38. Strickberger SA, Hummel JD, Bartlett TG, Frumin HI, Schuger CD, Beau SL, et al. Amiodarone versus implantable cardioverter-defibrillator: randomized trial in patients with nonischemic dilated cardiomyopathy and asymptomatic nonsustained ventricular tachycardia – AMIOVIRT. J Am Coll Cardiol. 2003;41(10):1707–12.
39. Kadish A, Dyer A, Daubert JP, Quigg R, Estes NA, Anderson KP, et al. Prophylactic defibrillator implantation in patients with nonischemic dilated cardiomyopathy. N Engl J Med. 2004;350(21):2151–8. doi:10.1056/NEJMoa033088.
40. Bardy GH, Lee KL, Mark DB, Poole JE, Packer DL, Boineau R, et al. Amiodarone or an implantable cardioverter-defibrillator for congestive heart failure. N Engl J Med. 2005;352 (3):225–37.

41. Goldberger JJ, Subacius H, Patel T, Cunnane R, Kadish AH. Sudden cardiac death risk stratification in patients with nonischemic dilated cardiomyopathy. J Am Coll Cardiol. 2014;63(18):1879–89. doi:10.1016/j.jacc.2013.12.021.
42. Goldberger JJ, Buxton AE, Cain M, Costantini O, Exner DV, Knight BP, et al. Risk stratification for arrhythmic sudden cardiac death: identifying the roadblocks. Circulation. 2011;123 (21):2423–30. doi:10.1161/CIRCULATIONAHA.110.959734.
43. Goldberger JJ, Basu A, Boineau R, Buxton AE, Cain ME, Canty Jr JM, et al. Risk stratification for sudden cardiac death: a plan for the future. Circulation. 2014;129(4):516–26. doi:10.1161/CIRCULATIONAHA.113.007149.
44. Deyell MW, Krahn AD, Goldberger JJ. Sudden cardiac death risk stratification. Circ Res. 2015;116(12):1907–18. doi:10.1161/CIRCRESAHA.116.304493.
45. Hohnloser SH, Kuck KH, Dorian P, Roberts RS, Hampton JR, Hatala R, et al. Prophylactic use of an implantable cardioverter-defibrillator after acute myocardial infarction. N Engl J Med. 2004;351(24):2481–8. doi:10.1056/NEJMoa041489.
46. Camm AJ, Pratt CM, Schwartz PJ, Al-Khalidi HR, Spyt MJ, Holroyde MJ, et al. Mortality in patients after a recent myocardial infarction: a randomized, placebo-controlled trial of azimilide using heart rate variability for risk stratification. Circulation. 2004;109(8):990–6. doi:10.1161/01.CIR.0000117090.01718.2A.
47. Huikuri HV, Raatikainen MJ, Moerch-Joergensen R, Hartikainen J, Virtanen V, Boland J, et al. Prediction of fatal or near-fatal cardiac arrhythmia events in patients with depressed left ventricular function after an acute myocardial infarction. Eur Heart J. 2009;30(6):689–98. doi:10.1093/eurheartj/ehn537.
48. Au-Yeung WT, Reinhall PG, Poole JE, Anderson J, Johnson G, Fletcher RD, et al. SCD-HeFT: use of R-R interval statistics for long-term risk stratification for arrhythmic sudden cardiac death. Heart Rhythm. 2015;12(10):2058–66. doi:10.1016/j.hrthm.2015.06.030.
49. Watanabe E, Tanabe T, Osaka M, Chishaki A, Takase B, Niwano S, et al. Sudden cardiac arrest recorded during Holter monitoring: prevalence, antecedent electrical events, and outcomes. Heart Rhythm. 2014;11(8):1418–25. doi:10.1016/j.hrthm.2014.04.036.
50. Cobb LA, Fahrenbruch CE, Olsufka M, Copass MK. Changing incidence of out-of-hospital ventricular fibrillation, 1980–2000. JAMA. 2002;288(23):3008–13.
51. Herlitz J, Andersson E, Bang A, Engdahl J, Holmberg M, Lindqvist J, et al. Experiences from treatment of out-of-hospital cardiac arrest during 17 years in Goteborg. Eur Heart J. 2000;21 (15):1251–8. doi:10.1053/euhj.2000.2150.
52. Kuisma M, Repo J, Alaspaa A. The incidence of out-of-hospital ventricular fibrillation in Helsinki, Finland, from 1994 to 1999. Lancet. 2001;358(9280):473–4. doi:10.1016/S0140-6736(01)05634-3.
53. Odemuyiwa O, Poloniecki J, Malik M, Farrell T, Xia R, Staunton A, et al. Temporal influences on the prediction of postinfarction mortality by heart rate variability: a comparison with the left ventricular ejection fraction. Br Heart J. 1994;71(6):521–7.
54. Shusterman V, Aysin B, Gottipaty V, Weiss R, Brode S, Schwartzman D, et al. Autonomic nervous system activity and the spontaneous initiation of ventricular tachycardia. ESVEM Investigators. Electrophysiologic Study Versus Electrocardiographic Monitoring Trial. J Am Coll Cardiol. 1998;32(7):1891–9.
55. Osaka M, Watanabe E, Murata H, Fuwamoto Y, Nanba S, Sakai K, et al. V-shaped trough in autonomic activity is a possible precursor of life-threatening cardiac events. Circ J. 2010;74 (9):1906–15.
56. Vybiral T, Glaeser DH, Goldberger AL, Rigney DR, Hess KR, Mietus J, et al. Conventional heart rate variability analysis of ambulatory electrocardiographic recordings fails to predict imminent ventricular fibrillation. J Am Coll Cardiol. 1993;22(2):557–65.
57. Makikallio TH, Koistinen J, Jordaens L, Tulppo MP, Wood N, Golosarsky B, et al. Heart rate dynamics before spontaneous onset of ventricular fibrillation in patients with healed myocardial infarcts. Am J Cardiol. 1999;83(6):880–4.

58. Wessel N, Ziehmann C, Kurths J, Meyerfeldt U, Schirdewan A, Voss A. Short-term forecasting of life-threatening cardiac arrhythmias based on symbolic dynamics and finite-time growth rates. Phys Rev E Stat Phys Plasmas Fluids Relat Interdiscip Top. 2000;61(1):733–9.
59. Kotecha D, Piccini JP. Atrial fibrillation in heart failure: what should we do? Eur Heart J. 2015;36:3250–7. doi:10.1093/eurheartj/ehv513.
60. Stein KM, Borer JS, Hochreiter C, Okin PM, Herrold EM, Devereux RB, et al. Prognostic value and physiological correlates of heart rate variability in chronic severe mitral regurgitation. Circulation. 1993;88(1):127–35.
61. Frey B, Heinz G, Binder T, Wutte M, Schneider B, Schmidinger H, et al. Diurnal variation of ventricular response to atrial fibrillation in patients with advanced heart failure. Am Heart J. 1995;129(1):58–65.
62. Yamada A, Hayano J, Sakata S, Okada A, Mukai S, Ohte N, et al. Reduced ventricular response irregularity is associated with increased mortality in patients with chronic atrial fibrillation. Circulation. 2000;102(3):300–6.
63. Watanabe E, Kiyono K, Hayano J, Yamamoto Y, Inamasu J, Yamamoto M, et al. Multiscale entropy of the heart rate variability for the prediction of an ischemic stroke in patients with permanent atrial fibrillation. PLoS ONE. 2015;10(9):e0137144. doi:10.1371/journal.pone.0137144.
64. Lip GY, Nieuwlaat R, Pisters R, Lane DA, Crijns HJ. Refining clinical risk stratification for predicting stroke and thromboembolism in atrial fibrillation using a novel risk factor-based approach: the euro heart survey on atrial fibrillation. Chest. 2010;137(2):263–72.
65. Bari V, Valencia JF, Vallverdu M, Girardengo G, Marchi A, Bassani T, et al. Multiscale complexity analysis of the cardiac control identifies asymptomatic and symptomatic patients in long QT syndrome type 1. PLoS ONE. 2014;9(4):e93808. doi:10.1371/journal.pone.0093808.
66. Porta A, Girardengo G, Bari V, George Jr AL, Brink PA, Goosen A, et al. Autonomic control of heart rate and QT interval variability influences arrhythmic risk in long QT syndrome type 1. J Am Coll Cardiol. 2015;65(4):367–74. doi:10.1016/j.jacc.2014.11.015.

Chapter 11
Heart Rate Variability and Neurological Disorders

Hisayoshi Oka

Abstract Heart rate variability (HRV) can generally be used to evaluate cardiac autonomic function in patients with neurological disorders. The low-frequency (LF) component of the RR interval on spectral analysis is influenced by both the sympathetic and parasympathetic nervous systems, whereas the high-frequency (HF) component of the RR interval mainly reflects cardiovagal tone. The ratio of the power at LF to that at HF (LF/HF) has been suggested to be an indicator of the sympathetic nervous system. LF and LF/HF were reduced even in early-stage de novo Parkinson's disease (PD) without orthostatic hypotension, despite no significant difference in HF. This finding indicated that sympathetic nerve dysfunction in the sinus node occurs in patients who have early-stage PD without parasympathetic nerve dysfunction. Cardiovascular function as evaluated by HRV was also associated with olfactory impairment. Reduced LF and HF components were found in multiple system atrophy. It seems that the impairment was more severe than that in PD. Although autonomic dysfunction in patients with progressive supranuclear palsy (PSP) has been controversial in previous studies, no clinically significant autonomic dysfunction was found in PSP patients who underwent evaluation of HRV. In dementias, the evaluation of HRV revealed reduced cardiac vagal denervation in dementia with Lewy bodies (DLB) and Parkinson's disease with dementia (PDD) and partial impairment of sympathovagal balance and abnormal postural heart modulation in Alzheimer disease (AD). It is thought that patients with ischemic stroke have depressed parasympathetic activity in the acute or chronic stage, while the impairment of sympathetic activity remains conflicting. HRV revealed that both sympathetic and parasympathetic functions were significantly decreased in patients with early-stage familial amyloid polyneuropathy (FAP), while sympathetic activity was more decreased in patients with advanced-stage FAP. Cardiac autonomic neuropathy (CAN) was evaluated by spectral analyses of the RR interval in diabetes mellitus. Parasympathetic dysfunction was apparently present in patients with mild autonomic dysfunction, while sympathetic dysfunction was observed in patients with moderate or severe autonomic neuropathy. A few

H. Oka, M.D., Ph.D. (✉)
Department of Neurology, Daisan Hospital, The Jikei University School of Medicine, 4-11-1 Izumihoncyo, Komae, Tokyo 201-8601, Japan
e-mail: h.oka@jikei.ac.jp

S. Iwase et al. (eds.), *Clinical Assessment of the Autonomic Nervous System*,
DOI 10.1007/978-4-431-56012-8_11

studies that used a 24-hour (24-h) heart rate power spectrum to evaluate autonomic dysfunction in Guillain-Barré syndrome have been reported. The HF component was significantly decreased at the height of the disease, and the LF/HF ratio increased as compared with the follow-up value after 1 year. Spectral analysis of HRV showed significantly reduced LF and HF components, and LF/HF was significantly greater in ALS patients than in healthy subjects. HRV is a useful marker of cardiovascular autonomic dysfunction in neurological disorders because it can be determined simply and noninvasively.

Keywords Parkinson's disease • Multiple system atrophy • Progressive supranuclear palsy • Orthostatic hypotension • Sympathetic and parasympathetic nervous system • Parkinson's disease with dementia • Dementia with Lewy bodies • Alzheimer disease • Ischemic stroke • Cardiac autonomic neuropathy • Diabetes mellitus • Familial amyloid polyneuropathy

11.1 Introduction

Heart rate variability (HRV) can generally be used to evaluate cardiac autonomic function in patients with neurological disorders. The high-frequency power (HF) of RR intervals is significantly affected by vagal gains and is quite independent of sympathetic gains [1–3], as demonstrated clinically and experimentally by autonomic maneuvers such as electrical vagal stimulation, muscarinic receptor blockade, and vagotomy. The HF of RR intervals is derived mainly from vagal activity. In contrast, the low-frequency power (LF) of RR intervals reflects both sympathetic activity and vagal activity. LF has been reported to be a quantitative marker of sympathetic modulations. The ratio of LF to HF (LF/HF) is considered to be influenced by the balance between sympathetic and parasympathetic activity. LF/HF is thought by some investigators to mirror sympathovagal balance or to reflect sympathetic modulations [4–6].

11.2 Neurodegenerative Disorders

11.2.1 Parkinson's Disease

Parkinson's disease (PD) is associated with resting tremor, rigidity, bradykinesia, and gait disturbance. PD is an important degenerative disease because of its high prevalence and non-motor symptoms such as autonomic dysfunction [7, 8]. Clinically significant dysfunction, such as orthostatic hypotension (OH) or postprandial hypotension, can occur in PD [8], although the autonomic dysfunction in PD is not as severe as that associated with multiple system atrophy [9–16]. OH occurs in 20–50 % of patients with PD and can contribute to falls and other accidental traumas [17–20]. Cardiac autonomic dysfunction develops early in Parkinson's disease before peripheral sympathetic dysfunction, as demonstrated by drops in blood pressure. Sympathetic failure in PD is common. OH is the most frequent symptom and the hallmark of sympathetic failure. Cardiac sympathetic denervation may occur very early and orthostatic dysregulation later in the course of PD progression. Decreased cardiac uptake of MIBG in Lewy body disease reflects cardiac sympathetic denervation, which precedes neuronal loss in sympathetic ganglia [21].

Parasympathetic dysfunction frequently occurs in PD and is manifested as constipation, bladder disturbance, and cardiac autonomic dysfunction. The pathogenesis of the parasympathetic nervous dysfunction associated with cardiac dysfunction in PD remains unclear. The cardiac parasympathetic nervous system might be relatively preserved in patients with PD. In fact, many studies have reported that the baroreceptor reflex of patients who have early-stage PD without OH is not severely impaired [22–25]. Cardiovagal efferent innervation, which produces reflex bradycardia, is mainly mediated by the nucleus ambiguus, which does not appear to suffer cell loss in patients with PD [26].

Quantitative autonomic testing of neurological patients has been reviewed recently [27], including Ewing's protocol [28] and HRV [29]. Spectral analysis of RR interval variations showed that RR-LF and LF/HF in early-stage de novo PD without OH were reduced as compared with the values in controls, despite no significant difference in RR-HF. The low RR-LF and LF/HF with preserved RR-HF indicate that sympathetic nerve dysfunction in the sinus node occurs in patients with early-stage PD without parasympathetic nerve dysfunction (Table 11.1) [30].

Sorensen et al. [31] documented cardiovascular autonomic dysfunction during wakefulness and sleep in patients with idiopathic rapid eye movement (REM) sleep behavior disorder (iRBD) and PD in whom HRV was evaluated. They reported that patients with iRBD had attenuated sympathetic nervous system activity as compared with controls, and this finding was more pronounced in patients with PD. As mentioned above in Oka's report [30], the cardiac parasympathetic nervous system may be relatively well preserved in patients with PD as well as iRBD, and sympathetic nervous dysfunction may progress in line with the postganglionic sympathetic nervous dysfunction seen in early PD.

Previous studies have reported that the autonomic dysfunction in PD is associated with disease characteristics such as progression and duration during wakefulness [32, 33]. Several studies have demonstrated a relation between autonomic dysfunction and PD severity by evaluating the circadian fluctuations of HRV in PD patients [34, 35]. Covassin et al. reported that PD severity was significantly related to nocturnal HRV indices, and such associations were restricted to REM [36]. HRV has also been used to assess therapy in PD. Liu et al. have examined whether subthalamic nucleus deep brain stimulation (STN-DBS) affects functions of the autonomic nervous system as assessed by spectral analysis of HRV. They reported that the power of LF and HF components was significantly activated in the stimulation "on" groups by STN-DBS [37].

Table 11.1 Hemodynamic autonomic function on the Valsalva maneuver and spectral analyses of RR intervals and blood pressure in control subjects and patients with de novo PD without OH

	Control $n = 20$	De novo PD without OH $n = 20$	P value
BRSII (ms/mmHg)	3.6 ± 2.2	2.7 ± 1.9	ns
BRSIV (ms/mmHg)	6.1 ± 4.0	4.0 ± 3.6	ns
PhaseII_E (mmHg)	20.5 ± 16.8	21.2 ± 13.6	ns
PhaseII_L (mmHg)	14.9 ± 12.7	9.5 ± 8.6	ns
PhaseIV	22.7 ± 9.5	11.6 ± 9.5	0.001
PRT	2.1 ± 1.1	4.6 ± 4.4	0.008
RR-LF (ms^2)	25.7 ± 21.8	12.7 ± 11.2	0.037
RR-HF (ms^2)	20.7 ± 17.4	22.0 ± 19.8	ns
LF/HF	1.55 ± 1.17	0.61 ± 0.29	0.003
SBP-LF ($mmHg^2$)	0.48 ± 0.35	0.30 ± 0.30	0.040

PD Parkinson's disease, *OH* orthostatic hypotension, *BRS II* baroreceptor reflex sensitivity in phase II, *BRS IV* baroreceptor reflex sensitivity in phase IV, *Phase II_E* systolic blood pressure (SBP) decrease in early phase II, *Phase II_L* SBP increase in late phase II, *Phase IV* the overshoot of blood pressure in phase IV, *PRT* blood pressure recovery time (time interval from the end of phase III to the complete return of SBP to the baseline value), *RR-LF* low-frequency component of RR interval, *RR-HF* high-frequency component of RR interval, *LF/HF* ratio of LF to HF, *SBP-LF* low-frequency component of systolic blood pressure

An association between non-motor symptoms and cardiovascular dysfunction as evaluated by HRV was recently reported in PD. The relations of olfactory function (odor stick identification test Japan, OSIT-J) to cardiovascular function as evaluated by HRV assessed on the basis of the coefficient of variation of RR intervals, cardiac ^{123}I-MIBG uptake of the heart, orthostatic hypotension, and other clinical variables were studied. The OSIT-J score was related to the cardiac ^{123}I-MIBG uptake, the fall in orthostatic blood pressure, and HRV, after adjustment for other clinical variables (Fig. 11.1) [38]. Kang et al. also reported that olfactory dysfunction in PD positively correlated with reduced HF bands on spectral analysis of RR intervals and suggested that olfactory involvement is associated with parasympathetic dysautonomia in cardiovascular systems [39].

11.2.2 Parkinson's Disease-Related Disorders

11.2.2.1 Multiple System Atrophy

It is well known that signs of dysautonomia such as orthostatic hypotension, abnormalities in bladder and bowel control, and temperature dysregulation are distinctive features in multiple system atrophy (MSA) [40]. A 24-h analysis of HRV was effective for characterizing cardiac autonomic status in MSA. Moreover, HRV variables negatively correlated with disease duration and severity [10]. Our previous study demonstrated reduced low RR-LF and HF components on spectral analysis of RR intervals (Table 11.2) [41].

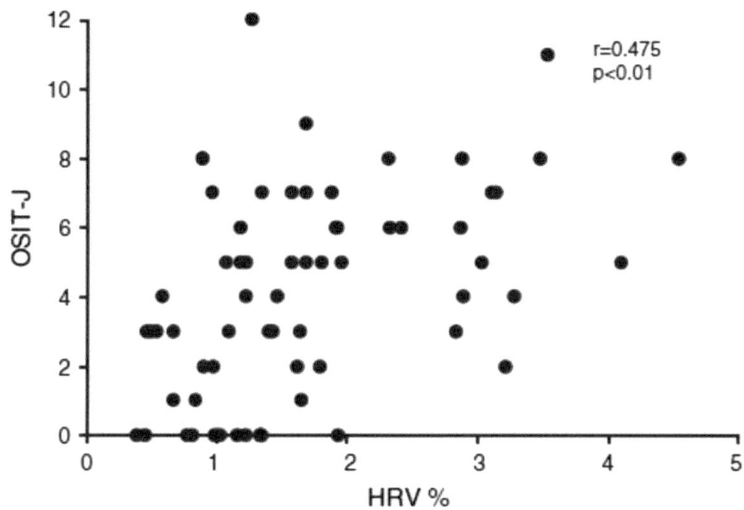

Fig. 11.1 Correlations between the OSIT-J score and HRV. A significant correlation between the OSIT-J score and HRV was found. HRV was assessed by the coefficient of variation for RR intervals. *OSIT-J* odor stick identification test Japan, *HRV* heart rate variability

Table 11.2 Comparison of spectral analysis of RR intervals between control and MSA

	Control ($n = 40$)	MSA ($n = 10$)	P value
RR-LF	280 ± 197	125 ± 99	0.015
RR-HF	213 ± 200	48 ± 37	0.001

Brisinda et al. [42] evaluated autonomic nervous system dysfunction in PD and MSA using the Cardiac Autonomic Nervous System Evaluation protocol (CANSEp), which consisted of Ewing's protocol and the analysis of HRV (HRVa). Ewing's protocol scores were higher in MSA than in PD and controls. HRVa revealed abnormal autonomic function in PD and MSA, as compared with controls. Markedly decreased LF and HF powers were found in PD and MSA, during both daily activity and sleep. In PD, depressed vagal tone was predominant during sleep, whereas in MSA, depressed sympathetic tone prevailed during daily activity.

11.2.3 Progressive Supranuclear Palsy

Autonomic dysfunction in progressive supranuclear palsy (PSP) remains controversial. Some studies found no significant autonomic dysfunction in PSP patients [43–46], whereas others reported autonomic cardiovascular abnormalities in PSP patients [14, 47–50]. Holmberg et al. [51] evaluated autonomic function in PD,

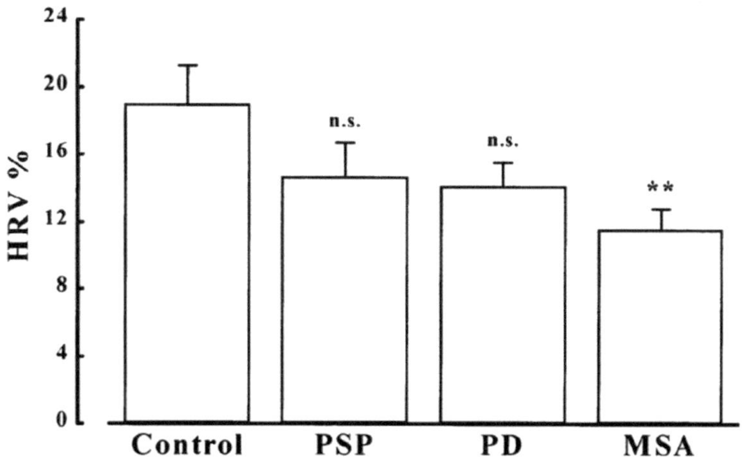

Fig. 11.2 Relative heart rate variability (HRV) during controlled deep breathing in the patient groups and healthy controls. Mean levels and S.E.M. are shown. HRV was significantly reduced in the multiple system atrophy (MSA) group (**$P < 0.01$), whereas no significant difference (not significant, n.s.) was found in the Parkinson's disease (PD) or progressive suparanuclear palsy (PSP) groups as compared with the controls

PSP, and MSA by means of laboratory autonomic tests, which included analysis of HRV at rest and during deep respiration, as well as blood pressure changes during 75° head-up tilt. Decreased HRV and severe hypotensive responses were seen in MSA patients regardless of age and disease duration, whereas this combination of findings was seen only in PD patients who were very elderly and had a prolonged disease duration. In PSP, only a few patients showed decreased HRV, and hypotensive responses were limited (Fig. 11.2).

11.3 Dementias

Resting HRV and heart rate changes to deep breathing (HRDB) and to standing (30:15 ratio), which indicate cardiac vagal denervation, are reduced in DLB. Cardiac vagal dysfunction is also found in dementia with PD (PDD) [52, 53]. Differences among dementia with Lewy bodies (DLB), PDD, and PD without dementia with respect to the involvement of the autonomic nervous system have been demonstrated in a study investigating cutaneous and cardiovascular autonomic functions in patients with Lewy body disease [53]. That study used the coefficient of variation of RR intervals (CVRR) to estimate cardiovascular function. Abnormally low values of CVRR were observed in half of the patients with DLB, as compared with 25 % of the patients with PDD and PD. The mean CVRR value was significantly lower in the patients with DLB (Fig. 11.3). It was also reported that sudomotor function may be severely affected in Lewy body disorders, while skin

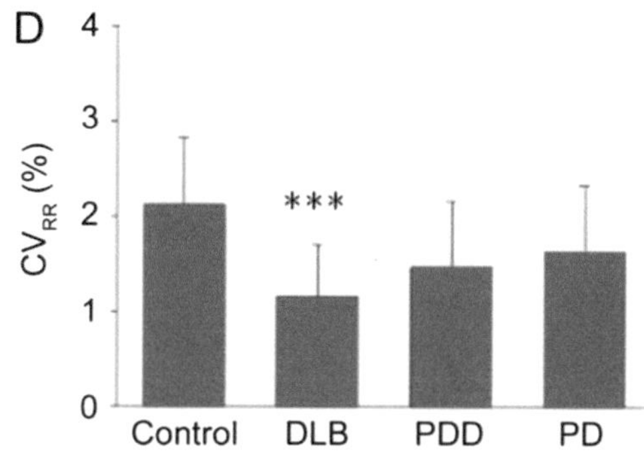

Fig. 11.3 Comparison of CV-RR according to the type of dementia. The mean CVRR value was significantly lower in the patients with DLB. *CV-RR* coefficient of variation of RR intervals

vasomotor function and the cardiovascular system may be more severely affected in DLB and PDD than in PD.

Allan et al. [52] used spectral analysis of HRV to assess cardiovascular autonomic function in patients with Alzheimer disease (AD), vascular dementia (VaD), DLB, and PDD. They reported that the severity of cardiovascular autonomic dysfunction differed significantly among these four types of dementia. These four groups of patients also showed significant differences in total spectral power as well as LF and HF components (Table 11.3). Mean RR interval, very LF component, and LF/HF ratio did not differ significantly among the four groups, while LF and HF components were lower in patients with PDD than in healthy controls. The LF and HF components were also reduced as compared with the AD patient group. The LF components in patients with DLB were significantly lower than those in the healthy controls. Allan et al. [52] also evaluated cardiac autonomic dysfunction using Ewing's protocol to perform autonomic function tests in patients with dementia. Ewing's protocol consisted of five tests, consisting of three predominantly parasympathetic tests and two predominantly sympathetic tests, to assess autonomic dysfunction at bedside [54]. The test results were used to classify autonomic dysfunction into five groups according to severity [55, 56].

According to Ewing's protocol, PDD was consistently associated with impairment of both parasympathetic and sympathetic function as compared with controls and AD. DLB showed impairment of parasympathetic function. PDD showed significantly higher impairment than DLB in terms of the Valsalva ratio, considered an indicator of parasympathetic function. Patients with VaD showed impairment based on the Valsalva ratio. Autonomic dysfunction occurs in all common dementias but is especially prominent in PDD and has important treatment implications [52]. Several studies have also described partial impairment of the sympathovagal balance and abnormal postural heart modulation in AD [57, 58], while recent reports indicate that cardiovagal dysfunction in AD correlates with cognitive decline [59], although resting HRV can be normal [60].

Table 11.3 Spectral analysis of HRV in patients with AD, VaD, DLB, and PDD

Heart rate variability						
Diagnosis (n)	Control (31/38)	AD (32/39)	VaD (27/30)	DLB (23/30)	PDD (38/40)	
Total power (ms^2)	1003 (575–1431)	820 (483–1158)	922 (332–1512)	617 (300–934)	628 (301–714)	AD vs DLB: 0.24 (0.18)
ANOVA: $p = 0.04$		0.49 (0.55)	0.27 (0.19)	0.08 (**0.05**)	0.003 (**0.003**)	**AD vs PDD: 0.02 (0.02)**
						DLB vs PDD: 0.38 (0.42)
Low-frequency power (ms^2)	2323 (169–477)	1324 (158–490)	389 (91.5–687)	261 (85.6–438)	171 (94.0–247)	AD vs DLB: 0.25 (0.13)
ANOVA: $p = 0.04$		0.41 (0.43)	0.32 (0.29)	0.059 (**0.02**)	0.007 (**0.002**)	**AD vs PDD: 0.06 (0.03)**
						DLB vs PDD: 0.61 (0.47)
High-frequency power (ms^2)	293 (81· 3–504)	165 (99· 1–232)	231 (54.4–407)	129 (46.5–212)	105 (44.0–166)	AD vs DLB: 0.13 (0.20)
ANOVA: $p = 0.003$		0.89 (0.83)	0.59 (0.39)	0.10 (0.23)	0.001 (**0.01**)	**AD vs PDD: 0.001 (0.01)**
						DLB vs PDD: 0.17 (0.25)

LF and HF components were reduced in patients with PDD as compared with healthy controls. LF and HF components were also reduced as compared with the AD patient group. LF components in patients with DLB were significantly reduced as compared with controls
P values in the left-hand column give the result of the univariate ANOVA across all groups
Columns 3–6 show the mean (95 % confidence intervals for the mean) for each heart rate variability test by diagnosis, with *p* values for that patient group in comparison with the control group in univariate ANOVA and in multivariate analyses in *brackets*
The results of other predefined contrasts are given in the right hand column
All significant results are shown in bold typeface
AD Alzheimer disease, *DLB* dementia with Lewy bodies, *PDD* Parkinson's disease dementia, *VaD* vascular dementia, *HRV* heart rate variability, *LF* low frequency, *HF* high frequency

11.4 Cerebrovascular Disease

Autonomic dysfunction is frequently seen in patients with ischemic stroke [61, 62]. Patients with ischemic stroke have depressed parasympathetic activity in the acute or chronic stages, while the impairment of sympathetic activity remains controversial [63]. Acute cerebrovascular diseases are thought to significantly decrease HRV as a result of cardiovascular autonomic dysregulation. Decreased HRV has been demonstrated in stroke by cardiovascular reflex tests [64], as well as by spectral analysis of HRV based on 24-h ECG recordings [65, 66]. The HF and LF spectral components of HRV seem to be persistently reduced in hemispheric lesions [67, 68]. It is also known that decreased HRV is an independent predictor of mortality in patients with acute ischemic stroke [69]. The underlying mechanisms of these findings in patients with ischemic stroke remain to be elucidated. Increased sympathetic activity has been reported in acute stroke in some studies [61], although Korpelainen et al. found no impairment of sympathetic activity in patients with acute ischemic stroke [63]. The increased blood pressure variability in patients with acute stroke has been attributed to impaired baroreceptor reflex sensitivity (BRS) associated with the central parasympathetic and sympathetic nervous systems [62, 69]. Impaired BRS has also been linked to increased long-term mortality after acute stroke.

Xiong et al. [70] have studied whether autonomic function is impaired during different phases in ischemic stroke, using Ewing's protocol of autonomic function tests as well as spectral analysis of HRV. Stroke patients in both the acute and chronic phases had significantly lower LF than did controls. On autonomic function testing by Ewing's protocol [53–55], patients with acute ischemic stroke showed impairment in two parasympathetic tests (Valsalva ratio, heart rate response to deep breathing), while those in the chronic stage showed impairment in all parasympathetic tests as compared with controls. They concluded that autonomic dysfunction in ischemic stroke persisted in the chronic stage (up to 6 months) after stroke and that parasympathetic dysfunction rather than sympathetic dysfunction predominates after ischemic stroke. The mechanism underlying the autonomic impairment in patients with stroke remains unclear. A few studies have assessed cardiovascular autonomic function in subtypes of ischemic stroke [71, 72], but the results were conflicting. One study reported that there was no significant association between impaired autonomic function and subtypes of acute ischemic stroke [73], whereas another study demonstrated that patients with acute atherosclerotic infarction of large arteries had lower parasympathetic activity and higher sympathetic activity than those with acute lacunar infarction [71]. Xiong et al. [73] evaluated cardiovascular autonomic function in ischemic stroke patients with large-artery atherosclerosis (LAA) and patients with small-vessel occlusion (SCO) by performing autonomic function tests by Ewing's protocol and spectral analysis of HRV. The results demonstrated that cardiovascular autonomic dysfunction was influenced by the subtype of ischemia and that patients with LAA had more severely impaired parasympathetic and sympathetic functions than those with SCO (Table 11.4).

Table 11.4 Comparisons of HRV among subtypes of stroke and control

Heart rate variability				
Diagnosis	Controls	Patients with LAA	Patients with SVO	*p**
Time domain				
Resting R-R interval (ms)	856.3 (816.3–896.3)[a]	865.2 (830.8–899.7)[a]	812.9 (762.7–863.1)[a]	0.189
Breathing frequency (cpm)	13 [12–15]	14 [12–16]	14 [13–16]	0.899
Frequency domain				
Very low-frequency power (ms^2)	94.9 (52.8–204.5)	65.4 (30.4–176.2)	56.0 (17.1–231.3)	0.249
Low-frequency power (ms^2)	141.7 (52.9–231.9)	54.8 (28.5–118.6)	57.8 (18.5–187.2)	0.019
High-frequency power (ms^2)	85.4 (35.1–167.9)	50.6 (27.5–165.7)	30.4 (15.5–171.4)	0.429
Total power (ms^2)	387.4 (175.6–532.2)	231.9 (95.5–526.8)	129.6 (42.4–636.8)	0.211
LF/HF	1.7 (0.8–2.9)	1.1 (0.4–2.0)	1.1 (0.4–2.4)	0.103

LF power was significantly reduced in the LAA and SVO patient groups as compared with the controls. There were no differences in mean RRI, respiratory frequencies, very LF power, HF power, or LF/HF ratio between the patients and controls

Data expressed as median (interquartile range) except for [a]where mean (95 % CI)

LF low frequency, *HF* high frequency

*Represents *P* values for comparison across all groups

Although the pathophysiologic basis of cardiovascular autonomic dysfunction remains unclear, it has been reported that damage of the cortical or subcortical structures or the neural pathways can induce dysregulation in the cardiovascular autonomic system. The insular cortex of the cerebral artery territory is considered the most important region for sympathetic and parasympathetic regulation of the cardiovascular autonomic system [74, 75]. The insular cortex has connections with the autonomic regulatory areas located in the subcortical limbic and forebrain regions [76]. The frequency of supraventricular tachycardia has been reported to be higher in patients with right hemispheric lesions than in those with left hemispheric lesions. Stimulation of the human right insula increases sympathetic cardiovascular tone, whereas parasympathetic activity increases more frequently during left insular stimulation [74, 77]. Left insular lesions may induce an increased incidence of new-onset atrial fibrillation [78]. The insular cortex appears to play an important role in regulating cardiovascular autonomic functions, and other regions of the brain, such as the forebrain or brain stem, may participate in cardiovascular regulation.

11.5 Peripheral Neuropathy

11.5.1 Familial Amyloid Polyneuropathy

Familial amyloid polyneuropathy (FAP) is a hereditary and polyneuropathic disease with long-term progression. Autonomic neuropathy occurs in most patients with early-onset FAP, while it is less prominent in late-onset FAP [79]. These patients often have cardiovascular, gastrointestinal, and urinary system dysfunction. Cardiovascular circulatory dysfunction is responsible for orthostatic hypotension. It can cause fatigue, blurred vision, or dizziness on standing. Postprandial diarrhea and constipation alternately occur. Erectile dysfunction is an early feature in men. Dysuria and urinary retention may develop in late stage.

Kinoshita et al. [80] evaluated autonomic dysfunction in patients with FAP by time and frequency domain analysis of HRV. The 24-h electrocardiograms of subjects in sinus rhythm and off medication were analyzed. The power spectra of LF and HF and the LF/HF ratio were calculated. Global measures of HRV, including the standard deviation of mean RR intervals (SDNN) and the standard deviation of 5-min mean RR intervals (SDANN), were decreased in patients with FAP. Specific vagal effects on HRV, including the proportion of RR intervals more than 50 ms different (pNN50) and the HF power on spectral analysis, were less in patients with FAP. LF power and LF/HF ratio were also more decreased in patients with FAP. They concluded that HRV was significantly decreased in patients with early-stage FAP, whereas sympathetic activity was more decreased in patients with advanced-stage disease. They also mentioned that the decrease in HRV is an indicator of FAP and that the power spectral analysis of HRV is useful for assessing the severity of autonomic dysfunction.

11.5.2 Diabetic Autonomic Neuropathy

Cardiovascular autonomic neuropathy (CAN) is frequently found in patients with diabetes mellitus and is a common form of autonomic dysfunction. Parasympathetic dysfunction in the heart induces a high resting heart rate and a fixed heart rate with advanced nerve dysfunction caused by vagal neuropathy, which results in increased sympathetic outflow [81, 82]. CAN has been evaluated by spectral analyses of the RR interval in patients with diabetes mellitus. Correlations between spectral analyses of the RR interval and various clinical features or the severity of autonomic neuropathy were assessed [83]. Figure 11.1 shows the relation between the severity of autonomic neuropathy and the spectral analysis of the RR interval in diabetic autonomic neuropathy. The severity of autonomic nervous dysfunction was scored by the Valsalva ratio, overshoot, cold pressor response, coefficient of variation of the RR interval (CV-RR), and orthostatic hypotension. The results of these tests were graded as normal (score 0) or abnormal (score 1). The subjects were classified into four groups according to the severity of autonomic dysfunction, which was based on the total scores of the tests: group I (score 0), group II (score 1–2), group III (score 3–4), and group IV (score 5). Normal controls were evaluated as group C. There were no significant differences in RR-LF between group C and group I or group II. The RR-LF was significantly smaller in groups III and IV than in groups C, I, and II. The RR-HF was essentially the same in groups I and C but differed markedly between groups II and C. The RR-HF was significantly smaller in groups III and IV than in group C. The results suggested that parasympathetic dysfunction was present in patients with mild autonomic dysfunction, while sympathetic dysfunction was observed in patients with moderate or severe autonomic neuropathy (Fig. 11.4).

CAN is associated with an increased risk of mortality, as well as orthostatic hypotension [84]. Impaired autonomic control of heart rate is known to be linked to an increased risk of mortality. Fatal arrhythmias in diabetes may be caused by reduced parasympathetic function or increased sympathetic activity [81]. Maser et al. [85] performed a meta-analysis examining the relation between CAN and the risk of mortality in individuals with diabetes. Fifteen studies published in English-language journals from 1966 to 2001 that included a baseline assessment of cardiovascular autonomic function and mortality follow-up were identified. CAN was found to be significantly associated with subsequent mortality when autonomic dysfunction was defined as the presence of abnormalities on two or more tests of HRV. Ziegler et al. [86] have also evaluated whether reduced HRV, prolonged corrected QT (QTc) interval, or increased QT dispersion (QTD) are predictors of mortality in the general diabetic and nondiabetic population. They reported that prolonged QTc interval, but not increased QTD, is an independent predictor of an

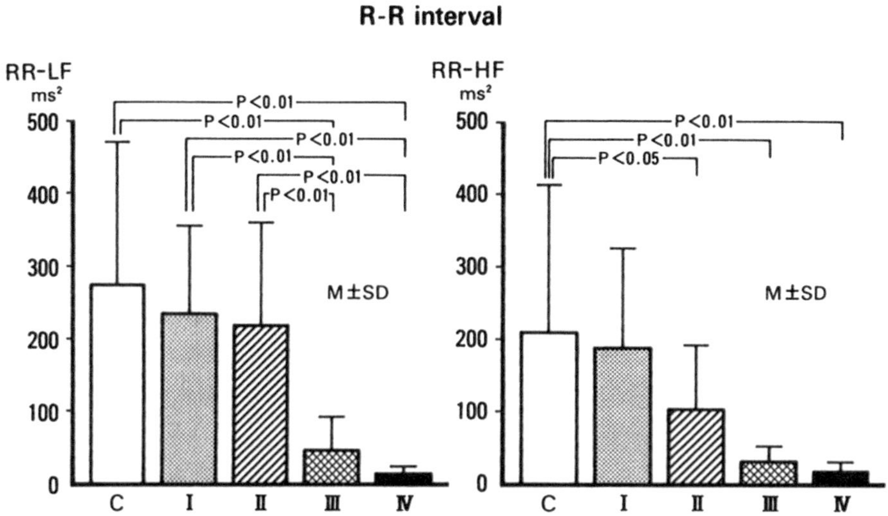

Fig. 11.4 Severity and spectral analysis data in diabetic autonomic neuropathy. *C* Control. *I* (score 0): patients without autonomic neuropathy. *II* (score 1–2): patients with mild autonomic neuropathy. *III* (score 3–4): patients with moderate autonomic neuropathy. *IV* (score 5): patients with severe autonomic neuropathy

increased risk of mortality in the nondiabetic and diabetic elderly general population and that low HRV during spontaneous breathing tends to be associated with excess mortality in the diabetic but not in the nondiabetic population [87].

11.5.3 Guillain-Barré Syndrome

Autonomic dysfunction frequently occurs as an important complication in approximately two-thirds of patients with Guillain-Barré syndrome (GBS). GBS is often associated with various types of autonomic dysfunction, including cardiovascular, vasomotor, or sudomotor dysfunction [88]. HRV has been analyzed in patients with GBS. Several studies used a 24-h heart rate power spectrum to evaluate autonomic dysfunction in GBS. Flachenecker et al. [89] estimated HRV by power spectrum analysis for up to 1 year in patients with GBS. At the peak of the disease, the HF component decreased significantly. The LF/HF ratio increased as compared with the follow-up value after 1 year. Both of these variables gradually returned to normal with time. In patients who presented with tachycardia, LF and HF power were strikingly decreased as compared with patients who had normal heart rates, while in patients showing vagal over-reactivity, the power of both spectral bands significantly increased. They concluded that spectral analysis of HRV is useful for investigating cardiovascular neural regulation in patients with GBS, and the sympathovagal balance clearly shifted to sympathetic predominance at the height of disease in patients with this disorder. They mentioned that 24-h heart rate power spectrum analysis is not only a powerful predictor of serious autonomic complications but also may help to identify patients with GBS who are at risk for potentially life-threatening arrhythmias [90, 91].

11.6 Amyotrophic Lateral Sclerosis

Autonomic dysfunction is frequently recognized in advanced amyotrophic lateral sclerosis (ALS) [92]. Sympathetic activity is considered to increase in the early stage [93, 94] and decrease in the late stage in ALS [95–98]. Sympathetic dysfunction is attributed to both central (preganglionic) and postganglionic involvement [99, 100]. Impaired HRV is expressed as depressed parasympathetic function [101, 102]. Pavlovic et al. [103] investigated autonomic cardiac control in patients with ALS according to Ewing's protocol, power spectrum analysis of HRV (LF, HF, and LF/HF), real-time beat-to-beat ECG signal monitoring with HRV analysis, and baroreflex function analysis. Short-term spectral analysis of HRV showed significantly reduced LF and HF in ALS patients. LF/HF was significantly greater in ALS patients than in healthy subjects. Continuous real-time, beat-to-beat ECG signal monitoring with HRV analysis at rest showed significantly increased sympathetic components and decreased parasympathetic components. Baroreflex sensitivity was lower in ALS patients than in controls [104]. The analysis of HRV revealed impaired cardiac autonomic control in ALS with marked parasympathetic dysfunction and sympathetic predominance.

References

1. Pagani M, Lombardi F, Guzzetti S, Rimoldi O, Furlan R, Pizzinelli P, et al. Power spectral analysis of heart rate and arterial pressure variabilities as a marker of sympatho-vagal interaction in man and conscious dog. Circ Res. 1986;59:178–93.
2. Pomeranz B, Macaulay RJ, Caudill MA, Kutz I, Adam D, Gordon D, et al. Assessment of autonomic function in humans by heart rate spectral analysis. Am J Physiol. 1985;248: H151–3.
3. Malliani A, Pagani M, Lombardi F, Cerutti S. Cardiovascular neural regulation explored in the frequency domain. Circulation. 1991;84:482–92.
4. Task Force of the European Society of Cardiology, North American Society of Pacing and Electrophysiology. Heart rate variability: standards of measurement, physiological interpretation, and clinical use. Circulation. 1996;93:1043–65.
5. Hayano J, Taylor JA, Yamada A, Mukai S, Hori R, Asakawa T, et al. Continuous assessment of hemodynamic control by complex demodulation of cardiovascular variability. Am J Physiol. 1993;264:H1229–38.
6. Cacioppo JT, Berntson GG, Binkley PF, Quigley KS, Uchino BN, Fieldstone A. Autonomic cardiac control. II. Noninvasive indices and basal response as revealed by autonomic blockades. Psychophysiology. 1994;31:586–98.

7. Rajput AH, Offord KP, Beard CM, Kurland LT. Epidemiology of parkinsonism: incidence, classification, and mortality. Ann Neurol. 1984;16:278–82.
8. Goldstein DS. Dysautonomia in Parkinson's disease: neurocardiological abnormalities. Lancet Neurol. 2003;2:669–76.
9. Bonuccelli U, Lucetti C, Del Dotto P, Ceravolo R, Gambaccini G, Bernardini S, et al. Orthostatic hypotension in de novo Parkinson disease. Arch Neurol. 2003;60:1400–4.
10. Jost WH. Autonomic dysfunctions in idiopathic Parkinson s disease. J Neurol. 2003;250 Suppl 1:I28–30.
11. Perani D, Bressi S, Testa D, Grassi F, Cortelli P, Gentrini S, et al. Clinical/metabolic correlations in multiple system atrophy. A fludeoxyglucose F 18 positron emission tomographic study. Arch Neurol. 1995;52:179–85.
12. Colosimo C, Albanese A, Hughes AJ, de Bruin VM, Lees AJ. Some specific clinical features differentiate multiple system atrophy (striatonigral variety) from Parkinson's disease. Arch Neurol. 1995;52:294–8.
13. Quinn N. Multiple system atrophy: the nature of the beast. J Neurol Neurosurg Psychiatry. 1989;52(Suppl):78–89.
14. Sandroni P, Ahlskog JE, Fealey RD, Low PA. Autonomic involvement in extrapyramidal and cerebellar disorders. Clin Auton Res. 1991;1:147–55.
15. Magalhaes M, Wenning GK, Daniel SE, Quinn NP. Autonomic dysfunction in pathologically confirmed multiple system atrophy and idiopathic Parkinson's disease – a retrospective comparison. Acta Neurol Scand. 1995;91:98–102.
16. Gilman S. Multiple system atrophy. In: Jankovic J, Tolosa E, editors. Parkinson's disease and movement disorders. 3rd ed. Baltimore: Williams & Wilkins; 1998. p. 245–62.
17. Micieli G, Martignoni E, Cavallini A, Sandrini G, Nappi G. Postprandial and orthostatic hypotension in Parkinson's disease. Neurology. 1987;37:386–93.
18. Seared JM, Raï S, Lapeyre-Mestre M, Brefel C, Rascol O, Rascol A, et al. Prevalence of orthostatic hypotension in Parkinson's disease. J Neurol Neurosurg Psychiatry. 1997;63:584–9.
19. Hillen ME, Wagner ML, Sage JI. "Subclinical" orthostatic hypotension is associated with dizziness in elderly patients with Parkinson disease. Arch Phys Med Rehabil. 1996;77:710–2.
20. Goldstein DS, Pechnik S, Holmes C, Eldadah B, Sharabi Y. Association between supine hypertension and orthostatic hypotension in autonomic failure. Hypertension. 2003;42:136–42.
21. Orimo S, Ozawa E, Oka T, Nakade S, Tsuchiya K, Yoshimoto M, et al. Different histopathology accounting for a decrease in myocardial MIBG uptake in PD and MSA. Neurology. 2001;57:1140–1.
22. Haensch CA, Lerch H, Jörg J, Isenmann S. Cardiac denervation occurs independent of orthostatic hypotension and impaired heart rate variability in Parkinson's disease. Parkinsonism Relat Disord. 2009;15:134–7.
23. Sachs C, Berglund B, Kaijser L. Autonomic cardiovascular responses in parkinsonism: effect of levodopa with dopa-decarboxylase inhibition. Acta Neurol Scand. 1985;71:37–42.
24. Ludin SM, Steiger UH, Ludin HP. Autonomic disturbances and cardiovascular reflexes in idiopathic Parkinson's disease. J Neurol. 1987;235:10–5.
25. Camerlingo M, Aillon C, Bottacchi E, Gambaro P, D'Alessandro G. Parasympathetic assessment in Parkinson's disease. Adv Neurol. 1987;45:267–9.
26. Benarroch EE, Schmeichel AM, Parisi JE. Preservation of branchimotor neurons of the nucleus ambiguus in multiple system atrophy. Neurology. 2003;60:115–7.
27. Weimer LH. Autonomic testing: common techniques and clinical applications. Neurologist. 2010;16:215–22.
28. Ewing DJ. Cardiovascular reflex and autonomic neuropathy. Clin Sci Mol Med. 1978;55:321–7.

29. Task force of the European Society of Cardiology, North American Society of pacing and Electrophysiology. Heart rate variability. Standards measurement, physiological interpretation, and clinical use. Eur Heart J. 1996;17:354–81.
30. Oka H, Toyoda C, Yogo M, Mochio S. Cardiovascular dysautonomia in de novo Parkinson's disease without orthostatic hypotension. Eur J Neurol. 2011;18:286–92.
31. Sorensen GL, Mehlsen J, Jennum P. Reduced sympathetic activity in idiopathic rapid-eye-movement sleep behavior disorder and Parkinson's disease. Auton Neurosci. 2013;179:138–41.
32. Mesec A, Sega S, Kiauta T. The influence of the type, duration, severity and levodopa treatment of Parkinson's disease on cardiovascular autonomic responses. Clin Auton Res. 1993;3:339–44.
33. Korchounov A, Kessler KR, Yakhno NN, Damulin IV, Schipper HI. Determinants of autonomic dysfunction in idiopathic Parkinson's disease. J Neurol. 2005;252:1530–6.
34. Haapaniemi TH, Pursiainen V, Korpelainen JT, Huikuri HV, Sotaniemi KA, Myllylä VV. Ambulatory ECG and analysis of heart rate variability in Parkinson's disease. J Neurol Neurosurg Psychiatry. 2001;70:305–10.
35. Niwa F, Kuriyama N, Nakagawa M, Imanishi J. Circadian rhythm of rest activity and autonomic nervous system activity at different stages in Parkinson's disease. Auton Neurosci. 2011;165:195–200.
36. Covassin N, Neikrug AB, Liu L, Maglione J, Natarajan L, Corey-Bloom J, et al. Relationships between clinical characteristics and nocturnal cardiac autonomic activity in Parkinson's disease. Auton Neurosci. 2012;171:85–8.
37. Liu KD, Shan DE, Kuo TB, Yang CC. The effects of bilateral stimulation of the subthalamic nucleus on heart rate variability in patients with Parkinson's disease. J Neurol. 2013;260:1714–23.
38. Oka H, Toyoda C, Yogo M, Mochio S. Olfactory dysfunction and cardiovascular dysautonomia in Parkinson's disease. J Neurol. 2010;257:969–76.
39. Kang P, Kloke J, Jain S. Olfactory dysfunction and parasympathetic dysautonomia in Parkinson's disease. Clin Auton Res. 2012;22:161–6.
40. Albanese A, Colosimo C, Bentivoglio AR, Fenici R, Melillo G, Tonali P. Multiple system atrophy presenting as parkinsonism: clinical features and diagnostic criteria. J Neurol Neurosurg Psychiatry. 1995;59:144–51.
41. Oka H, Mochio S, Katayama K, Inoue K. Autonomic dysfunction in multiple system atrophy by spectral analyses of RR interval and systolic blood pressure. Auton Nerv Syst (Tokyo). 1997;34:322–6.
42. Brisinda D, Sorbo AR, Di Giacopo R, Venuti A, Bentivoglio AR, Fenici R. Cardiovascular autonomic nervous system evaluation in Parkinson disease and multiple system atrophy. J Neurol Sci. 2014;336:197–202.
43. Kimber J, Mathias CJ, Lees AJ, Bleasdale-Barr K, Chang HS, Churchyard A, et al. Physiological, pharmacological and neurohormonal assessment of autonomic function in progressive supranuclear palsy. Brain. 2000;123:1422–30.
44. Kikkawa Y, Asahina M, Suzuki A, Hattori T. Cutaneous sympathetic function and cardiovascular function in patients with progressive supranuclear palsy and Parkinson's disease. Parkinsonism Relat Disord. 2003;10:101–6.
45. Deguchi K, Sasaki I, Tsukaguchi M, Kamoda M, Touge T, Takeuchi H, et al. Abnormalities of rate-corrected QT intervals in Parkinson's disease-a comparison with multiple system atrophy and progressive supranuclear palsy. J Neurol Sci. 2002;199:31–7.
46. Brefel-Courbon C, Thalamas C, Rascol O, Montastruc JL, Senard JM. Lack of autonomic nervous dysfunction in progressive supranuclear palsy, a study of blood pressure variability. Clin Auton Res. 2000;10:309–12.
47. Gutrecht JA. Autonomic cardiovascular reflexes in progressive supranuclear palsy. J Auton Nerv Syst. 1992;39:29–35.

48. Sasaki I, Takeuchi H, Deguchi K, Yamada A, Nishioka M, Sakamoto H. Autonomic nervous function in progressive supranuclear palsy–comparison with Parkinson's disease and healthy controls. Clin Neurol (Rinsho Shinkeigaku). 1994;34:975–9.
49. Wenning GK, Scherfler C, Granata R, Bösch S, Verny M, Chaudhuri KR, et al. Time course of symptomatic orthostatic hypotension and urinary incontinence in patients with postmortem confirmed parkinsonian syndromes: a clinicopathological study. J Neurol Neurosurg Psychiatry. 1999;67:620–3.
50. Van Dijk JG, Haan J, Koenderink M, Roos RA. Autonomic nervous function in progressive supranuclear palsy. Arch Neurol. 1991;48:1083–4.
51. Holmberg B, Kallio M, Johnels B, Elam M. Cardiovascular reflex testing contributes to clinical evaluation and differential diagnosis of parkinsonian syndromes. Mov Disord. 2001;16:217–5.
52. Allan LM, Ballard CG, Allen J, Murray A, Davidson AW, McKeith IG, et al. Autonomic dysfunction in dementia. J Neurol Neurosurg Psychiatry. 2007;78:671–7.13.
53. Akaogi Y, Asahina M, Yamanaka Y, Koyama Y, Hattori T. Sudomotor, skin vasomotor and cardiovascular reflexes in 3 clinical forms of Lewy body disease. Neurology. 2009;73:59–65.
54. Ewing DJ, Borsey DQ, Bellavere F, Clarke BF. Cardiac autonomic neuropathy in diabetes: comparison of measures of R–R interval variation. Diabetologia. 1981;21:18–24.
55. Ewing DJ, Clarke BF. Diagnosis and management of diabetic autonomic neuropathy. Br Med J (Clin Res Ed). 1982;285(6346):916–8.
56. Ewing DJ, Martyn CN, Young RJ, Clarke BF. The value of cardiovascular autonomic function tests: 10 years experience in diabetes. Diabetes Care. 1985;8:491–8.
57. Idiaquez J, Sandoval E, Seguel A. Association between neuropsychiatric and autonomic dysfunction in Alzheimer's disease. Clin Auton Res. 2002;12:43–6.
58. De Vilhena Toledo MA, Junqueira Jr LF. Cardiac sympathovagal modulation evaluated by short-term heart interval variability is subtly impaired in Alzheimer's disease. Geriatr Gerontol Int. 2008;8:109–18.
59. Toledo MA, Junqueira Jr LF. Cardiac autonomic modulation and cognitive status in Alzheimer's disease. Clin Auton Res. 2010;20:11–7.
60. Allan LM, Kerr SR, Ballard CG, Allen J, Murray A, McLaren AT, et al. Autonomic function assessed by heart rate variability is normal in Alzheimer's disease and vascular dementia. Dement Geriatr Cogn Disord. 2005;19:140–4.
61. Dutsch M, Burger M, Dorfler C, Schwab S, Hilz MJ. Cardiovascular autonomic function in post stroke patients. Neurology. 2007;69:2249–55.
62. McLaren A, Kerr S, Allan L, Steen IN, Ballard C, Allen J, et al. Autonomic function is impaired in elderly stroke survivors. Stroke. 2005;36:1026–30.
63. Korpelainen JT, Sotaniemi KA, Huikuri HV, Myllya VV. Abnormal heart rate variability as a manifestation of autonomic dysfunction in hemispheric brain infarction. Stroke. 1996;27:2059–63.
64. Tokgozoglu SL, Batur MK, Top uoglu MA, Saribas O, Kes S, Oto A. Effects of stroke localization on cardiac autonomic balance and sudden death. Stroke. 1999;30:1307–11.
65. Naver HK, Blomstrand C, Wallin BG. Reduced heart rate variability after right-sided stroke. Stroke. 1996;27:247–51.
66. Barron SA, Rogovski Z, Hemli J. Autonomic consequences of cerebral hemisphere infarction. Stroke. 1994;25:113–6.
67. Korpelainen JT, Huikuri HV, Sotaniemi KA, Myllylä VV. Abnormal heart rate variability reflecting autonomic dysfunction in brainstem infarction. Acta Neurol Scand. 1996;94:337–42.
68. Colivicchi F, Bassi A, Santini M, Caltagirone C. Cardiac autonomic derangement and arrhythmias in right-sided stroke with insular involvement. Stroke. 2004;35:2094–8.
69. Robinson TG, Dawson SL, Eames PJ, Panerai RB, Potter JF. Cardiac baroreceptor sensitivity predicts long-term outcome after acute ischemic stroke. Stroke. 2003;34:705–12.

70. Xiong L, Leung HH, Chen XY, Han JH, Leung TW, Soo YO, et al. Comprehensive assessment for autonomic dysfunction in different phases after ischemic stroke. Int J Stroke. 2013;8:645–51.
71. Chen PL, Kuo TB, Yang CC. Parasympathetic activity correlates with early outcome in patients with large artery atherosclerotic stroke. J Neurol Sci. 2012;314:57–61.
72. Kwon DY, Lim HE, Park MH, Oh K, Yu SW, Park KW, et al. Carotid atherosclerosis and heart rate variability in ischemic stroke. Clin Auton Res. 2008;18:355–7.
73. Xiong L, Leung HW, Chen XY, Leung WH, Soo OY, Wong KS. Autonomic dysfunction in different subtypes of post-acute ischemic stroke. J Neurol Sci. 2014;337:141–6.
74. Oppenheimer SM, Gelb A, Girvin JP, Hachinski VC. Cardiovascular effects of human insular cortex stimulation. Neurology. 1992;42:1727–32. 44.
75. Kedem G, Martin WM. Left-insular cortex lesions perturb cardiac autonomic tone in humans. Clin Auton Res. 1996;6:131–40.
76. Cechetto DF, Saper CB. Role of the cerebral cortex in autonomic function. In: Loewy AD, Spyer KM, editors. Central regulation of autonomic functions. New York: Oxford University Press; 1990. p. 208–23.
77. Lane RD, Wallace JD, Petrosky PP, Schwartz GE, Gradman AH. Supraventricular tachycardia in patients with right hemisphere strokes. Stroke. 1992;23:362–6.
78. Vingerhoets F, Bogousslavsky J, Regli F, van Malle G. Atrial fibrillation after stroke. Stroke. 1993;24:20–6.
79. Planté-Bordeneuve V, Said G. Familial amyloid polyneuropathy. Lancet Neurol. 2011;10:1086–97.
80. Kinoshita O, Hongo M, Saikawa Y, Katsuyama T, Tanaka M, Takeda M, et al. Heart rate variability in patients with familial amyloid polyneuropathy. PACE. 1997;20:2949–53.
81. Maser RE, Lenhard MJ. Cardiovascular autonomic neuropathy due to diabetes mellitus: clinical manifestations, consequences, and treatment. J Clin Endocrinol Metab. 2005;90:5896–903.
82. Callaghan BC, Cheng HT, Stables CL, Smith AL, Feldman EL. Diabetic neuropathy: clinical manifestations and current treatments. Lancet Neurol. 2012;11:521–34.
83. Oka H, Mochio S, Sato K, Sato H, Katayama K, Watanabe S, et al. Spectral analyses of R-R interval and systolic blood pressure in diabetic autonomic neuropathy. J Auton Nerv Syst. 1995;52:203–11.
84. Vinik AI, Maser RE, Mitchell BD, Freeman R. Diabetic autonomic neuropathy. Diabetes Care. 2003;26:1553–79.
85. Maser RE, Mitchell BD, Vinik AI, Freeman R. The association between cardiovascular autonomic neuropathy and mortality in individuals with diabetes: a meta-analysis. Diabetes Care. 2003;26:1895–901.
86. Ziegler D, Zentai CP, Perz S, Rathmann W, Haastert B, Döring A, et al. Prediction of mortality using measures of cardiac autonomic dysfunction in the diabetic and nondiabetic population. Diabetes Care. 2008;31:556–61.
87. Tahrani AA, Dubb K, Raymond NT, Begum S, Altaf QA, Sadiqi H, et al. Cardiac autonomic neuropathy predicts renal function decline in patients with type 2 diabetes: a cohort study. Diabetologia. 2014;57:1249–56.
88. Hartung HP, Pollard JD, Harvey GK, Toyka KV. Immunopathogenesis and treatment of the Guillain-Barré syndrome – part I. Muscle Nerve. 1995;18:137–53.
89. Flachenecker P, Hartung HP, Reiners K. Power spectrum analysis of heart rate variability in Guillain-Barre syndrome. A longitudinal ‘study. Brain. 1997;120:1885–94.
90. Flachenecker P, Reiners K. Twenty-four-hour heart rate power spectrum for evaluation of autonomic dysfunction in Guillain-Barré syndrome. J Neurol Sci. 1999;165:144–53.
91. Flachenecker P, Lem K, Müllges W, Reiners K. Detection of serious bradyarrhythmias in Guillain-Barré syndrome: sensitivity and specificity of the 24-hour heart rate power spectrum. Clin Auton Res. 2000;10:185–91.

92. Shimizu T, Hayashi H, Hayashi M, Kato S, Tanabe H. Hyposensitivity of peripheral alpha-receptors in respiratory-dependent amyotrophic lateral sclerosis assessed by intravenous norepinephrine infusion. Clin Auton Res. 1995;5:165–9.
93. Shindo K, Tsunoda S, Shiozawa Z. Microneurographic analysis of muscle sympathetic nerve activity in amyotrophic lateral sclerosis. Clin Auton Res. 1993;3:131–5.
94. Shindo K, Tsunoda S, Shiozawa Z. Increased sympathetic outflow to muscles in patients with amyotrophic lateral sclerosis: a comparison with other neuromuscular patients. J Neurol Sci. 1995;134:57–60.
95. Karlsborg M, Andersen EB, Wiinberg N, Gredal O, Jorgensen L, Mehlsen J. Sympathetic dysfunction of central origin in patients with ALS. Eur J Neurol. 2003;10:229–34.
96. Hecht MJ, Bron CM, Mittelhamm F, Werner D, Heuss D, Neundorfer B, et al. Increased hypoxic blood pressure response in patients with amyotrophic lateral sclerosis. J Neurol Sci. 2003;213:47–53.
97. Baltadzhieva R, Gurevich T, Korczyn AD. Autonomic impairment in amyotrophic lateral sclerosis. Curr Opin Neurol. 2005;18:487–93.
98. Sachs C, Conradi S, Kaijser L. Autonomic function in amyotrophic lateral sclerosis: a study of cardiovascular responses. Acta Neurol Scand. 1985;71:373–8.
99. Oey PL, Vos PE, Wieneke GH, Wokke JH, Blankestijn PJ, Karemaker JM. Subtle involvement of the sympathetic nervous system in amyotrophic lateral sclerosis. Muscle Nerve. 2002;25:402–8.
100. Kennedy P, Duchen L. A quantitative study of intermedio-lateral column cells in amyotrophic lateral sclerosis and Shy-Drager. J Neurol Neurosurg Psychiatry. 1985;48:1103–6.
101. Pisano F, Miscio G, Mazzuero G, Lanfranchi P, Colombo R, Pinelli P. Decreased heart rate variability in amyotrophic lateral sclerosis. Muscle Nerve. 1995;18:1225–31.
102. Merico A, Cavinato M. Autonomic dysfunction in the early stage of ALS with bulbar involvement. Amyotroph Lateral Scler. 2011;12:363–7.
103. Pavlovic S, Stevic Z, Milovanovic B, Milicic B, Rakocevic-Stojanovic V, Lavrnic D, et al. Impairment of cardiac autonomic control in patients with amyotrophic lateral sclerosis. Amyotroph Lateral Scler. 2010;11:272–6.
104. Linden D, Diehl RR, Berlit P. Reduced baroreflex sensitivity and cardiovascular regulation in amyotrophic lateral sclerosis. Electroencephalogr Clin Neurophysiol. 1998;109:387–90.

Part III
Cardiac MIBG Scintigraphy

Chapter 12
An Introduction to MIBG Cardiac Scintigraphy

Satoshi Orimo

Abstract *Meta*-iodobenzylguanidine (MIBG), a guanethidine analog, has an uptake and storage mechanism similar to that of noradrenaline. Using this agent, we can noninvasively assess the pathophysiology of postganglionic cardiac sympathetic nerve endings. Cardiac scintigraphy using MIBG radiolabeled with iodine-123 (^{123}I-MIBG) has been applied to patients with pheochromocytoma and various types of heart disease as well as neurological disorders, including Parkinson's disease (PD) and dementia with Lewy bodies (DLB). Reduced cardiac MIBG uptake is specific to Lewy body disease. Therefore, ^{123}I-MIBG cardiac scintigraphy is a useful imaging tool to differentiate PD and DLB from other types of parkinsonism and dementia. Reduced cardiac MIBG uptake is associated with cardiac sympathetic nerve degeneration that is specific to Lewy body disease and is also found in early PD, which accounts for the reduced cardiac MIBG uptake in Lewy body disease. Currently, it is thought that reduced cardiac MIBG uptake is a potential biomarker for the presence of Lewy bodies. Recently, the cross-calibration phantom method has helped convert institutional heart-to-mediastinum (H/M) ratios to standard H/M ratios. This standardization will contribute to multi-center studies as well as data comparisons with other institutions.

Keywords MIBG cardiac scintigraphy • Pathological background • Lewy bodies • Standardization

12.1 *Meta*-Iodobenzylguanidine (MIBG)

Meta-iodobenzylguanidine (MIBG) is an analog of guanethidine, an adrenergic blocking agent. Its uptake and storage mechanism is similar to that of noradrenaline (also known as norepinephrine) [1, 2]. It is actively taken up by postganglionic presynaptic nerve endings in the adrenergic nervous system, using an energy-dependent noradrenaline transporter mechanism. MIBG is also actively taken up

S. Orimo (✉)
Division of Neurology, Kanto Central Hospital of the Mutual Aid Association of Public School Teachers, 6-25-1 Kami-Yoga, Setagaya, Tokyo 158-8531, Japan
e-mail: sorimo0307@nifty.com

S. Iwase et al. (eds.), *Clinical Assessment of the Autonomic Nervous System*,
DOI 10.1007/978-4-431-56012-8_12

by noradrenaline vesicles and the ATPase-dependent proton pump via a vesicle monoamine transporter [3–5]. Therefore, we can perform noninvasive assessment of the pathophysiology of organs that are innervated by the sympathetic nervous system by using this agent. Noradrenaline and MIBG kinetics in the sympathetic nervous system are described in Chap. 13.

12.2 A Historical Perspective of MIBG Cardiac Scintigraphy

In 1980, the tissue distributions of three kinds of radiolabeled iodobenzylguanidine (e.g., ^{125}I-*meta*-, ^{125}I-*para*-, and ^{131}I-*para*-iodobenzylguanidine) were determined in dogs by Wieland et al. for the first time [1]. The high affinity and retention of these agents were observed in the adrenal medulla; high myocardial concentrations were also observed at early time intervals. Images of the adrenal medullae were obtained with ^{131}I-*para*-iodobenzylguanidine. In 1981, ^{123}I-MIBG was used to image the heart in five normal men [6]. The left ventricle was visualized 1–2 min after MIBG injection. The mean cardiac uptake was 0.63 % of the injected dose after 5 min and 0.76 % at 2 h. In 1988, ^{123}I-MIBG cardiac scintigraphy was used in patients with idiopathic congestive cardiomyopathy and myocardial infarctions to determine the function and damage of the cardiac sympathetic nerve endings [7, 8]. Since then, this imaging tool has been further developed to evaluate various kinds of heart disease, such as heart failure, ischemic heart disease, and cardiomyopathy. A detailed history of MIBG development is described in Chap. 13. Technical considerations for MIBG cardiac scintigraphy are described in the Chap. 14. Findings using MIBG cardiac scintigraphy in cardiac diseases, diabetes mellitus (DM), and other disorders are described in Chap. 15.

12.3 A Historical Perspective of MIBG Cardiac Scintigraphy in Japan

In Japan in 1987, the Daiichi Radioisotope Laboratory (currently, the FUJIFILM RI Pharma Co., Ltd.) performed a multicenter clinical trial of ^{123}I-MIBG (named MyoMIBG) to image the heart. In 1991, the results of this multicenter clinical trial using ^{123}I-MIBG were reported [9]. In 1992, the Japanese Ministry of Welfare (currently known as the Ministry of Health, Labor and Welfare) approved the clinical use of ^{123}I-MIBG in cardiology practice. Since then, we have accumulated extensive clinical experience dealing with various heart diseases, including ischemic heart disease, arrhythmia, idiopathic dilated cardiomyopathy, hypertrophic cardiomyopathy, and cardiomyopathies secondary to DM, renal failure, and other metabolic disorders. In 1994, ^{123}I-MIBG cardiac scintigraphy was applied to patients with neurological disorders, such as Parkinson's disease (PD) and familial amyloidotic neuropathy, to assess systemic autonomic dysfunction for the first time [10, 11]. Reduced cardiac MIBG uptake was observed in patients with PD, not only with orthostatic hypotension (OH) but also without OH [10]. Seki reported on a patient with PD with reduced cardiac MIBG uptake and the possible usefulness of ^{123}I-MIBG cardiac scintigraphy to evaluate disturbances in the sympathetic nervous system in PD [12]. Yoshita and Orimo reported that cardiac MIBG uptake was reduced in PD [13, 14] and that the degree of reduced cardiac MIBG uptake correlated with motor symptom severity [14]. After these reports, research focusing on ^{123}I-MIBG cardiac scintigraphy in PD progressed in Japan and Europe. Many studies have reported that this imaging tool is useful to differentiate PD from other types of parkinsonism, such as multiple system atrophy (MSA), progressive supranuclear palsy (PSP), corticobasal degeneration (CBD), and vascular parkinsonism together with essential tremor. Goldstein reported that reduced cardiac fluorodopamine uptake was observed in PD, but not in MSA, using positron emission tomography with ^{18}F-fluorodopamine [15]. In 2001, Watanabe and Yoshita reported that cardiac MIBG uptake was reduced in dementia with Lewy bodies (DLB) and that ^{123}I-MIBG cardiac scintigraphy was a useful tool to differentiate DLB from AD [16, 17]. Findings from MIBG cardiac scintigraphy in PD and related disorders and DLB and related dementia are described in Chaps. 15 and 16, respectively. The physiological background of reduced cardiac MIBG uptake is described in Chap. 17.

12.4 Pathological Background of Cardiac MIBG Uptake

In 2001, to clarify the pathological basis of reduced cardiac MIBG uptake in PD, Orimo et al. examined cardiac tissue from pathologically confirmed patients with PD and MSA using an antibody against tyrosine hydroxylase (TH), the rate-limiting enzyme in catecholamine biosynthesis [18]. This study demonstrated a near-complete depletion of TH-immunoreactive nerve fibers in the epicardial nerve fascicles in PD, but not in MSA and control; this accounted for the reduced cardiac MIBG uptake in PD, but not in MSA. Subsequently, pathological studies focused on the pathophysiological mechanism of cardiac MIBG uptake, and the degeneration mechanism of the cardiac sympathetic nerve in Lewy body disease had extensively progressed [19–23]. A series of studies demonstrated that (1) - TH-immunoreactive fibers in the heart were decreased in pathologically confirmed Lewy body disease, but not in other related disorders, supporting the findings of reduced cardiac MIBG uptake in Lewy body disease, (2) degeneration of the cardiac sympathetic nerve began early in PD, (3) α-synuclein aggregates accumulated in neurons and the nerve fibers of the cardiac sympathetic nervous system centripetally in PD, and (4) degeneration of the cardiac sympathetic nerve was closely related to the presence of Lewy bodies in a wide range of neurodegenerative processes, indicating that reduced cardiac MIBG uptake is a potential biomarker for the presence of Lewy bodies. The pathological background of reduced cardiac MIBG uptake and others are described in Chap. 18, while the clinical implications of reduced cardiac MIBG are described in Chap. 19.

12.5 Standardization

The heart-to-mediastinum (H/M) ratio is a comparison of the myocardial uptake (counts/pixel) and mediastinal uptake (counts/pixel) of region of interests (ROIs) in anterior planar imaging, which estimates MIBG accumulation to the myocardium. This index is considerably influenced by the location and size of ROIs, and an average count is reduced from lower to upper, according to mediastinal ROI. The H/M ratio is influenced by the choice of collimetor, which is attached to a gamma camera. Therefore, the H/M ratio can be variable depending on the differences of the collimetor and scinticameras used at each institute as well as the setting place and ROI. Inter-institutional differences in MIBG quantification using the H/M ratio have hampered multicenter comparisons of H/M ratios, such that the results of a single center study cannot easily be extrapolated to studies from other facilities [24]. Recently, Nakajima et al. developed a novel software named "smart MIBG" to standardize H/M ratios, which is cooperated by FUJIFILM RI Pharma Co., Ltd. Furthermore, they performed a study to standardize the MIBG H/M ratio using a calibration phantom method [25, 26]. This standardization is described in detail in Chap. 14.

References

1. Wieland DM, Wu J-I, Brown LE, Mangner TJ, Swanson DP, Beierwaltes WH. Radiolabeled adrenergic neuron-blocking agents: adrenomedullary imaging with [^{131}I]iodobenzylguanidine. J Nucl Med. 1980;21:349–53.
2. Wieland DM, Wu J-I, Brown LE, Mangner TJ, Swanson DP, Beierwaltes WH. Imaging the primate adrenal medulla with [^{123}I] and [^{131}I] meta-iodobenzylguanidine: concise communication. J Nucl Med. 1980;21:349–53.
3. Wieland DM, Brown LE, Rogers WL, Worthington KC, Wu JL, Clinthorne NH, et al. Myocardial imaging with a radioiodinated norepinephrine storage analog. J Nucl Med. 1981; 22(1):22–31.
4. Tobes MC, Jaques Jr S, Wieland DM, Sisson JC. Effect of uptake-one inhibitors on the uptake of norepinephrine and meta-iodobenzylguanidine. J Nucl Med. 1985;26(8):897–907.
5. Sisson JC, Shapiro B, Meyers L, Mallette S, Mangner TJ, Wieland DM, et al. Meta-iodobenzylguanidine to map scintigraphically the adrenergic nervous system in man. J Nucl Med. 1987;28:1625–36.
6. Kline RC, Swanson DP, Wieland DM, Thrall JH, Gross MD, Pitt B, et al. Myocardial imaging in man with I-123 meta-iodobenzylguanidine. J Nucl Med. 1981;22(2):129–32.
7. Schofer J, Spielmann R, Schuchert A, Weber K, Schluter M. Iodine-123 meta-iodobenzylguanidine scintigraphy: a noninvasive method to demonstrate myocardial adrenergic nervous

system disintegrity in patients with idiopathic dilated cardiomyopathy. J Am Coll Cardiol. 1988;12:1252–8.
8. Henderson EB, Kahn JK, Corbett JR, Jansen DE, Pippin JJ, Kulkarni P, et al. Abnormal I-123 meta-iodobenzylguanidine myocardial washout and distribution may reflect myocardial adrenergic derangement in patients with congestive cardiomyopathy. Circulation. 1988;78:1192–9.
9. Morozumi T, Ishida Y, Tani A, Sato H, Matsuyama T, Ozaki H, et al. Clinical significance of I-123 MIBG myocardial scintigraphy for evaluating the severity of congestive heart failure. Kaku Igaku [in Japanese]. 1991;28(3):271–80.
10. Hakusui S, Yasuda T, Yanagi T, Tohyama J, Hasegawa Y, Koike Y, et al. A radiological analysis of heart sympathetic functions with meta-[^{123}I]iodobenzylguanidine in neurological patients with autonomic failure. J Auton Nerv Syst. 1994;49:81–4.
11. Ando Y, Obayashi K, Tanaka Y, Tsuji T, Uchino M, Takahashi M, et al. Radiolabelled meta-iodobenzylguanidine in assessment of autonomic dysfunction. Lancet. 1994;343:984–5.
12. Seki K, Tanaka M, Kiyono Y. A trial of 123I-MIBG SPECT in Parkinson's disease as an autonomic function test. Niigata Kakuigakukonwakai. 1994;18:19–22.
13. Yoshita M, Matubara S, Tada A. Reduced MIBG uptake on^{123}I-meta-iodobenzyl-guanidine cardiac scintigraphy in Parkinson's disease. Sinkeinaika [in Japanese]. 1996;45:221–5.
14. Orimo S, Ozawa E, Nakade S. ^{123}I-meta-iodobenzylguanidine myocardial scintigraphy in Parkinson's disease. Igakunoayumi [in Japanese]. 1996;179:1728–9.
15. Goldstein DS, Holmes C, Cannon III RO, Eisenhofer G, Kopin IJ. Sympathetic cardioneuropathy in dysautonomias. New Eng J Med. 1997;336:696–702.
16. Watanabe H, Ieda T, Katayama T, Takeda A, Aiba I, Doyu M, et al. Cardiac ^{123}I-meta-iodobenzylguanidine (MIBG) uptake in dementia with Lewy bodies: comparison with Alzheimer's disease. J Neurol Neurosurg Psychiatry. 2001;70:781–3.
17. Yoshita M, Taki J, Yamada M. A clinical role for [^{123}I]MIBG myocardial scintigraphy in the distinction between dementia of the Alzheimer's-type and dementia with Lewy bodies. J Neurol Neurosurg Psychiatry. 2001;71:583–8.
18. Orimo S, Ozawa E, Oka T, Nakade S, Tsuchiya K, Yoshimoto M, et al. Different histopathology accounting for a decrease in myocardial MIBG uptake in PD and MSA. Neurology. 2001;57: 1140–1.
19. Orimo S, Oka T, Miura H, Tsuchiya K, Mori F, Wakabayashi K, et al. Sympathetic cardiac denervation in Parkinson's disease and pure autonomic failure but not in multiple system atrophy. J Neurol Neurosurg Psychiatry. 2002;73:776–7.
20. Amino T, Orimo S, Ihoh Y, Takahashi A, Uchihara H, Mizusawa H. Profound cardiac sympathetic denervation occurs in Parkinson disease. Brain Pathol. 2005;15:29–34.
21. Orimo S, Amino T, Itoh Y, Takahashi A, Kojo T, Uchihara T, et al. Cardiac sympathetic denervation precedes neuronal loss in the sympathetic ganglia in Lewy body disease. Acta Neuropathol. 2005;109:583–8.
22. Orimo S, Takahashi A, Uchihara T, Mori F, Kakita A, Wakabayashi K, et al. Degeneration of cardiac sympathetic nerve begins in the early disease process of Parkinson's disease. Brain Pathol. 2007;17:24–30.
23. Orimo S, Uchihara T, Nakamura A, Mori F, Kakita A, Wakabayashi K, et al. Axonal α-synuclein aggregates herald centripetal degeneration of cardiac sympathetic nerve in Parkinson's disease. Brain. 2008;131:642–50.
24. Okuda K, Nakajima K, Hosoya T, Ishikawa T, Konishi T, Matsubara K, et al. Semi-automated algorithm for calculating heart-to-mediastinum ratio in cardiac Iodine-123 MIBG imaging. J Nucl Cardiol. 2011;18:82–8.
25. Nakajima K, Okuda K, Matsuo S, Yoshita M, Taki J, Yamada M, et al. Standardization of meta-iodobenzylguanidine heart to mediastinum ratio using a calibration phantom: effects of correction on normal databases and a multicentre study. Eur J Nucl Med Mol Imaging. 2012; 39:113–9.
26. Nakajima K, Okuda K, Yoshimura M, Matsuo S, Wakabayashi H, Imanishi Y, et al. Multicenter cross-calibration of I-123 meta-iodobenzylguanidine heart-to mediastinum ratios to overcome camera-collimator variations. J Nucl Cardiol. 2014;21:970–8.

Chapter 13
Noradrenaline and ^{123}I-*Meta*-Iodobenzylguanidine Kinetics in the Sympathetic Nervous System

Tomohiko Nakamura and Gen Sobue

Abstract Noradrenaline is a neurotransmitter released from sympathetic nerve endings and is synthesized from the amino acid tyrosine through a series of enzymatic reactions. When sympathetic nerves are excited, noradrenaline is released to increase the heart rate and cardiac contractility through beta-1 adrenergic receptors. Only a fraction of the released noradrenaline binds to receptors, and most of the remainder (80–95 %) undergoes active reuptake into nerve endings (uptake-1 mechanism) to be restored in vesicles. ^{123}I-*meta*-iodobenzylguanidine (*MIBG*), an analog of guanethidine (an adrenergic neuron-blocking agent), is a radioactive tracer of noradrenaline and is used to evaluate presynaptic sites of the adrenergic system. MIBG can determine the distribution, activity, and disorders of postsynaptic cardiac sympathetic nerves and is widely used to evaluate various kinds of heart diseases, diabetes, and autonomic disorders, such as neurodegenerative disorders. MIBG is transported into the nerve by noradrenaline monoamine transporter 1, trapped in storage vesicles, and released by a mechanism similar to noradrenaline but is not catabolized like noradrenaline, and this is a necessary characteristic as sympathetic nerve-imaging agent. Understanding MIBG kinetics is important in recognizing the characteristics and the difference between early and delayed image of MIBG scintigraphy.

Keywords Noradrenaline • MIBG • Kinetics • Sympathetic nerve

T. Nakamura (✉)
Department of Neurology, Nagoya University Graduate School of Medicine, 65 Tsurumai-cho, Showa-ku, Nagoya, Aichi 466-8550, Japan
e-mail: tomohiko@med.nagoya-u.ac.jp

G. Sobue
Nagoya University Graduate School of Medicine, 65 Tsurumai-cho, Showa-ku, Nagoya, Aichi 466-8550, Japan

S. Iwase et al. (eds.), *Clinical Assessment of the Autonomic Nervous System*,
DOI 10.1007/978-4-431-56012-8_13

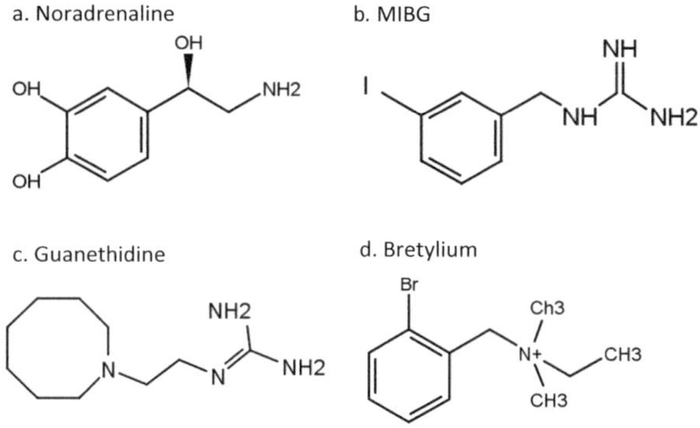

Fig. 13.1 Structural formula of (**a**) noradrenaline, (**b**) *meta*-iodobenzylguanidine (MIBG), (**c**) guanethidine, and (**d**) bretylium. MIBG was synthesized on the basis of the structural formulas of guanethidine and bretylium

13.1 Introduction

Noradrenaline plays an important role in the regulation of blood pressure and heart rate. Upon excitation, noradrenaline is released from sympathetic nerve terminals and stimulates the sinoatrial or atrioventricular node to increase the heart rate. Noradrenaline also increases cardiac contractility through beta-1 adrenergic receptors.

^{123}I-*meta*-iodobenzylguanidine (*MIBG*), a chemical analog of noradrenaline, was developed based on the chemical structure of guanethidine, a sympathetic blocker (Fig. 13.1a–c). The uptake and storage mechanisms of MIBG in sympathetic nerve terminals are similar to noradrenaline. MIBG scintigraphy can determine the distribution, activity, and disorders of postsynaptic cardiac sympathetic nerves. It is widely used to evaluate various kinds of heart diseases, diabetes, and autonomic disorders, such as neurodegenerative disorders. Here we discuss the role and kinetics of noradrenaline in cardiac sympathetic nerves and the development of MIBG and its kinetics.

13.2 The Role of Noradrenaline in Cardiac Sympathetic Nerves

Autonomic control of the heart consists of sympathetic and parasympathetic nerves [1]. The neurotransmitter of the sympathetic system is noradrenaline. The neurotransmitter of the parasympathetic system is acetylcholine. Working together, these systems stimulate or inhibit the heart via adrenergic and muscarinic receptors [1]. Sympathetic output is controlled by regulatory centers in the brain that integrate input signals from other brain sites and receptors throughout the body, such as the carotid sinus and aortic sinus [1]. The heart contains densely distributed postsynaptic sympathetic fibers where noradrenaline functions as a neurotransmitter.

Efferent sympathetic nerves descend in the spinal cord, synapse with preganglionic fibers, synapse with paravertebral stellate ganglia, and innervate the right ventricle and anterior/lateral left ventricle. In the heart, sympathetic nerves follow coronary arteries in the subepicardium, penetrating into the myocardium to regulate cardiac function [1, 2]. Sympathetic nerves consist of a cell body, long axon, and adrenergic postganglionic fibers that branch to form nerve endings. These adrenergic postganglionic fibers control the heart, with the sinoatrial node, atrioventricular node, and atrium receiving the most innervation. The ventricles receive less innervation and Purkinje fibers the least innervation [3]. Sympathetic nerve fibers contain a number of varicose structures, so-called varicosities. The transmission of nerve impulses to the nerve terminal produces nerve terminal depolarization. Synaptic vesicles stored in these varicosities aggregate and integrate into the cell membrane releasing noradrenaline. This is called exocytosis, and noradrenaline is thus released into the synapse between nerve endings and cardiomyocytes, binds to receptors on effector organs, and enhances adenyl cyclase activity through intermediary G proteins [4–6]. Lastly, the impulse is transmitted to cardiac myocytes to modulate physiological functions, such as increasing heart rate and cardiac contraction.

13.3 Kinetics of Noradrenaline

13.3.1 Noradrenaline Synthesis

The synthesis of noradrenaline begins with the synthesis of dopamine. Once dopamine is synthesized from the amino acid tyrosine, which is taken up by active transport from the blood, it is stored in synaptic vesicles (Fig. 13.2). The enzyme dopamine beta-hydroxylase further hydroxylates dopamine into noradrenaline. Tyrosine is converted to levodopa via the enzyme tyrosine hydroxylase and converted to dopamine by DOPA decarboxylase. The reaction catalyzed by tyrosine hydroxylase is the rate-limiting *step* in the production of catecholamines. The end product, noradrenaline itself, controls noradrenaline synthesis by feedback to the rate-limiting enzyme, and the released noradrenaline promotes increased tyrosine hydroxylase activity. Dopamine beta-hydroxylase, which catalyzes the conversion of noradrenaline from dopamine, is localized in tissues that synthesize catecholamines, such as noradrenergic neurons and chromaffin cells, and is confined within vesicles [3].

13.3.2 Noradrenaline Release, Uptake, and Inactivation

Relatively little noradrenaline released from nerve endings binds to receptors, and most of the remaining noradrenaline (80–95 %) undergoes active reuptake into nerve endings (uptake-1 mechanism). This mechanism is facilitated by the specific

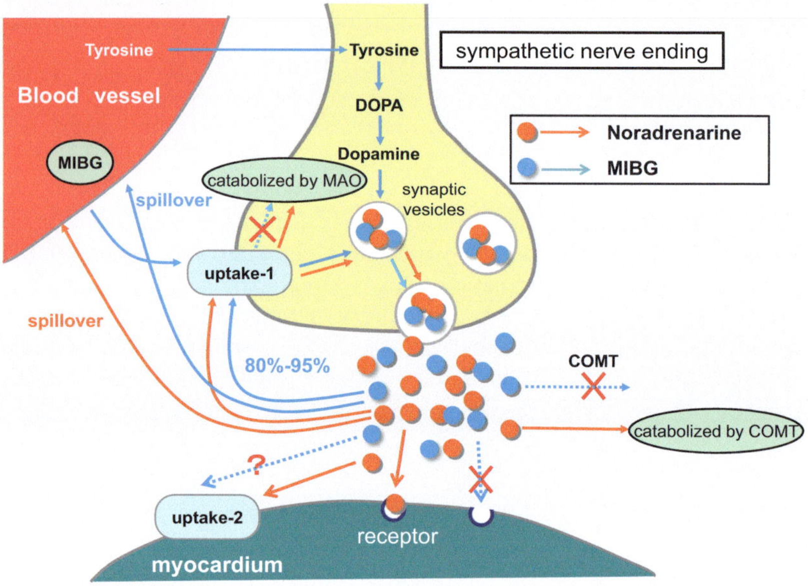

Fig. 13.2 Kinetics of noradrenaline and MIBG in cardiac sympathetic nerve ending (Reprinted and adapted with permission from FUJIFILM RI Pharma Co., Ltd.)

noradrenaline transporter (NAT, also known as the norepinephrine transporter, or NET) protein [7]. This uptake mechanism, via NET, is a "specific" sodium-dependent process characterized by high affinity, low capacity, saturability, and temperature and ouabain sensitivity [8]. It is an important inactivation mechanism to inhibit excessive noradrenaline increments in the interstitial myocardium.

Noradrenaline taken into nerve endings is inactivated by metabolism, but this process is limited, and most noradrenaline is repackaged into noradrenaline-storing vesicles by another active transport process, facilitated by the specific vesicular monoamine transporter (vMAT) protein, and reused in exocytosis [7]. On the other hand, noradrenaline not binding to receptors or undergoing reuptake is taken up by nonneuronal postsynaptic cells, probably by sodium-independent passive diffusion. Such "nonspecific" low-level, energy-independent, unsaturable, extraneuronal uptake takes place in all cells and is sometimes called uptake-2 [7, 9, 10]. Under ideal conditions in vitro, the "specific" uptake process is ~50 times more efficient than passive uptake [11]. The remaining noradrenaline spills over into the blood. Finally, metabolic enzymes deactivate noradrenaline. If it does not undergo reuptake or is not taken up by passive diffusion, it is catabolized by catechol-O-methyltransferase (COMT) outside nerve cells. If it undergoes reuptake, but is not repackaged in noradrenaline-storing vesicles, it is catabolized by monoamine oxidase (MAO). The final product of these pathways is vanillylmandelic acid.

13.4 History of the Development of MIBG

13.4.1 Cardiac Imaging Before MIBG

MIBG, an analog of guanethidine (an adrenergic neuron-blocking agent), is a radioactive tracer of noradrenaline and used to evaluate presynaptic sites of the adrenergic system. It is virtually the only radiotracer used for scintigraphy studies of the cardiac autonomic nervous system [7]. The history of its development dates to the 1970s. At the time, noradrenaline labeled with ^{3}H showed rapid localization to the heart. Myocardial imaging with noradrenaline was used often, and noradrenaline labeled with the positron-emitting nuclide ^{11}C was developed. However, this method required a hospital cyclotron and a positron emission tomography (PET) device and thus was not suitable for implementation in conventional facilities. Cardiac tracer development was an unintended result of efforts by Dr. William H. Beierwaltes, then Chief of Nuclear Medicine at the University of Michigan, to develop scintigraphic imaging for adrenal diseases, particularly neoplasms associated with the medulla [1, 12]. It is impossible to fully appreciate the development of MIBG as a cardiac radiotracer without discussing the events leading to its conception as a noninvasive imaging tool for clinical endocrinology [12]. Initially, their group was developing scintigraphic imaging for adrenal disease. They were exploring iodine (I)-labeled analogs of the adrenergic-blocking antiarrhythmic drug bretylium (Fig. 13.1d) and found that the *ortho*-iodo-bretylium analog was effectively taken up by the heart [13]. However, in later research, they found that the *para*-iodo-bretylium analog effectively imaged the adrenal medulla and concluded that the *para*-iodo-bretylium analog was better suited for portraying the adrenal medulla [14].

13.4.2 Usability of MIBG

The efforts of Dr. Beierwaltes were continued by Dr. Donald M. Wieland. In 1978, he synthesized five ^{125}I-labeled bretylium analogs, including *para*- and *ortho*-iodo-bretylium analogs. Biodistribution studies demonstrated that the *ortho*-iodo-bretylium analog had higher uptake levels in the adrenal medulla than *para*-iodo compounds, which was different from the results of previous studies. In the summer of 1978, when reviewing adrenergic neuron-blocking agents such as guanethidine, he found an article on the adrenergic neuron-blocking activity of benzylguanidines written by Short and Darby [15]. Struck by the structural similarity between benzylguanidines and the iodo-bretylium he had recently worked on, he decided to synthesize and study ^{125}I-*ortho*-iodobenzylguanidine (OIBG) and ^{125}I-*para*-iodobenzylguanidine (PIBG) as potential adrenal medulla imaging agents [12] (Fig. 13.3). In January 1979, the first biodistribution studies of ^{125}I-labeled OIBG and PIBG were performed in dogs. The results showed that the *ortho*-substituted

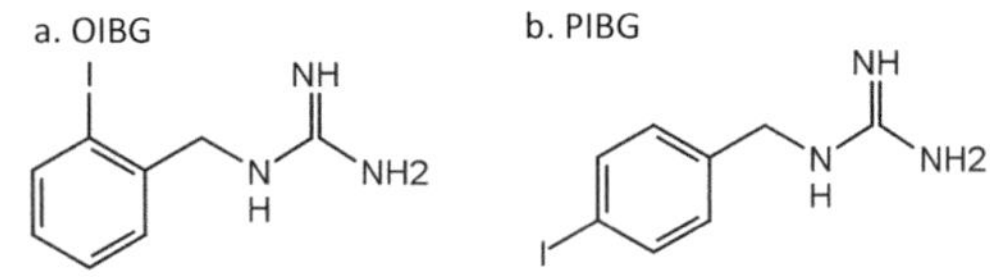

Fig. 13.3 Structural formula of (**a**) *ortho*-iodobenzylguanidine (OIBG) and (**b**) *para*-iodobenzylguanidine (PIBG)

analog OIBG had significantly lower uptake in the adrenal medulla than *para*-substituted PIBG, in contrast to the previous findings with iodo-bretylium analogs [12]. Since the results with the *ortho*- and *para*-iodobenzylguanidines were opposite from what had been observed with the iodo-bretyliums, he felt that a thorough assessment of the structure-activity relationships for iodobenzylguanidines was necessary. He immediately began synthesizing the third possible ring-iodinated structure, MIBG [12]. In 1979, biodistribution studies of MIBG were performed and showed uptake levels comparable with PIBG. After a detailed evaluation, it was concluded that although PIBG had a somewhat higher uptake in the adrenal medulla, MIBG was more metabolically stable, undergoing less deiodination in vivo. Thus MIBG proved to be the optimal radioiodinated benzylguanidine for adrenomedullary imaging [16]. He also performed research on three isomers (*para*, *ortho*, *meta*) and found high early accumulation in the heart [17]. In these studies, the *para*- and *meta*-isomers showed good accumulation in the adrenal medulla and had high affinity for sympathetic nerves. However, thyroid uptake was significantly increased with *para*- and *ortho*-isomers, while thyroid and liver uptake was low with the *meta*-isomer. Thus the *meta*-isomer was considered to be the optimal radioiodinated benzylguanidine for adrenomedullary and cardiac imaging. Wieland et al. studied the biodistribution in animals, such as dogs, and found that MIBG was useful as an adrenomedullary and cardiac-imaging agent [17, 18]. In 1981, he succeeded in performing cardiac imaging of dogs and monkeys using ^{131}I and ^{123}I-MIBG [19].

13.4.3 Human and Clinical Application of MIBG

The first cardiac-imaging assessment in humans was a study of five normal male volunteers conducted by Kline et al. in 1981 [20]. They administered MIBG intravenously and reported that the left ventricular myocardium could be visualized after 1–2 min [20]. They also concluded that MIBG is a new agent for human myocardial imaging that may provide quantitative information on myocardial catecholamine content, and with its ^{123}I label, this radiopharmaceutical can be used without special equipment in a typical nuclear medicine laboratory [20]. When first developed, MIBG was labeled with ^{131}I and used to detect various tumors [9]. In the United States, only ^{131}I-MIBG has been clinically approved for neuroendocrine tumor imaging and occasionally for therapy of these tumors [21]. However, ^{123}I-MIBG was preferred because ^{131}I gives off relatively high-energy (365 keV) gamma emissions, emits beta particles, and has a relatively long half-life (approximately 8.02 days). ^{123}I emits predominantly gamma photons, with energies of 159 keV, and has a half-life of 13.2 h [9, 21]. It is easily imaged and well tolerated. Later, MIBG imaging was speculated to be useful in evaluating heart failure and identifying patients with autonomic neuropathies predisposed to arrhythmias and sudden cardiac death [22–24]. However, the efforts at Michigan shifted toward PET compounds, resulting in clinical MIBG studies being performed mostly in Europe and Japan [1].

13.5 MIBG Kinetics in the Heart [25]

13.5.1 Uptake of MIBG by Sympathetic Nerve Endings Through NET-1

In 1981, Wieland et al. reported that blocking studies in dogs using reserpine, generally thought to selectively block the vesicular uptake of noradrenaline in adrenergic neurons, caused a 30 % decrease in the myocardial concentration of MIBG [19]. This result indicated that a considerable portion of myocardial MIBG accumulation is by entrapment within neuronal storage vesicles in adrenergic nerves through NET-1 [19].

In 1985, Tobes et al. evaluated the mechanisms underlying the uptake of MIBG and noradrenaline using cultured bovine adrenomedullary cells as an in vitro model [26]. Selective uptake-1 inhibitors, desmethylimipramine and cocaine, demonstrated that 52.3 % of MIBG uptake takes place through uptake-1 mechanisms and 47.7 % through passive diffusion [26].

In 1987, Sisson et al. compared the movement of ^{125}I-MIBG and ^{3}H-noradrenaline in the heart of rats. They observed that 6-hydroxydopamine, a drug designed to impair adrenergic nerve terminals, inhibited the uptake of MIBG and ^{3}H-noradrenaline (0.31 and 0.12 of control values, respectively) [27]. Additionally, intraperitoneal injection of desmethylimipramine reduced MIBG uptake to a lesser extent (0.50 of control value) than uptake of ^{3}H-noradrenaline (0.06 of control value), indicating that distribution to extraneural tissue through passive diffusion is greater for MIBG than noradrenaline [27]. This extraneural uptake of MIBG can reach significant levels in animal testing, but in humans the uptake of MIBG in extraneural tissue is considered very limited because patients receiving heart transplants show an absence of MIBG uptake in early and delayed images [28–30].

13.5.2 MIBG Release from Nerve Endings by Exocytosis

Injection of phenylpropanolamine, an indirectly acting sympathomimetic drug that displaces noradrenaline from nerve endings, depleted MIBG and noradrenaline from the left ventricle, leaving a residual of 0.47 and 0.63 of the control values, respectively. This showed that MIBG and noradrenaline are qualitatively released by similar mechanisms [27].

13.5.3 MIBG Does Not Show Physiological Activity nor Bind Sympathetic Receptors

Wieland et al. showed that administration of reserpine markedly lowers the MIBG content of the dog adrenal medulla, but the adrenergic-blocking agents phenoxybenzamine and propranolol have no effect on the uptake of MIBG [18]. This indicates that MIBG does not bind to sympathetic receptors nor show physiological activity [18]. This result also indicates that MIBG imaging is not a visualization of sympathetic receptors.

13.5.4 MIBG Is Not Catabolized by COMT or MAO

Tissue distribution studies in dogs by Wieland et al. showed that concentrations (% kg dose/g) in the heart at 30 min and 2 h decreased from 0.74 to 0.39 after injection of ^{3}H-noradrenaline and increased from 0.47 to 0.63 after injection of MIBG [19]. Radio thin-layer chromatography of 24 h urine collection from two dogs administered MIBG showed that >95 % of the radioactivity was unchanged MIBG [19]. These results indicated MIBG is not catabolized by COMT or MAO [9, 19], and this is also a necessary characteristic as sympathetic nerve-imaging agent.

Based on the above findings, it was proved that MIBG is a chemical analog of noradrenaline, taken up by NET-1 into the nerve, trapped in noradrenaline storage vesicles, and released into the synaptic gap by a mechanism similar to noradrenaline (Fig. 13.2). After intravenous administration, MIBG diffuses into the synaptic gap followed immediately by myocardial uptake, achieving a stable level [7] and allowing left ventricle visualization within 1–2 min [20]. Mean myocardial uptake is 0.63 % of the injected dose at 5 min and 0.76 % at 120 min [20]. The intravesicular percentage of the total cardiac tissue concentration reached a plateau 4 h after intravenous injection. This indicates that at 4 h after injection, MIBG myocardial imaging best represents the neuronal accumulation of the tracer and may best provide useful information on the state of adrenergic neurons in various pathological conditions of the human heart [31].

At present, extraneural tissue uptake of MIBG is considered very limited. Thus, as early uptake of MIBG occurs specifically within sympathetic nerves, early images (15–20 min after administration) reflect uptake-1 mechanisms, i.e., reflecting presynaptic integrity and distribution of sympathetic nerves. Delayed images (3–4 h after administration) are considered the result of exocytosis, reuptake, partial spill over to the blood, and urine excretion of incorporated MIBG, thus reflecting cardiac sympathetic functional activity.

13.6 MIBG Kinetics and Metabolism in the Whole Body

13.6.1 Pharmacokinetics

Blood clearance of MIBG is rapid. After 5 min, the administered dose is reduced by about 10 %, 50 % at 15 min, and 80 % in an hour [20]. Ten to fifteen percent of the injected activity accumulates in cells with neuroendocrine receptors [8]. Retention is especially prolonged in highly adrenergically innervated tissues, such as the adrenal medulla, heart, and salivary glands [8]. During the first hour, MIBG accumulates in the lungs and then in the heart and reaches the highest heart concentrations after 2–3 h [8]. Uptake in the liver is always higher than the heart, and the heart-to-hepatic ratio is 0.87 after 15 min and remains at 0.3–0.6 until 24 h [20, 30, 32]. Uptake in the lung is always lower than the heart, and the heart-to-pulmonary ratio is 1.17 after 5 min, 1.44 after 2 h, and 2.9 after 24 h [20, 30].

13.6.2 Metabolism and Elimination of MIBG

Between 85 and 90 % of the administered dose is excreted through the kidneys as an unchanged compound [8]. In patients with normal renal function, about 50 % of the injected radioactivity was recovered in the urine during the first 24 h, and about 90 % was recovered in the urine after 4 days as unchanged MIBG [8]. According to biodistribution studies, MIBG undergoes little metabolism. There are de-iodinated metabolites, such as m-iodohippuric acid, m-iodobenzoic acid, 4-hydroxy-3-iodobenzylguanidine, and free radioiodine, but this is less than 10 % [8]. The pharmacologic activity of these metabolites is not well known [8].

13.7 Factor Impacting MIBG Kinetics

A number of drugs are known to, or considered to, have an impact on the uptake or storage of MIBG [33]. The details are covered elsewhere (Chap. 17):

1. Inhibition of sodium-dependent uptake systems (uptake-1) from synaptic clefts (cocaine, tricyclic antidepressant, labetalol)
2. Inhibition of uptake by active transport into vesicles (reserpine, desmethylimipramine)
3. Competition for transport into vesicles (noradrenaline, serotonin, guanethidine)
4. Depletion of storage vesicle (reserpine, labetalol, sympathomimetic drugs)
5. Other non-explained mechanisms (pH, osmotic gradients) [7]

It is also recommended that foods containing vanilla and catecholamine-like ingredients (such as chocolate and blue cheese that have a high tyramine; tyramine acts as a catecholamine-releasing agent) should be avoided, as some might influence the uptake of MIBG [7].

References

1. Travin MI. Cardiac autonomic imaging with SPECT tracers. J Nucl Cardiol. 2013;20 (1):128–43; quiz 46. doi:10.1007/s12350-012-9655-1.
2. Carrio I. Cardiac neurotransmission imaging. J Nucl Med. 2001;42(7):1062–76.
3. Goldstein DS. Noradrenergic neurotransmission. In: Robertson D, editor. Primer on the autonomic nervous system. 2nd ed. Amsterdam: Elsevier; 2004. p. 44–9.
4. Travin MI. Cardiac neuronal imaging at the edge of clinical application. Cardiol Clin. 2009;27 (2):311–27. doi:10.1016/j.ccl.2008.12.007.
5. Verrier RL, Antzelevitch C. Autonomic aspects of arrhythmogenesis: the enduring and the new. Curr Opin Cardiol. 2004;19(1):2–11.
6. Flotats A, Carrio I. Cardiac neurotransmission SPECT imaging. J Nucl Cardiol. 2004;11 (5):587–602. doi:10.1016/j.nuclcard.2004.07.007.
7. Teresinska A. Metaiodobenzylguanidine scintigraphy of cardiac sympathetic innervation. Nucl Med Rev Cent East Eur. 2012;15(1):61–70. doi:10.5603/nmr-18732.
8. Vallabhajosula S, Nikolopoulou A. Radioiodinated metaiodobenzylguanidine (MIBG): radiochemistry, biology, and pharmacology. Semin Nucl Med. 2011;41(5):324–33. doi:10.1053/j.semnuclmed.2011.05.003.
9. Chirumamilla A, Travin MI. Cardiac applications of 123I-mIBG imaging. Semin Nucl Med. 2011;41(5):374–87. doi:10.1053/j.semnuclmed.2011.04.001.
10. Sisson JC, Wieland DM. Radiolabeled meta-iodobenzylguanidine: pharmacology and clinical studies. Am J Physiol Imaging. 1986;1(2):96–103.
11. Mairs RJ, Cunningham SH, Russell J, Armour A, Owens J, McKellar K, et al. No-carrier-added iodine-131-MIBG: evaluation of a therapeutic preparation. J Nucl Med. 1995;36 (6):1088–95.
12. Raffel DM, Wieland DM. Development of mIBG as a cardiac innervation imaging agent. JACC Cardiovasc Imag. 2010;3(1):111–6. doi:10.1016/j.jcmg.2009.09.015.
13. Counsell RE, Yu T, Ranade VV, Buswink AA, Carr Jr EA, Carroll M. Radioiodinated bretylium analogs for myocardial scanning. J Nucl Med. 1974;15(11):991–6.
14. Korn N, Buswink A, Yu T, Carr Jr EA, Carroll M, Counsell RE. A radioiodinated bretylium analog as a potential agent for scanning the adrenal medulla. J Nucl Med. 1977;18(1):87–9.
15. Short JH, Darby TD. Sympathetic nervous system blocking agents. 3. Derivatives of benzylguanidine. J Med Chem. 1967;10(5):833–40.
16. Wieland DM, Mangner TJ, Inbasekaran MN, Brown LE, Wu JL. Adrenal medulla imaging agents: a structure-distribution relationship study of radiolabeled aralkylguanidines. J Med Chem. 1984;27(2):149–55.

17. Wieland DM, Wu J, Brown LE, Mangner TJ, Swanson DP, Beierwaltes WH. Radiolabeled adrenergi neuron-blocking agents: adrenomedullary imaging with [131I]iodobenzylguanidine. J Nucl Med. 1980;21(4):349–53.
18. Wieland DM, Brown LE, Tobes MC, Rogers WL, Marsh DD, Mangner TJ, et al. Imaging the primate adrenal medulla with [123I] and [131I] meta-iodobenzylguanidine: concise communication. J Nucl Med. 1981;22(4):358–64.
19. Wieland DM, Brown LE, Rogers WL, Worthington KC, Wu JL, Clinthorne NH, et al. Myocardial imaging with a radioiodinated norepinephrine storage analog. J Nucl Med. 1981;22(1):22–31.
20. Kline RC, Swanson DP, Wieland DM, Thrall JH, Gross MD, Pitt B, et al. Myocardial imaging in man with I-123 meta-iodobenzylguanidine. J Nucl Med. 1981;22(2):129–32.
21. Link JM, Caldwell JH. Diagnostic and prognostic imaging of the cardiac sympathetic nervous system. Nat Clin Pract Cardiovasc Med. 2008;5 Suppl 2:S79–86. doi:10.1038/ncpcardio1150.
22. Sisson JC, Shapiro B, Meyers L, Mallette S, Mangner TJ, Wieland DM, et al. Metaiodobenzylguanidine to map scintigraphically the adrenergic nervous system in man. J Nucl Med. 1987;28(10):1625–36.
23. Sisson JC, Lynch JJ, Johnson J, Jaques Jr S, Wu D, Bolgos G, et al. Scintigraphic detection of regional disruption of adrenergic neurons in the heart. Am Heart J. 1988;116(1 Pt 1):67–76.
24. Rabinovitch MA, Rose CP, Schwab AJ, Fitchett DH, Honos GN, Stewart JA, et al. A method of dynamic analysis of iodine-123-metaiodobenzylguanidine scintigrams in cardiac mechanical overload hypertrophy and failure. J Nucl Med. 1993;34(4):589–600.
25. Ishida Y. ^{123}I-MIBG cardiac sympathetic nervous function imaging. In: Inoue H, editor. Cardiac diseases and autonomic nerve function. Tokyo: Igakushoin; 2001. p. 110–36.
26. Tobes MC, Jaques Jr S, Wieland DM, Sisson JC. Effect of uptake-one inhibitors on the uptake of norepinephrine and metaiodobenzylguanidine. J Nucl Med. 1985;26(8):897–907.
27. Sisson JC, Wieland DM, Sherman P, Mangner TJ, Tobes MC, Jacques Jr S. Metaiodobenzylguanidine as an index of the adrenergic nervous system integrity and function. J Nucl Med. 1987;28(10):1620–4.
28. Dae MW, De Marco T, Botvinick EH, O'Connell JW, Hattner RS, Huberty JP, et al. Scintigraphic assessment of MIBG uptake in globally denervated human and canine hearts--implications for clinical studies. J Nucl Med. 1992;33(8):1444–50.
29. Fagret D, Wolf JE, Vanzetto G, Borrel E. Myocardial uptake of metaiodobenzylguanidine in patients with left ventricular hypertrophy secondary to valvular aortic stenosis. J Nucl Med. 1993;34(1):57–60.
30. Glowniak JV, Turner FE, Gray LL, Palac RT, Lagunas-Solar MC, Woodward WR. Iodine-123 metaiodobenzylguanidine imaging of the heart in idiopathic congestive cardiomyopathy and cardiac transplants. J Nucl Med. 1989;30(7):1182–91.
31. Nakajo M, Shimabukuro K, Yoshimura H, Yonekura R, Nakabeppu Y, Tanoue P, et al. Iodine-131 metaiodobenzylguanidine intra- and extravesicular accumulation in the rat heart. J Nucl Med. 1986;27(1):84–9.
32. Farahati J, Bier D, Scheubeck M, Lassmann M, Schelper LF, Grelle I, et al. Effect of specific activity on cardiac uptake of iodine-123-MIBG. J Nucl Med. 1997;38(3):447–51.
33. Solanki KK, Bomanji J, Moyes J, Mather SJ, Trainer PJ, Britton KE. A pharmacological guide to medicines which interfere with the biodistribution of radiolabelled meta-iodobenzylguanidine (MIBG). Nucl Med Commun. 1992;13(7):513–21.

Chapter 14
Technical Considerations for MIBG Cardiac Scintigraphy

Hirohisa Watanabe and Gen Sobue

Abstract The most commonly used means of imaging to assess cardiac sympathetic denervation is cardiac ^{123}I-metaiodobenzylguanidine (^{123}I-MIBG) scintigraphy. The heart-to-mediastinum (H/M) ratio generated by planar imaging is used widely to assess sympathetic nerve involvement. The H/M ratio is a simple and reliable index, but its values can be influenced profoundly by differences in collimators. Region of interest (ROI) setting methods are also an important problem to generate reliable data. More recently, multicenter cross-calibration of ^{123}I-MIBG H/M ratios to overcome camera-collimator variations, with the use of reference H/M ratios assessed by the phantom method and of a conversion coefficient for each camera-collimator system, demonstrated that H/M ratios under various conditions can be converted to standard H/M ratios for a range of ratios from normal to low. In addition, a standardized method for semiautomatic ROI setting in MIBG was developed. These technical advancements enable us to promote multicenter studies. We also discuss the imaging technique, analysis, and interpretation in this section.

Keywords Collimator • Phantom • Planar • Regions of interest • Heart-to-mediastinum ratio

14.1 Introduction

The most widely used radiotracer for single-photon emission computed tomography (SPECT) to evaluate cardiac sympathetic innervation is ^{123}I-metaiodobenzylguanidine (MIBG) [1–5]. MIBG is a noradrenaline analog developed initially to image neuroendocrine tumors; however, pilot studies quickly identified the uptake of MIBG in the myocardium [6].

At sympathetic nerve terminals, MIBG shows similar kinetics to noradrenaline and is taken up from the blood before cardiac sympathetic nerve-ending synapses by a mechanism called uptake-1, which is a transporter protein-mediated and

H. Watanabe (✉) • G. Sobue
Brain and Mind Research Center, Nagoya University, Nagoya University Graduate School of Medicine, 65 Tsurumai-cho, Showa-ku, Nagoya, Aichi 466-8550, Japan
e-mail: nabe@med.nagoya-u.ac.jp

S. Iwase et al. (eds.), *Clinical Assessment of the Autonomic Nervous System*,
DOI 10.1007/978-4-431-56012-8_14

sodium- and energy-dependent process [1, 7]. However, it does not bind to receptors of the heart muscle membranes and is not metabolized by enzymes. Therefore, MIBG has the potential to stay in sympathetic nerve terminals for a long period of time and has excellent properties as an imaging preparation. Nowadays, ^{123}I-MIBG is a useful method that can evaluate cardiac sympathetic nerve function objectively (see the chapter "Noradrenaline and ^{123}I-MIBG Kinetics in the Sympathetic Nervous System" for more information).

When we evaluate the implementation of MIBG, various points should be considered. Several common medications, such as tricyclic antidepressants, labetalol, reserpine, and calcium antagonists, are known to affect the accumulation of MIBG [1–5, 8, 9]. Although the heart-to-mediastinum (H/M) ratio is used widely as an indicator for MIBG imaging equipment, its value varies greatly and is particularly affected by differences in the collimator used [10–12]. Therefore, differences of reference by facilities have occurred as a problem. The method for determining the region of interest (ROI) is also an important problem [13, 14].

In this section, we focus on the technical considerations of the advanced imaging technique, analysis, and interpretation of MIBG to obtain more reliable results and to promote multicenter studies.

14.2 Patient Information and Preparation and Tracer Administration

When we use MIBG in women of childbearing age, with respect to pregnancy or lactation, a cautious clinical decision is necessary to consider its merits and demerits. Although it is still undetermined whether MIBG is excreted into human milk, Iodine-123 (^{123}I) can be excreted into human milk. Since the physical half-life of ^{123}I is 13.2 h, the EANM Cardiovascular Committee and the European Council of Nuclear Cardiology recommended that breastfeeding should be discontinued for 48 h after MIBG administration in order to minimize the risk to a nursing infant [5].

Cardiac ^{123}I-MIBG scintigraphy should be conducted with the patient resting in a supine position for at least 5 min before administration and under conditions of nil per os. Prior to administration, patients should be questioned for a history of reactions to iodine, an iodine-containing contrast agent, or other products containing iodine. MIBG should be administered by slow (over 1–2 min), secure peripheral intravenous injection and flushed with saline, in accordance with local radiation protection practices [5]. As initial images are acquired a few minutes later, the tracer is best administered while the patient is under the camera or in close proximity to it.

It is recommended to stop temporarily some medications and substances known to hamper the mechanism of noradrenaline uptake, including opioids, tricyclic antidepressants, isoproterenol, labetalol, reserpine, calcium channel blockers, phenothiazines, and foods containing vanillin and catecholamine-like compounds such as chocolate and blue cheese [1–5, 8, 9]. According to more recent critical review of the literature on drug interactions with MIBG uptake, tricyclic antidepressants and labetalol should be withdrawn because they have moderate to strong inhibitory effect on MIBG accumulation. If reserpine and phenylephrine are still used, they also should be considered the withholding prior to MIBG imaging since they have strong MIBG-inhibitory effect. Based on the mechanism of the drugs, we should consider the cessation of selective serotonin reuptake inhibitors (SSRIs) and serotonin and norepinephrine reuptake inhibitors (SNRIs) if possible although the level of evidence is not high. Beta blockers except for labetalol do not require discontinuing. Generally, only a small number of drugs have strong evidence for inhibition of MIBG uptake [15].

There are conflicting opinions about premedication with thyroid-blocking agents. Historically, such blockade was undertaken to protect the thyroid from unnecessary irradiation; however, the amount of these compounds is minimal with present production procedures, and many clinicians and researchers feel that this pretreatment can be unnecessary. For the moment, pretreatment should be determined according to local and institutional regulations. More recently, Friedman et al. examined the effectiveness of thyroid blockade in subjects undergoing MIBG cardiac scintigraphy and estimated the relative contribution of bound and unbound radioiodine to imaging findings [16]. They demonstrated that the mean visual score on late images was significantly higher for non-blocked than blocked subjects ($P < 0.001$), and the net thyroid counts of ROI analyses were also significantly higher on the late images of non-blocked subjects ($P < 0.0001$). Compared to early and delayed images, 87 % of blockade subjects showed decreased or unchanged counts, but 75 % of non-blocked subjects had increased net thyroid activity. Thus, they concluded that thyroid blockade pretreatment in MIBG cardiac scintigraphy could have a potential to hamper the increase of thyroid activity over time due to the uptake of unbound ^{123}I. Nevertheless, the administration dose is 111–148 MBq over 1 min for the diagnosis of Parkinson's disease in Japan, but 370 MBq have been used in some western studies [17]. We should also consider the differences of these doses.

Adverse reactions to ^{123}I-MIBG are rare [1, 5]. However, if it is administered too quickly, palpitations, heat sensations, short-term hypertension, shortness of breath, and abdominal cramps may be induced more easily than with the standard protocol. An anaphylactic reaction is also very rare but possible. The highest exposure is to the bladder, liver, spleen, gallbladder, heart, and adrenals, and the absorbed dose may be higher in patients with severe renal failure. Patients are encouraged to void frequently to facilitate the excretion of the radiopharmaceutical and to minimize the radiation dose to the bladder.

14.3 Imaging Technique, Analysis, and Interpretation

14.3.1 General Consideration

Considerable data from ^{123}I-MIBG imaging have been generated based on the analysis of planar images, mostly with a standard anterior view. Planar images of the thorax are acquired at approximately 15–30 min (early images) and 3–4 h (late images) after injection for 10 min. SPECT images can also be acquired using standard perfusion-imaging methods. In the analysis of ^{123}I-MIBG cardiac scintigraphy, semiquantitative analyses of global uptake with the H/M ratio and differences in tracer uptake/retention in early and late images with the washout rate (WR) are the most common measures.

The H/M ratio on MIBG cardiac scintigraphy is used as a semiquantitative index [1–5]. The early H/M ratio probably reflects the receptor density, the integrity of presynaptic nerve terminals, and the uptake-1 function. The delayed H/M ratio combines information on neuronal function from uptake to release through storage vesicles at nerve terminals. Cardiac ^{123}I-MIBG washout has been shown to be an important measure of cardiac sympathetic innervation; early and delayed planar images are used for this calculation. The WR is thought to reflect the turnover of catecholamines, which relates to the degree of sympathetic drive [1–5]. Increased sympathetic activity, reflecting worsening heart failure or cardiac sympathetic nerve involvement, is associated with diminished retention of cardiac ^{123}I-MIBG on delayed images and thus a greater washout of cardiac ^{123}I-MIBG. A normal washout value in control subjects is reported to be $10 \pm 9\,\%$. Nevertheless, more studies are needed to establish the differences between the early H/M ratio, delayed H/M ratio, and WR.

14.3.2 Collimators

In nuclear medicine, a radiopharmaceutical is administered to a patient in order to detect the gamma (γ) rays emitted from the radioactive isotope, which can reflect functional information of a target organ. Gamma radiation-emitting radioisotopes are imaged by using a gamma camera. A gamma camera consists of a lead collimator and an array of detectors such as photomultiplier tubes, light guide, and crystal (Fig. 14.1) [18]. The collimator has a unique structure to identify the direction of the gamma rays emitted from the single-photon emission nuclear species by shielding gamma rays emitted from other directions and allowing detectors to receive gamma rays from a specific direction. Gamma rays produce light flashes in the scintillators, and the light output is converted to an electrical signal by photomultipliers. Finally, a computer constructs an image from the detector output.

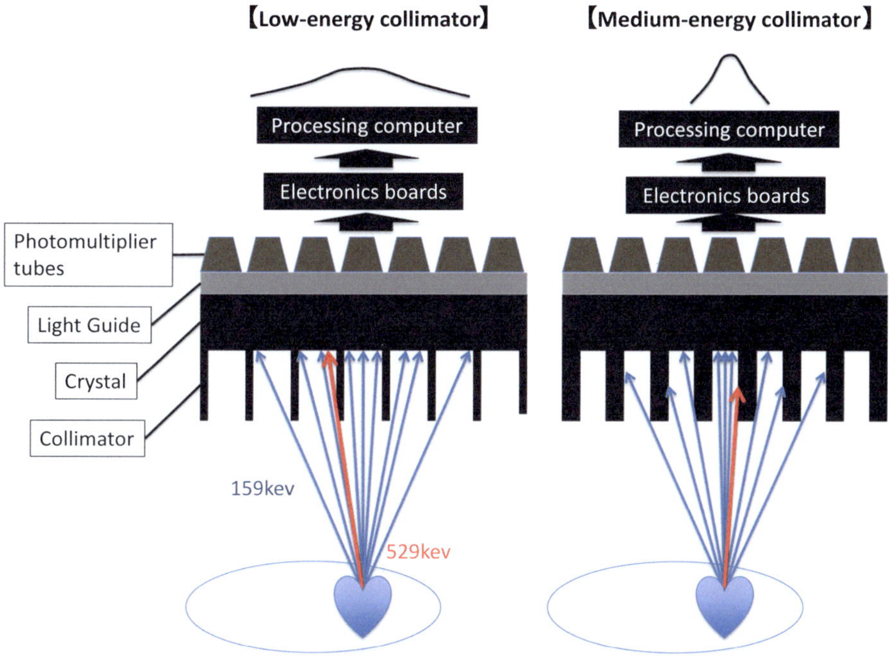

Fig. 14.1 Structures of gamma camera and the effect of high-energy photons penetration between low-energy collimator and medium-energy collimator on the results of heart-to-mediastinum ratio. ^{123}I emits not only 159 keV photons but also high-energy photons of more than 400 keV [~2.87 %, main contributor 529 keV (1.28 %)], which have a potential to penetrate the thinner collimator septa resulting scatter detected in the 159 keV energy windows. Thus, high-energy photons penetration will more profoundly affect the heart-to-mediastinum (H/M) ratio assessed by low-energy (LE) collimator, which has thinner septa than that by medium-energy (ME) collimator. Although the use of ME provides lower spatial resolution than LE, H/M ratios are assessed from the counts in relatively large regions compared to brain imaging. Thus, semiquantitative cardiac ^{123}I-MIBG scintigraphy can be best performed using ME collimators.

The collimator, which is an important part that determines the quality of SPECT imaging, has continued to be used since Anger first developed a gamma camera in the 1950s. The most common collimator used in nuclear medicine is a parallel-hole collimator, which has developed into a myriad of holes aligned in parallel with each other with a thin lead plate (height number cm) as a partition wall. The shape of the holes, e.g., hexagonal, round, and square, is determined by the production method of the collimator.

Collimators are classified into two major groups: low energy (LE) and medium energy (ME). The thickness of the partition wall (septa) of collimators is designed so that the ratio of the gamma rays passing through it is less than 5 %. Therefore, ME collimators have a thicker partition wall than LE collimators [5] to prevent the penetration of higher-energy gamma rays. The increased septal thickness of ME collimators contributes to increase image resolution and decrease scatter in the images compared to LE collimators but also degrades count sensitivity and spatial resolution. Moreover, nuclear laboratories are not always equipped with ME collimators, but use LE collimators instead because they are suitable for technetium (^{99m}Tc), which is used most commonly in clinical practice.

Actually, in addition to LE and ME collimators, various types of collimators are also available, depending on the purpose of the study, in order to achieve good balance among sensitivity, resolution, and appropriate energy extent: low-energy high resolution (LEHR), low-energy high sensitivity, low-energy general purpose (LEGP), medium-energy general purpose, high-energy general purpose, and high-energy pinhole collimators [17, 19]. With respect to ^{123}I-MIBG cardiac scintigraphy, most clinical and published studies have used LE collimators, such as LEGP or LEHR collimators.

Iodine-123 predominantly emits energy of 159 keV (97 %), which is an optimal imaging energy for an LE collimator, but also emits high-energy photons of more than 400 keV (approximately 2.9 %), with the main contributing photons at 529 keV. Higher-energy ^{123}I photons, mostly at 529 keV, can penetrate the thinner septal wall and contaminate 159 keV imaging data [5, 12, 20] (Fig. 14.1). Thus, an LE collimator, which has a thinner septal wall, can show a lower H/M ratio than an ME collimator. On the basis of these findings, recent guidelines on MIBG imaging from the European Council of Nuclear Cardiology recommend using an ME collimator, which has a thicker septal wall than an LE collimator, for MIBG imaging to minimize scattered radiation noise from high-energy emissions [5], but this recommendation is not usually followed in clinical practice. In addition, a lot of previous studies have used an LE collimator.

Clinically, it is a considerable problem that differences in the collimator can influence the H/M ratio profoundly. To compensate for the resulting corruption of image quantitation related to LE collimator penetration by higher-energy photons, Chen et al. developed a mathematical technique (iterative reconstruction with deconvolution of septal penetration) that appears to improve the quantitative accuracy of cardiac ^{123}I-MIBG uptake in reference to a phantom standard [21]. However, clinical applications have yet to be determined.

On the contrary, Nakajima et al. have endeavored to standardize the heart-to-mediastinum (H/M) ratio assessed by using planar images. They designed a simple and reproducible phantom standard for planar image acquisition [20, 21]. The phantom consisted of multiple slices of the heart, lung, mediastinum, and liver containing a radioisotope as a single component. Using these two phantoms, four H/M ratios (anterior and posterior views for each) were obtained to calculate the regression equation. A linear regression equation was analyzed using the formula $y - 1 = Ki \times (x - 1)$. The first step was to convert the H/M ratio generated by the LE collimator to the reference value based on mathematical theory using the conversion coefficient Ki. The second step was to convert this H/M ratio to a standardized H/M ratio with the conversion coefficient Ks, which was defined as the average K value for a typical ME collimator. They verified the initial standardization efforts in ten centers, suggesting that this phantom method could be applied to calibrate the results from ME collimators and LE collimators [22].

More recently, much larger Japanese multicenter studies including 84 institutions for the measurement of planar H/M ratios using standard nuclear cameras from a variety of vendors and collimators also confirmed and extended these findings [19]. On the basis of phantom studies, a conversion coefficient of 0.88 was determined to integrate H/M ratios from all acquisition conditions. Using the reference H/M ratio and conversion coefficients for the system can convert an H/M ratio under various conditions converted to the standard one, irrespective of the collimator used. They also demonstrated that two H/M ratios from one phantom could be comparable to one from two phantoms because the conversion coefficient showed a significantly high correlation with an R^2 of 0.997. Standardization of the H/M ratio for ^{123}I-MIBG among various scinticamera-collimator combinations will be useful for not only clinical practice but also multicenter studies.

14.3.3 ROI Setting

Cardiac MIBG uptake is semiquantitatively obtained by calculating an H/M ratio, after drawing ROIs over the heart and the upper mediastinum above the lung apices, but below the thyroid gland, in the planar anterior view. Average counts per pixel in the myocardium are divided by average counts per pixel in the mediastinum, thus generating the H/M ratio.

At least three methods have been used to acquire an H/M ratio [1]. Firstly, square or rectangular ROIs are drawn in the center of the heart and upper mediastinum, and the count per pixel ratio is calculated [23]. Secondly, an ROI is drawn around the epicardial border and the valve plane, including the left ventricular cavity [24]. Thirdly, an ROI encompassing the myocardium alone, tracing the epicardial and endocardial borders, excluding the valve plane and cavity, has also been used [25].

The H/M ratio is a simple measure and the above three measuring methods appear to give similar results in previous reports when they were performed at a single center. However, it is prone to variation among setters with respect to the position of the ROI [13, 26]. As for the mediastinum, differences in ROI location sensitivity can influence the results; therefore, it is desirable to provide a certain set reference.

Okuda, Nakajima, and colleagues developed novel software for semiautomatically measuring the H/M ratio (standardized method for automatic ROI setting in MIBG, smartMIBG) [27]. In this semiautomatic measuring system, the examiner manually sets a circular contour to the cardiac region by checking and modifying the center point and circular ROI displayed on the screen. The mediastinum ROI is set up automatically by using four steps:

1. To determine the trunk right border of the body, after determining the heart ROI manually, the right border of the liver is determined from the count profile curve of the bottom of the ROI as the position of the right border of the trunk.

2. To determine the trunk left border, after the upper boundary corresponding to the apex of the lungs is estimated as three times the vertical cardiac diameter from the lower border consistent with the center point of the cardiac ROI, a chest counter profile curve is created to detect the minimal count point on the mediastinum. Subsequently, it is folded back the distance between this point and the trunk right border and defined as the left border.
3. To determine the superior mediastinal edge, the apex of the lung is determined by vertically detecting the minimum value on the upper mediastinum, applying thyroid uptake as a reference. If thyroid uptake is too low to analyze, three times the vertical cardiac diameter is employed.
4. Finally, the rectangular mediastinal ROI is placed in the upper 30 % part of the mediastinum, and the horizontal size of the mediastinal ROI is defined as 10 % of body width.

The test-retest reliability of the H/M ratio calculated using this automated method is better than that for manual analysis. Interobserver agreement was also good using the semiautomated method. The H/M ratios generated with the method were higher than those obtained manually, because the count point of the mediastinum ROI is minimal. By using the semiautomatic method to set an ROI, it is possible to reduce differences between setters and between facilities.

14.3.4 Washout Rate

Washout of cardiac ^{123}I-MIBG has been shown to be an important measure of cardiac sympathetic innervation. The WR is thought to reflect the turnover of catecholamines, which relates to the degree of sympathetic drive, although it represents several mechanisms (at least vesicular exocytosis and extravesicular diffusion in the nerve terminals) [28], and the mechanisms responsible for this phenomenon have still not been completely explained. Increased sympathetic activity is associated with diminished cardiac ^{123}I-MIBG retention on delayed images, resulting in a greater degree of cardiac ^{123}I-MIBG washout.

The WR of MIBG from the heart is most often expressed as a percentage difference between the average number of counts within the heart after 15–30 min and 3–4 h and the average number of counts within the heart after 15–30 min [29, 30]. A normal washout value in control subjects is reported to be 10 ± 9 %. However, there is a wide variety of calculation methods with or without standardization of H/M counts, with or without background correction by subtracting mediastinum counts from heart counts, and with or without correction for radioisotope decay over time [5, 13, 31]. For example, the WR could be overestimated by 15–20 % if correction for radioisotope decay over time is not considered. The timing of late images, i.e., whether they are acquired after 3 or 4 h, can also influence the results. Calculation method proposed by the recent European Association of Nuclear Medicine guidelines is shown in Fig. 14.2.

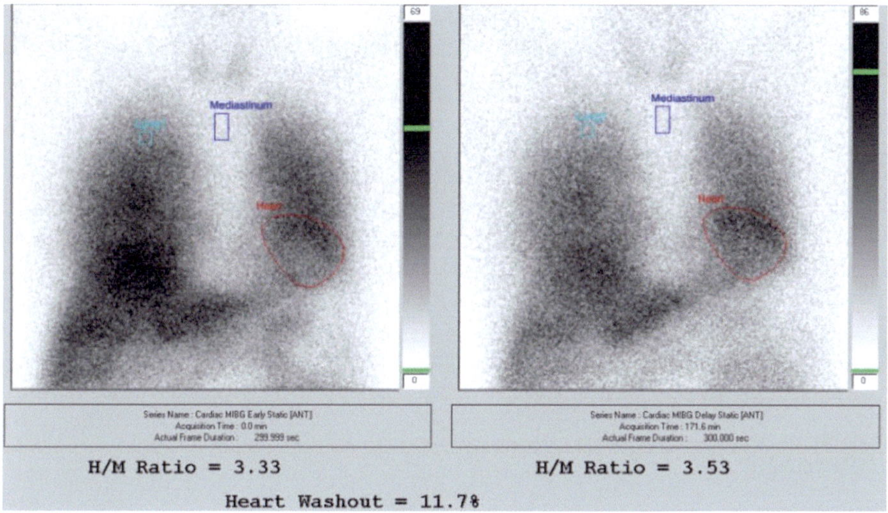

$$H/M = \frac{\{H\}}{\{M\}}$$

$$WR = \frac{\{H\}_{early} - (\{H\}_{delay} \times 1.21^{*})}{\{H\}_{early}} \times 100$$

$$WR = \frac{(\{H\}_{early} - \{M\}_{early}) - ((\{H\}_{delay} - \{M\}_{delay}) \times 1.21^{*})}{(\{H\}_{early} - \{M\}_{early})} \times 100$$

{} = mean counts per pixel $\quad$ *= ^{123}I decay correction for 3h and 45 min (1÷0.8213)

Fig. 14.2 Calculation methods of hear-to-mediastinum ratio and washout ratio. This figure shows normal result of ^{123}I-metaiodobenzylguanidine scintigraphy in subject. Calculation methods of heart-to-mediastinum ratio and washout rate recommended by European Association of Nuclear Medicine guidelines are also shown: H, heart; M, mediastinum.

14.3.5 SPECT

In ^{123}I-MIBG cardiac scintigraphic studies of patients with neurodegenerative disorders, planar images are generally used, but SPECT is not always performed, even though it has an advantage for detecting the localization, severity, and extent of the decreased uptake of the radiopharmaceutical area and better results could be expected to be obtained with quantitative SPECT imaging.

Some programs include manual steps such as the identification of the apex and base in short-axis slices, resulting in considerable interoperator variability. Completely automated programs have the advantage of low interoperator variability, but there can be considerable differences between different automated programs, which should be taken into account, especially if more than one program is used at the same site. In addition, SPECT inter-site reproducibility has not yet been shown to be superior to that of a simple planar method [5, 32].

For SPECT image interpretation, tomographic slices should be assessed scaled to the planar appearance [33]. All three planes of image, i.e., the short axis, horizontal long axis, and vertical long axis, should be investigated. The early and delayed images should be aligned identically in such a way that the tomograms are displayed carefully with anatomically corresponding the early and delayed slices under each other, from apex to base.

Normal cardiac MIBG distribution includes a relatively low uptake in the inferior wall, which is more pronounced in the elderly subjects. In addition, there may be substantial MIBG uptake in the liver, which overlaps with the inferior LV wall [34, 35]. Moreover, scattering from the lung field to the lateral LV wall may also occur. We should also consider that variables such as patient age, acquisition, and processing protocols, as well as the compensation algorithms applied, can potentially influence cardiac sympathetic images.

References

1. Travin MI. Cardiac autonomic imaging with SPECT tracers. J Nucl Cardiol. 2013;20 (1):128–43; quiz 46.
2. Goldstein DS. Sympathetic neuroimaging. Handb Clin Neurol. 2013;117:365–70.
3. Teresinska A. Metaiodobenzylguanidine scintigraphy of cardiac sympathetic innervation. Nucl Med Rev Cent East Eur. 2012;15(1):61–70.
4. Chirumamilla A, Travin MI. Cardiac applications of 123I-mIBG imaging. Semin Nucl Med. 2011;41(5):374–87.
5. Flotats A, Carrio I, Agostini D, Le Guludec D, Marcassa C, Schafers M, et al. Proposal for standardization of 123I-metaiodobenzylguanidine (MIBG) cardiac sympathetic imaging by the EANM Cardiovascular Committee and the European Council of Nuclear Cardiology. Eur J Nucl Med Mol Imaging. 2010;37(9):1802–12.
6. Wieland DM, Brown LE, Rogers WL, Worthington KC, Wu JL, Clinthorne NH, et al. Myocardial imaging with a radioiodinated norepinephrine storage analog. J Nucl Med. 1981;22(1):22–31.
7. Sisson JC, Wieland DM, Sherman P, Mangner TJ, Tobes MC, Jacques Jr S. Metaiodobenzylguanidine as an index of the adrenergic nervous system integrity and function. J Nucl Med. 1987;28(10):1620–4.
8. Solanki KK, Bomanji J, Moyes J, Mather SJ, Trainer PJ, Britton KE. A pharmacological guide to medicines which interfere with the biodistribution of radiolabelled meta-iodobenzylguanidine (MIBG). Nucl Med Commun. 1992;13(7):513–21.
9. Wafelman AR, Hoefnagel CA, Maes RA, Beijnen JH. Radioiodinated metaiodobenzylguanidine: a review of its biodistribution and pharmacokinetics, drug interactions, cytotoxicity and dosimetry. Eur J Nucl Med. 1994;21(6):545–59.
10. Dobbeleir AA, Hambye AS, Franken PR. Influence of high-energy photons on the spectrum of iodine-123 with low- and medium-energy collimators: consequences for imaging with 123I-labelled compounds in clinical practice. Eur J Nucl Med. 1999;26(6):655–8.
11. Inoue Y, Suzuki A, Shirouzu I, Machida T, Yoshizawa Y, Akita F, et al. Effect of collimator choice on quantitative assessment of cardiac iodine 123 MIBG uptake. J Nucl Cardiol. 2003;10 (6):623–32.
12. Verberne HJ, Feenstra C, de Jong WM, Somsen GA, van Eck-Smit BL, Busemann SE. Influence of collimator choice and simulated clinical conditions on 123I-MIBG heart/mediastinum ratios: a phantom study. Eur J Nucl Med Mol Imaging. 2005;32(9):1100–7.

13. Verberne HJ, Habraken JB, van Eck-Smit BL, Agostini D, Jacobson AF. Variations in 123I-metaiodobenzylguanidine (MIBG) late heart mediastinal ratios in chronic heart failure: a need for standardisation and validation. Eur J Nucl Med Mol Imaging. 2008;35(3):547–53.
14. Van der Veen L, Scholte A, Stokkel M. Mathematical methods to determine quantitative parameters of myocardial 123I-MIBG studies: a review of the literature. Nucl Med Commun. 2010;31(7):617–28.
15. Jacobson AF, Travin MI. Impact of medications on mIBG uptake, with specific attention to the heart: comprehensive review of the literature. J Nucl Cardiol. 2015;15. [Epub ahead of print].
16. Friedman NC, Hassan A, Grady E, Matsuoka DT, Jacobson AF. Efficacy of thyroid blockade on thyroid radioiodine uptake in 123I-mIBG imaging. J Nucl Med. 2014;55(2):211–5.
17. Nakajima K, Nakata T. Cardiac 123I-MIBG imaging for clinical decision making: 22-year experience in Japan. J Nucl Med. 2015;56 Suppl 4:11S–9.
18. Peterson TE, Furenlid LR. SPECT detectors: the Anger Camera and beyond. Phys Med Biol. 2011;56(17):R145–82.
19. Nakajima K, Okuda K, Yoshimura M, Matsuo S, Wakabayashi H, Imanishi Y, et al. Multicenter cross-calibration of I-123 metaiodobenzylguanidine heart-to-mediastinum ratios to overcome camera-collimator variations. J Nucl Cardiol. 2014;21(5):970–8.
20. Nakajima K, Matsubara K, Ishikawa T, Motomura N, Maeda R, Akhter N, et al. Correction of iodine-123-labeled meta-iodobenzylguanidine uptake with multi-window methods for standardization of the heart-to-mediastinum ratio. J Nucl Cardiol. 2007;14(6):843–51.
21. Chen J, Garcia EV, Galt JR, Folks RD, Carrio I. Improved quantification in 123I cardiac SPECT imaging with deconvolution of septal penetration. Nucl Med Commun. 2006;27(7):551–8.
22. Nakajima K, Okuda K, Matsuo S, Yoshita M, Taki J, Yamada M, et al. Standardization of metaiodobenzylguanidine heart to mediastinum ratio using a calibration phantom: effects of correction on normal databases and a multicentre study. Eur J Nucl Med Mol Imaging. 2012;39(1):113–9.
23. Agostini D, Belin A, Amar MH, Darlas Y, Hamon M, Grollier G, et al. Improvement of cardiac neuronal function after carvedilol treatment in dilated cardiomyopathy: a 123I-MIBG scintigraphic study. J Nucl Med. 2000;41(5):845–51.
24. Yamada T, Shimonagata T, Fukunami M, Kumagai K, Ogita H, Hirata A, et al. Comparison of the prognostic value of cardiac iodine-123 metaiodobenzylguanidine imaging and heart rate variability in patients with chronic heart failure: a prospective study. J Am Coll Cardiol. 2003;41(2):231–8.
25. Gerson MC, Craft LL, McGuire N, Suresh DP, Abraham WT, Wagoner LE. Carvedilol improves left ventricular function in heart failure patients with idiopathic dilated cardiomyopathy and a wide range of sympathetic nervous system function as measured by iodine 123 metaiodobenzylguanidine. J Nucl Cardiol. 2002;9(6):608–15.
26. Nakajima K. Normal values for nuclear cardiology: Japanese databases for myocardial perfusion, fatty acid and sympathetic imaging and left ventricular function. Ann Nucl Med. 2010;24(3):125–35.
27. Okuda K, Nakajima K, Hosoya T, Ishikawa T, Konishi T, Matsubara K, et al. Semi-automated algorithm for calculating heart-to-mediastinum ratio in cardiac Iodine-123 MIBG imaging. J Nucl Cardiol. 2011;18(1):82–9.
28. Wakabayashi T, Nakata T, Hashimoto A, Yuda S, Tsuchihashi K, Travin MI, et al. Assessment of underlying etiology and cardiac sympathetic innervation to identify patients at high risk of cardiac death. J Nucl Med. 2001;42(12):1757–67.
29. Morozumi T, Kusuoka H, Fukuchi K, Tani A, Uehara T, Matsuda S, et al. Myocardial iodine-123-metaiodobenzylguanidine images and autonomic nerve activity in normal subjects. J Nucl Med. 1997;38(1):49–52.
30. Somsen GA, Verberne HJ, Fleury E, Righetti A. Normal values and within-subject variability of cardiac I-123 MIBG scintigraphy in healthy individuals: implications for clinical studies. J Nucl Cardiol. 2004;11(2):126–33.

31. Agostini D, Carrio I, Verberne HJ. How to use myocardial 123I-MIBG scintigraphy in chronic heart failure. Eur J Nucl Med Mol Imaging. 2009;36(4):555–9.
32. Slomka PJ, Mehta PK, Germano G, Berman DS. Quantification of I-123-meta-iodobenzylguanidine heart-to-mediastinum ratios: not so simple after all. J Nucl Cardiol. 2014;21(5):979–83.
33. Cerqueira MD, Weissman NJ, Dilsizian V, Jacobs AK, Kaul S, Laskey WK, et al. Standardized myocardial segmentation and nomenclature for tomographic imaging of the heart. A statement for healthcare professionals from the Cardiac Imaging Committee of the Council on Clinical Cardiology of the American Heart Association. Circulation. 2002;105(4):539–42.
34. Tsuchimochi S, Tamaki N, Tadamura E, Kawamoto M, Fujita T, Yonekura Y, et al. Age and gender differences in normal myocardial adrenergic neuronal function evaluated by iodine-123-MIBG imaging. J Nucl Med. 1995;36(6):969–74.
35. Estorch M, Serra-Grima R, Flotats A, Mari C, Berna L, Catafau A, et al. Myocardial sympathetic innervation in the athlete's sinus bradycardia: is there selective inferior myocardial wall denervation? J Nucl Cardiol. 2000;7(4):354–8.

Chapter 15
Findings of ^{123}I-MIBG Cardiac Scintigraphy: Parkinson's Disease and Related Disorders and Others (RBD, Cardiac Diseases, DM, etc.)

Makiko Yogo and Masahiko Suzuki

Abstract A wide range of autonomic dysfunctions has been described in Parkinson's disease (PD). Since its description, myocardial iodine-123-*meta*-iodobenzylguanidine (^{123}I-MIBG) cardiac scintigraphy has attracted attention as a useful tool for the diagnosis of PD. There is a general agreement that the H/M ratio in ^{123}I-MIBG cardiac scintigraphy is significantly low in both early and delayed images of PD patients. Reduced cardiac MIBG uptake has been reported in not only patients with PD but also in dementia with Lewy bodies (DLB), rapid eye movement behavior disorder (RBD), and pure autonomic failure (PAF). In this regard, ^{123}I-MIBG cardiac scintigraphy can be useful for differentiating Lewy body disorders from other parkinsonian syndromes and dementias.

In this chapter ^{123}I-MIBG cardiac scintigraphy findings are described from various perspectives of PD, and we also review RBD, PAF, and other related disorders. Finally, we describe comorbidities associated with reduced cardiac MIBG uptake for careful interpretation.

Keywords ^{123}I-MIBG cardiac scintigraphy • Parkinson's disease • Lewy body disease • Parkinsonism

15.1 Parkinson's Disease

Parkinson's disease (PD) is the most common neurodegenerative parkinsonism characterized by degeneration of both dopaminergic and non-dopaminergic neurons with intracytoplasmic eosinophilic inclusions known as Lewy bodies. The cardinal clinical motor signs of PD are resting tremor, rigidity, and bradykinesia and response to dopaminergic drugs. Postural instability is specific to the advanced

M. Yogo • M. Suzuki (✉)
Katsushika Medical Center, Department of Neurology, The Jikei University School of Medicine, 6-41-2 Aoto, Katsushika-ku, Tokyo 125-8506, Japan
e-mail: suzukimdphd@gmail.com

S. Iwase et al. (eds.), *Clinical Assessment of the Autonomic Nervous System*,
DOI 10.1007/978-4-431-56012-8_15

stage of PD. Other non-motor signs are frequently present, including autonomic dysfunction, cognitive and psychiatric changes, sensory symptoms, sleep disturbances, and hyposmia, although the frequency of these features varies from one patient to another.

15.1.1 ^{123}I-MIBG Cardiac Scintigraphy Findings in Parkinson's Disease

A wide range of autonomic dysfunctions has been described in PD such as orthostatic and postprandial hypotension, gastrointestinal dysfunction especially constipation, urinary disturbance, sexual dysfunction, and sweating abnormalities. Since its description, cardiac dysautonomia has attracted attention as a useful tool for the diagnosis of PD. Iodine-123-*meta*-iodobenzylguanidine (^{123}I-MIBG) cardiac scintigraphy is a useful diagnostic tool for cardiac dysautonomia. Reduced cardiac MIBG uptake is observed in 80–90 % of patients with PD, and most images show no accumulation of MIBG in the heart. Hakusui et al. [1] were the first to report reduced cardiac MIBG uptake in patients with PD. Since the first description, reduced cardiac MIBG uptake in patients with PD has become one of the main characteristics of PD, and many groups reported that the reduced uptake could be implemented in the differential diagnosis of PD from other atypical parkinsonisms.

The threshold value of the H/M ratio can be used for the differentiation of PD from other neurodegenerative parkinsonisms. In this regard, Shin et al. [2] reported that the threshold values of the early and delayed H/M ratios, which distinguished PD patients from the controls, were 1.57 (sensitivity, 95.0 %, specificity, 88.9 %) and 1.56 (sensitivity, 95.0 %, specificity, 100 %), respectively. There was no difference in the discrimination power of the early and delayed H/M ratios ($p = 0.188$, 95 % confidence interval [95 %CI] = −0.014 to 0.07). The same study found that the threshold values of the early and delayed H/M ratio that distinguished PD patients from MSA patients were 1.38 (sensitivity, 65.7 %; specificity, 95.7 %) and 1.36 (sensitivity, 80.0 %; specificity, 100 %), respectively. Furthermore, the delayed H/M ratio was significantly better than the early H/M ratio ($p = 0.068$, 95 % CI= −0.005 to 0.138).

However, Nagayama et al. [3] reported a lower specificity than other studies. In their study, patients showing one or more parkinsonian-like symptoms were enrolled. The sensitivity and specificity of ^{123}I-MIBG cardiac scintigraphy for the diagnosis of PD were 87.7 % and 37.4 %, respectively; these values were due to the fact that nearly half of their patients with senile dementia of Alzheimer type (SDAT) had low MIBG uptake. In their study, the patients with SDAT showed motor signs, and they could have had Lewy body disorder in addition to the Alzheimer pathology.

On the other hand, Orimo et al. [4] conducted meta-analysis study of 13 studies and reported that the pooled sensitivity and specificity of the early H/M ratio to differentiate PD from other neurodegenerative parkinsonisms were 82.6 % and 89.2 %, respectively, and those of the delayed H/M ratio were 89.7 % and 82.6 %, respectively. When PD was limited to that of early stage (Hoehn and Yahr (H-Y) stage 1 or 2), the pooled sensitivity and specificity by the delayed H/M ratio were 94.1 % and 80.2 %, respectively.

Considered together, we conclude that ^{123}I-MIBG cardiac scintigraphy can be used with high sensitivity and specificity to differentiate PD from other neurodegenerative parkinsonisms using both early and delayed imaging phases, despite potential interinstitutional differences in the H/M ratio based on differences in collimators and scinticameras used at each institution. Standardization of the H/M ratio is continuously carried out in Japan and standard values will be presented in the near future. This standardization process is described in detail in Chap. 14.

Postmortem examination of the anterior walls of the left ventricles demonstrated markedly low density of nerve fibers immunoreactive to tyrosine hydroxylase (TH), a marker of sympathetic axons, in patients with PD, compared to those with multiple system atrophy (MSA), progressive supranuclear palsy (PSP), corticobasal degeneration (CBD), and control subjects [5–8]. These results suggest postganglionic involvement in PD [5]. More details about the pathophysiological mechanism of the reduced cardiac MIBG uptake in PD are described in Chap. 18.

15.1.2 ^{123}I-MIBG Cardiac Scintigraphy and Clinical Features of PD

The correlation between ^{123}I-MIBG cardiac scintigraphy findings and age at disease onset, disease duration, different clinical phenotypes, disease severity, non-motor symptoms, and other tests are described below.

15.1.2.1 Age at Onset and Disease Duration

Age at onset correlates negatively with the H/M ratio [9–11]. Previous studies also reported that the H/M ratio correlated with disease duration [12], although others reported no such correlation [9, 13, 14].

15.1.2.2 Disease Severity

Several studies have demonstrated a significant negative correlation between cardiac MIBG uptake and H-Y stage [3, 9, 13]. One study of 34 patients with PD reported a significant correlation between early H/M ratio and the Unified Parkinson's Disease Rating Scale (UPDRS) score [10]. However, no correlation between the H/M ratio and disease severity in UPDRS III motor score was found in patients with early PD H-Y stages I–III [15] and those with advanced stage. Another study showed no significant difference in the H/M ratio in 24 patients with PD H-Y stages III, IV, and V [12]. The same trend was noted also in another study of 40 patients with H-Y stage III or IV [14] with no significant correlation between the H/M ratio and disease severity in UPDRS III motor score.

15.1.2.3 Clinical Phenotype

PD patients can be classified according to the clinical phenotype into the tremor-dominant type (TDT), akinetic/rigid type (ART), mixed type (MT), and postural instability/gait difficulty (PIGD) dominant type. One study of 102 patients with PD reported that cardiac MIBG uptake was significantly higher in TDT patients than ART and MT patients [16], while another study of 34 patients showed higher uptake in TDT patients than PIGD [10]. In another study of 143 patients with PD, cardiac MIBG uptake correlated negatively with bradykinesia, but not with tremor, rigidity, or postural instability [11]. Another study of 102 patients with PD showed a significant correlation between cardiac MIBG uptake and severity of hypokinesia as well as rigidity, but not with severity of resting or postural tremor irrespective of H-Y stage [16]. In contrast, in another study of 53 patients with PD, cardiac MIBG uptake was significantly lower in TDT patients compared with ART patients [15]. The study concluded that the discrepant results were probably due to different disease duration. The same results were later confirmed in another study of 37 patients with PD [17].

Generally, the presence of bradykinesia and rigidity tends to correlate with reduced cardiac MIBG uptake compared with the tremor type. These results suggest that cardiac sympathetic degeneration correlates with hypokinetic rigidity symptoms in PD, but not with tremor. Admittedly, these results remain controversial mainly due to population heterogeneity. Nevertheless, ^{123}I-MIBG cardiac scintigraphy could be used to predict the rate of progress of rigidity and axial symptoms in UPDRS III motor score in the subsequent 3–8 years. Such prediction is not possible using other motor symptoms, such as resting tremor, postural tremor, and bradykinesia [18].

15.1.2.4 Non-motor Symptoms

Clinical Symptoms of Dysautonomia

The relation between cardiac MIBG uptake and symptoms of autonomic failure in PD remains poorly defined. Most clinical studies found that reduced cardiac MIBG uptake is independent of such symptoms, as represented by orthostatic hypotension (OH) [19, 20]. However, Orimo et al. [21] reported significantly lower cardiac MIBG uptake in PD patients with OH than those without it. Interestingly, they also reported no significant difference in cardiac MIBG uptake between patients with and without constipation. Another study showed a significant difference in cardiac MIBG uptake between patients with and without bladder dysfunction [14]. These results indicate that not all dysautonomia-related clinical symptoms correlate with reduced cardiac MIBG uptake.

Hyposmia

The prodromal stage of PD occurs several years in advance of typical motor symptoms. Early diagnosis is important for early implementation of disease-modifying therapy. Several studies have indicated that hyposmia precedes the motor symptoms of PD [22, 23]. Furthermore, Jennings et al. [24] reported ^{123}I-FP-CIT ^{23}I-N-3-fluoropropyl-2beta-carbomethoxy-3beta-4-iodophenyl tropane (^{123}I-FP-CIT) single-photon emission computed tomography (SPECT) deficit in 11 % of 203 hyposmic subjects compared with 1 % of 100 normosmic subjects.

In their study of 23 non-demented patients with early-stage PD, Mizutani et al. [25] demonstrated that hyposmia correlates with cardiac MIBG uptake within 2 years from the onset of motor symptoms. The results also highlighted the potential utility of ^{123}I-MIBG cardiac scintigraphy in subjects with hyposmia and prodromal PD. Hyposmia and reduced MIBG uptake correlate well in the early stage of PD, suggesting that degenerations in olfactory nerves and cardiovascular sympathetic nerves represent similar processes. Therefore, it could be postulated that the prodromal stage of PD encompasses cardiovascular sympathetic nerve degeneration and reduced cardiac MIBG uptake.

Hallucination

Psychosis is a disabling non-motor complication of PD. Hallucination is one of the common psychotic symptoms in both demented and non-demented PD, occurring in approximately 40–50 % of PD patients [26, 27]. Uchiyama and colleagues [28] reported that in 97 patients with Lewy body disease (LBD) (90 patients with PD, seven with DLB), early and delayed H/M ratios independently correlated with hallucinations. In another study of 95 patients with PD [29], analysis of covariance, adjusted for the age of the patients as covariate, demonstrated that early and delayed H/M ratios were significantly lower in PD patients with hallucinations but no dementia, as well as PD patients with dementia, than in PD patients with no hallucinations or dementia. These findings indicate that reduced cardiac MIBG uptake may be associated with hallucination in PD patients.

15.1.3 Comparison with Other Tests

15.1.3.1 ^{123}I-FP-CIT SPECT

^{123}I-FP-CIT SPECT is helpful in the evaluation of patients with nigrostriatal dopaminergic deficit. Several studies have demonstrated that ^{123}I-FP-CIT SPECT is highly accurate in differentiating patients with neurodegenerative parkinsonisms from those with non-neurodegenerative parkinsonisms. In particular, ^{123}I-FP-CIT scanning offers help in the diagnosis of PD at early stages. However, only few studies have evaluated the relationship between ^{123}I-FP-CIT and cardiac MIBG uptake. In a study of 18 PD patients with early stage (limited only to H-Y stage I), Spiegel et al. [30] found a strong relationship between striatal binding on the impaired site and the H/M ratio of ^{123}I-MIBG cardiac scintigraphy. They concluded that the functional loss of nigrostriatal dopaminergic neurons is closely related to cardiac sympathetic dysfunction [30, 31]. In contrast to the above study, Chiaravalloti et al. [17] reported no relationships between ^{123}I-FP-CIT SPECT and early and delayed images of ^{123}I-MIBG cardiac scintigraphy in 37 patients with PD (H-Y stage I, $n = 15$; stage IV, $n = 8$; stage II, $n = 7$; stage III, $n = 7$). They also reported no significant relationship between subtypes TDT and ART of the single PD phenotype. The results suggest that the rate of sympathetic neurodegeneration is not related to the rate of nigrostriatal degeneration and vice versa.

A possible explanation of these discrepancies can be based on the different rates of disease severity and differences in PD phenotypes. Differences in PD phenotypes do not seem to explain the results of Chiaravalloti et al. [17]. In fact, the weak correlation between early cardiac MIBG uptake and ^{123}I-FP-CIT uptake in ART in the ipsilateral striatum is hampered by the medium-size effect on their statistical analysis. Further research is needed to determine the relationship between sympathetic and dopaminergic system in PD.

Only a few studies have compared the use of ^{123}I-FP-CIT SPECT and ^{123}I-MIBG cardiac scintigraphy. These two techniques were compared in 68 patients with suspected LBD [32]. The overall sensitivity, specificity, accuracy, and positive and negative predictive values for ^{123}I-MIBG cardiac scintigraphy were 83 %, 79 %, 82 %, 86 %, and 76 %, respectively, and 93 %, 41 %, 73 %, 71 %, and 80 %, respectively, for ^{123}I-FP-CIT SPECT. While these values were significantly different between the two methods in patients without LBD, they were not in patients with LBD. ^{123}I-FP-CIT SPECT has a high sensitivity in the diagnosis of LBD, while ^{123}I-MIBG cardiac scintigraphy may have a complementary role in the differential diagnosis of PD and other parkinsonisms [32].

15.1.3.2 Brain Perfusion SPECT

Hypoperfusion of the occipital lobe is one of the characteristics of PD [33, 34]. Nagamachi et al. [35] compared absolute regional cerebral blood flow (rCBF) values of ^{99m}Tc-HMPAO between 49 patients with PD and 28 patients with other neurodegenerative parkinsonisms. In the correlation analysis, reduced rCBF of occipital lobe correlated positively with reduced H/M ratio. With regard to the diagnostic ability, neither specificity nor accuracy improved by adding occipital lobe hypoperfusion to reduced cardiac MIBG uptake findings. However, the sensitivity improved by accounting for occipital hypoperfusion compared with reduced cardiac MIBG uptake findings alone. Adding analysis of occipital lobe rCBF to ^{123}I-MIBG cardiac scintigraphy is recommended.

15.1.3.3 Transcranial Sonography (TCS) of the Substantia Nigra (SN)

Becker et al. [36] first described SN hyperechogenicity as a typical sign for PD in 1995. About the pathophysiology of SN hyperechogenicity, Berg et al. [37] reported in a rat experimental model that iron may be the cause of the increased echogenicity of the SN. The same group also performed postmortem studies in 20 patients without extrapyramidal disorders during lifetime and showed that the echogenicity of the SN correlated with its iron content [38]. Another group also reported positive correlation between SN hyperechogenicity and iron and ferritin levels and negative correlation with neuromelanin content in postmortem brains from normal subjects [39]. In this regard, Berg et al. [40] also reported that SN echogenicity in 33 brains correlated with microglia activation, after correction for iron and neuromelanin content.

With regard to the correlation between SN hyperechogenicity and reduced cardiac MIBG uptake, Kajimoto et al.[41] reported that in 30 patients with PD whose midbrain was adequately displayed by TCS (46.2 % of the study group), no significant correlation was found between the area of SN echogenicity and the early H/M ratio. However, when the cutoff values were set at mean + 1 SD for TCS and mean − 2 SD for ^{123}I-MIBG cardiac scintigraphy, 29 patients (97 %) were identified as abnormal by the combination of TCS and ^{123}I-MIBG cardiac scintigraphy. Behnke et al. [42] analyzed the relation between the findings of TCS and ^{123}I-MIBG cardiac scintigraphy in 42 patients with PD. They demonstrated no correlation between the extent of SN hyperechogenicity (which was contralateral to the clinically more affected body side) and reduced cardiac MIBG uptake. The sensitivity of TCS in the diagnosis of PD was 79 %, compared with 81 % for ^{123}I-MIBG cardiac scintigraphy. On the other hand, the sensitivity of the combination of both was 95 %. Based on the above two studies, it seems that the combination of TCS of SN and ^{123}I-MIBG cardiac scintigraphy could improve the diagnosis of PD.

15.1.3.4 Microneurography

Shindo et al. [43] recorded muscle sympathetic nerve activity (MSNA) from the peroneal nerve in 14 patients with PD. No significant correlation was found between MSNA and H/M ratio and washout ratio of ^{123}I-MIBG cardiac scintigraphy.

15.1.3.5 Pupillary Sympathetic and Parasympathetic Sensitivity

Pupillary postganglionic autonomic dysfunction (pupillary sensitivity) is tested by assessing changes in pupil diameter using eye drops of parasympathomimetic [0.05 % pilocarpine hydrochloride (PL)] and sympathomimetic agents [0.02 % dipivefrine hydrochloride (DPE)]. Hori et al. [44] reported that in 40 patients with PD, pupillary supersensitivity to PL and DPE was found to be significantly greater in PD patients than in control subjects. Pupillary sensitivity to PL and DPE did not correlate with delayed H/M ratio. On the other hand, Yamashita et al. [45] reported in 40 patients with PD a weak and inverse correlation between delayed H/M ratio and pupillary sympathetic sensitivity to DPE. Further studies are required to explore this issue.

15.1.3.6 Apparent Diffusion Coefficient (ADC)

Kollensperger et al. [46] measured the apparent diffusion coefficient (ADC) in the putamina to differentiate PD from MSA parkinsonism (MSA-P). The sensitivity and specificity of ADC were higher than ^{123}I-MIBG cardiac scintigraphy. Their data suggest that ADC is superior to ^{123}I-MIBG cardiac scintigraphy in the differential diagnosis of PD versus MSA-P. However, the number of subjects in their study (nine with PD and nine with MSA-P) was small and more than half of the patients with MSA-P showed reduced cardiac MIBG uptake. Further studies are required to determine the diagnostic value of ADC.

15.1.4 Prodromal PD and Incidental Lewy Body Disease (iLBD)

15.1.4.1 Prodromal PD or DLB

The prodromal stage of PD is somewhat mentioned in non-motor features of this chapter. Again, detecting biomarkers of neurodegeneration that precede the apparent clinical symptoms of PD is important for early implementation of disease-modifying therapy. Braak et al. [47] proposed that prodromal pathological changes in PD first appear in the dorsal motor nuclei of the glossopharyngeal and vagal nerves and the anterior olfactory nucleus. Autonomic failure (constipation and postural hypotension), depression/anxiety, rapid eye movement behavior disorder (RBD), hyposmia, and mild memory disorder are known to be the prodromal symptoms of PD or DLB

Sakakibara et al. [48] studied 254 patients with memory complaints and found reduced cardiac MIBG uptake in 13 of 44 amnestic MCI cases (30 %). None of the 13 patients had the core clinical features of DLB. They also reported that among 1600 outpatients, only five had constipation and reduced cardiac MIBG uptake without apparent motor disorders, indicating LBD, nor apparent neurologic diseases other than LBD. Some patients also showed RBD, hallucinations, occipital hypoperfusion by SPECT, and other prodromal biomarkers. These studies suggest that reduced cardiac MIBG uptake can be a biomarker for Lewy body pathology in patients with amnestic MCI and constipation. Other prodromal biomarkers are discussed elsewhere in this chapter.

15.1.4.2 Incidental Lewy Body Disease

Incidental Lewy body disease (iLBD) [49, 50] is a term used to describe the presence of Lewy bodies at routine postmortem examination in individuals free of any clinical signs of PD or dementia in life. Orimo et al. [51] compared cardiac tissues and paravertebral sympathetic ganglia of patients with iLBD and PD and demonstrated earlier accumulation of α-synuclein aggregates in the distal axons of the cardiac sympathetic nervous system compared with neuronal somata or neurites in the paravertebral sympathetic ganglia, a finding that heralds centripetal degeneration of the cardiac sympathetic nerve in LBD, such as iLBD and PD. These results suggest that the reduced cardiac MIBG uptake in iLBD subjects represents degeneration of the cardiac sympathetic nerves that might predate the damage of striatal presynaptic dopaminergic terminals.

At present, ^{123}I-MIBG cardiac scintigraphy is not recommended for patients free of parkinsonism or dementia, because it is mainly used to differentiate LBD from other parkinsonisms or dementia. No postmortem studies have directly investigated the relationship between the findings of antemortem ^{123}I-MIBG cardiac scintigraphy and cardiac sympathetic denervation in cases of iLBD. Reduced cardiac MIBG uptake was reported in an autopsy case of MSA with Lewy bodies in neurons of the dorsal vagal nucleus, locus coeruleus, and the basal nucleus of Meynert [52]. Reduced cardiac MIBG uptake was also reported in an autopsy case of CBD with Lewy bodies in the sympathetic ganglia [53]. Although these were not ideal iLBD cases, the presence of reduced cardiac MIBG uptake could be considered to represent incidental Lewy body pathology during the lifetime.

15.1.5 ^{123}I-MIBG Cardiac Scintigraphy Findings in Familial PD

To date, 22 familial PD mutations have been identified. It seems that cardiac MIBG uptake varies according to the type of mutations.

15.1.5.1 Autosomal-Dominant Forms of PD: PARK1, PARK4, and PARK8

Point mutations in the gene for *α-synuclein*, non-A4 component of amyloid precursor (*SNCA*), duplications and triplications of the entire gene, and mutations in the leucine-rich repeat kinase 2 (*LRRK2*) gene cause autosomal-dominant forms of PD.

PARK1

The point mutation E46K in *SNCA* gene (PARK1) [54] causes an aggressive form of PD with relatively early onset. Most cases have been identified in families with multiple affected individuals. Patients with E46K substitution in the *SNCA* gene both with and without autonomic symptoms showed complete lack of cardiac MIBG uptake in life and showed a complete absence of TH-immunoreactive nerve fibers in the cardiac tissues on postmortem examination [55] and normal ^{123} I-FP-CIT SPECT in a subject with asymptomatic carrier [56].

PARK4

***SNCA* Gene Duplication and Triplication and Iowa Kindred** *SNCA* gene duplication and triplication causes typical late-onset PD (PARK4) [57]. Incomplete penetrance of *SNCA* gene duplication may result in a negative family history. The H/M ratio of ^{123}I-MIBG cardiac scintigraphy is reported to be less than that of the normal control [58]. Orimo et al. [59] used immunohistochemistry to examine cardiac tissues from three patients with PD linked to *SNCA* duplication. They found severe degeneration of the cardiac sympathetic nerve and sparse α-synuclein aggregates in the epicardial nerve fascicles of all three patients, a finding similar to idiopathic PD. In addition, scattered swollen phosphorylated neurofilament-immunoreactive axons were especially found in two of three patients, suggesting possible impairment of axonal flow. These results suggest that cardiac sympathetic denervation is closely related to the presence of Lewy bodies in not only sporadic Lewy body disorders but also in familial PD with duplication of *SNCA* and that it is associated with reduced cardiac MIBG uptake [59].

SNCA gene triplication causes parkinsonism in a well-characterized family called the "Iowa kindred" [60]. The results of cardiac PET scanning using the sympathoneural imaging agent, 6-[^{18}F]fluorodopamine, in affected members of this kindred showed a clear loss of cardiac sympathetic innervation [61].

PARK8

Mutations in the *LRRK2* gene are a common cause of dominant PD (PARK8) [62, 63]. To date, six mutations are known to be pathogenic (N1437H, R1441C, R1441G, Y1699C, G2019S, and I2020T). Overall, *LRRK2* mutations account for 5–15 % of dominant familial PD cases [64] and 1–3 % of sporadic PD cases [65]. Clinically, *LRRK2*-associated PD is indistinguishable from sporadic typical PD. A proportion of *LRRK2* patients with G2019S and R1441G mutations present with lower cardiac MIBG uptake compared with the control and relatively higher uptake than PD patients, while the remaining of the patients with these mutations showed normal cardiac MIBG uptake [66, 67].

I2020T mutation is described to have a single-founder effect in Japanese patients. One patient with the mutation had normal cardiac MIBG uptake, while the remaining two patients had reduced cardiac MIBG uptake [68]. These results highlight the heterogeneity of this disease.

15.1.5.2 Autosomal-Recessive Forms of PD: PARK2, PARK6, and PARK7

PARK2

Homozygous or compound heterozygous mutations in *parkin* gene are the most common familial PD (PARK2) [69]. Up to half of familial PD cases with a disease onset under the age of 45 and a recessive form of inheritance are caused by *parkin* mutations. Similarly, *parkin* mutations underlie about 15 % of sporadic PD cases with disease onset before the age of 45. The clinical course of patients with *parkin* mutations is overall benign. Motor fluctuations and levodopa-induced dyskinesia are frequent, whereas marked cognitive or autonomic disturbances are rare. Except for one patient, the reported cases of *parkin* mutations had normal cardiac MIBG uptake [67, 70–72]. In one study, postmortem examination showed preservation of TH-immunoreactive nerve fibers in the epicardial nerve fascicle [71]. These results suggest that PARK2 is a distinct disease entity from PD.

PARK6 and PARK7

Homozygous or compound heterozygous mutations in *PINK1*(PARK6) [73] and homozygous or compound heterozygous mutations in *DJ-1* genes (PARK7) [74]

Table 15.1 Summary of cardiac MIBG uptake in familial PD

Locus	Inheritance	Gene	Cardiac MIBG uptake	Lewy bodies
PARK1	AD	SNCA mutation	Reduced	+
PARK2	AR	Parkin	Normal	−
PARK4	AD	SNCA multiplication	Reduced	+
PARK6	AR	PINK1	Reduced or normal	+
PARK7	AR	DJ-1	Reduced or normal	Not published
PARK8	AD	LRRK2	Reduced or normal	+ or −

Cardiac MIBG uptake results may vary due to the relatively small number of subjects
AD autosomal dominant, *AR* autosomal recessive

are less common, accounting for only 1–8 % and 1–2 % of early-onset and sporadic cases, respectively [75–77]. The phenotype associated with *PINK1* and *DJ-1* mutations is basically indistinguishable from that associated with mutation in *parkin*. One previous study reported preservation of cardiac MIGB uptake in one of two patients with *DJ-1* mutations and in one of two brothers with *PINK1* mutations [67] (Table 15.1).

15.2 Gaucher Disease (GD)

Mutations in the gene encoding the lysosomal enzyme glucocerebrosidase are associated with GD, the most common autosomal-recessive lysosomal storage disease. It is now clear that a subset of patients with GD develop parkinsonism. Postmortem brain tissue examination of patients with GD associated with parkinsonism has consistently shown classic PD pathology. Itokawa et al. [78] reported a patient with type I GD who had near-normal cardiac MIBG uptake, with partially defective cardiac MIBG uptake considered to be due to old myocardial infarction confirmed on Tc-99 m SPECT. On the other hand, Lebouvier et al. [79] reported a case with type I GD and abnormal cardiac MIBG uptake, with reduced early and delayed H/M ratios. Another study that sequenced the entire coding exons and exon/intron boundaries reported that 27 of 144 families (18.8 %) of index patients with Japanese familial PD were heterozygous for known GD mutations and presented with reduced cardiac MIBG uptake [80].

15.3 ^{123}I-MIBG Cardiac Scintigraphy Findings in Other Lewy Body Diseases

Cardiac sympathetic denervation has been reported to develop in patients with other Lewy body diseases, such as DLB and pure autonomic failure (PAF). In this regard, ^{123}I-MIBG cardiac scintigraphy can be useful for differentiating Lewy body diseases from other parkinsonian syndromes and dementias. Furthermore, RBD is observed in α-synucleinopathy, especially in relation to Lewy body disease.

For this reason, we discuss PAF and RBD in this chapter, while DLB is described in Chap. 16.

15.3.1 Pure Autonomic Failure

PAF is considered a rare clinical manifestation of Lewy body diseases and is characterized by autonomic failure without any sign of parkinsonism. Previous studies reported reduced cardiac MIBG uptake in patients with PAF [81, 82].

15.3.2 Rapid Eye Movement (REM) Sleep Behavior Disorder (RBD)

RBD is characterized by dream-enactment behavior resulting from muscle activity during REM sleep. The development of RBD may be one of the first manifestations of α-synucleinopathy including PD, DLB, and MSA. While the exact mechanism of RBD is still unclear, it is speculated that disturbances in regions of the brainstem that control REM sleep play a pathological role. A few studies reported significantly reduced cardiac MIBG uptake in patients with idiopathic RBD similar to PD with and without RBD [83, 84]. The results suggest that most of RBD indicate Lewy body diseases. Furthermore, Koyama et al. [85] reported the presence of hyposmia, impaired facial expression recognition, and reduced cardiac MIBG uptake in one RBD patient free of parkinsonism. Impaired facial expression recognition may reflect the dysfunction of the amygdala. In RBD patients, neurodegeneration may occur more diffusively than the brainstem alone.

15.4 ^{123}I-MIBG Cardiac Scintigraphy Findings in Other Neurodegenerative Parkinsonism and Related Disorders

The clinical feature of atypical neurodegenerative parkinsonism, such as MSA, PSP, and CBD, resembles those of PD, especially in the early stages. Despite the presence of clinical consensus criteria for PD and other parkinsonisms, accurate diagnosis of these disorders remains a challenge for neurologists. ^{123}I-MIBG cardiac scintigraphy is a useful imaging tool for differentiating PD from other parkinsonisms.

15.4.1 Neurodegenerative Parkinsonism

15.4.1.1 Multiple System Atrophy

MSA is a late-onset neurodegenerative disease characterized by progressive autonomic failure, parkinsonism, and cerebellar and pyramidal tract symptoms. Neurodegeneration and the formation of glial cytoplasmic inclusions immunostained with α-synuclein are the hallmark of the disease. ^{123}I-MIBG cardiac scintigraphy is significantly higher in MSA than PD or within the normal range [86]. However, cardiac MIBG uptake is not necessarily preserved in patients with MSA, and approximately 30 % of patients with MSA have reduced cardiac MIBG uptake, and these levels do not correlate with disease duration or severity [87]. The mean value of the H/M ratio in striatonigral degeneration (SND), a subtype of MSA, with OH is lower than SND without OH [86].

It is known that central or preganglionic lesions, such as those in the cardiovascular autonomic center of the ventrolateral medulla or intermediolateral cell column of the spinal cord, contribute to the dysautonomia of MSA. Cohen et al. [88] hypothesized that in MSA, preganglionic lesions cause dysfunction of the postganglionic sympathetic fibers through the transsynaptic effects. Postmortem examination of a MSA patient with a slightly reduced cardiac MIBG uptake demonstrated no obvious neuronal loss in the peripheral sympathetic ganglia. Furthermore, TH-immunoreactive epicardial nerve fibers in the anterior wall of the left ventricle of the heart were preserved in patients with MSA [5]. Based on the above findings, it was proposed that postganglionic sympathetic nerves are not involved in MSA; rather, the central and preganglionic lesions account for the slightly reduced cardiac MIBG uptake in MSA [5]. The same group also reported mild or moderate decrease in the number of TH-immunoreactive epicardial nerve fibers of six of 15 patients with MSA, of whom four showed a decrease in TH immunoreactivity in the neuronal somata in the sympathetic ganglia. These pathological changes supported the hypothesis that preganglionic lesions cause dysfunction of the postganglionic sympathetic fibers through the transsynaptic effects in MSA.

15.4.1.2 Progressive Supranuclear Palsy

PSP is characterized by parkinsonisms, gaze palsy, pseudobulbar palsy, dysarthria, axial rigidity, frontal lobe dysfunction, and dementia and thus partially resembles PD. However, the clinical spectrum of PSP is known to be wider than originally described. The typical pathology is abnormal tau deposition that affects neurons and glial cells represented by tufted astrocytes. The mean value of the H/M ratio in patients with PSP is significantly higher than in PD, but the majority of patients have normal to slightly reduced cardiac MIBG uptake compared with normal control subjects [3, 83, 86, 89, 90]. Orimo et al. [6] reported that

TH-immunoreactive epicardial nerve fibers were well preserved in five pathologically confirmed patients with PSP.

15.4.1.3 Corticobasal Degeneration

CBD is characterized by asymmetric symptoms including rigidity and tremor, thus partially resembling PD. However, some patients with CBD present with clinical symptoms that do not resemble those of PD, such as dystonia, myoclonus, apraxia, cortical sensory deficits, and alien limb phenomena. In addition, some suffer from various cognitive and language deficits. CBD is pathologically characterized by abnormal tau deposition in neurons and glial cells, including astrocytic plaques. Only a few studies examined ^{123}I-MIBG cardiac scintigraphy in patients with CBD. The H/M ratio of CBD is not different from that of normal control subjects [8, 89, 90].

Another case report of postmortem histopathological examination of a CBD patient demonstrated well-preserved TH-immunoreactive epicardial nerve fibers, similar to the control subjects [8].

15.4.2 ^{123}I-MIBG Cardiac Scintigraphy Findings in Other Related Disorders

15.4.2.1 Vascular Parkinsonism

Orimo et al. [21] examined ^{123}I-MIBG cardiac scintigraphy in 11 patients with vascular parkinsonism and found it was not significantly different from normal and disease control subjects. Furthermore, the mean H/M ratio of the PD patients was significantly lower than patients with vascular parkinsonism. The same findings were confirmed in six patients [12] and 19 patients [91] with vascular parkinsonism.

15.4.2.2 Essential Tremor (ET)

ET is characterized by postural tremor affecting the hands, head, and other parts of the body. By definition, patients with ET should not have other clinical signs of parkinsonism. In some cases, it is difficult to differentiate ET from PD, especially in the early stages of the disease.

Orimo et al. [21] indicated that cardiac MIBG uptake measured in five patients with ET was comparable to that in normal and disease control subjects. Another group [92] also examined ^{123}I-MIBG cardiac scintigraphy in 20 patients with ET and reported significantly higher mean H/M ratio in patients with ET than in TDT or early PD [rated at H-Y of I and II with recent diagnosis (symptom duration 2 years)]. However, the mean H/M ratio of the ET group was not significantly different from that of the control group. In the same study [92], the H/M ratio was higher in patients with ET than two standard deviations above the range of the ratio in patients with early PD or TDT. Another study of 16 patients with rest and postural tremor (mixed tremor), together with mild extrapyramidal features and abnormal striatal ^{123}I-FP-CIT SPECT, reported reduced cardiac MIBG uptake in delayed images in half of the patients [93]. The results suggest that the combined use of both ^{123}I-FP-CIT SPECT and ^{123}I-MIBG cardiac scintigraphy in patients with mixed tremors and extrapyramidal features can help distinguish patients with ET from those with PD and parkinsonism.

15.4.2.3 Drug-Induced Parkinsonism

Drug-induced parkinsonism (DIP) is a heterogeneous clinical syndrome, but almost all patients with DIP have normal cardiac MIBG uptake. All patients show improvement or complete resolution of parkinsonism after withdrawal of the offending drug. However, some DIP patients with significantly reduced cardiac MIBG uptake develop persistent and worsening parkinsonism or PD after discontinuation of the offending drug. ^{123}I-MIBG cardiac scintigraphy may be a useful tool for detecting DIP unrelated to PD and to identify DIP patients with subclinical PD [94–96] (Fig. 15.1).

15.4.2.4 Others

Spinocerebellar Ataxia Type 2

SCA2 is an autosomal-dominant neurodegenerative disorder associated with expanded CAG trinucleotide repeat in the *ATXN2* gene. It is clinically characterized by gait and limb ataxia, dysarthria, supranuclear ophthalmoplegia, peripheral neuropathy, sleep disorders, postural tremor, chorea, myoclonus, parkinsonism, pyramidal signs, and dementia. Most cases exhibit the cerebellar phenotype, though few present with parkinsonism as the predominant phenotype instead of cerebellar ataxia. The levodopa-responsive parkinsonism is considered a rare clinical presentation in SCA2.

Koyano et al. [97] found reduced cardiac MIBG uptake in SCA2 parkinsonian phenotype with homozygous SCA2 expansion (36/38). Another group [98] reported the postmortem findings of a SCA2 patient with parkinsonian phenotype and reduced cardiac MIBG uptake. They found atrophy of the olivopontocerebellar system and substantia nigra. Both findings were compatible with SCA2. In addition, they also found Lewy body pathology in the SN, the locus coeruleus, dorsal motor nuclei of vagus, and cardiac sympathetic nerves. De Rosa et al. [99] performed ^{123}I-MIBG cardiac scintigraphy in nine patients with SCA2 free of parkinsonism. The early and delayed H/M ratios were significantly lower in patients with SCA2 than the control subjects, though less marked than in PD patients. Another study reported

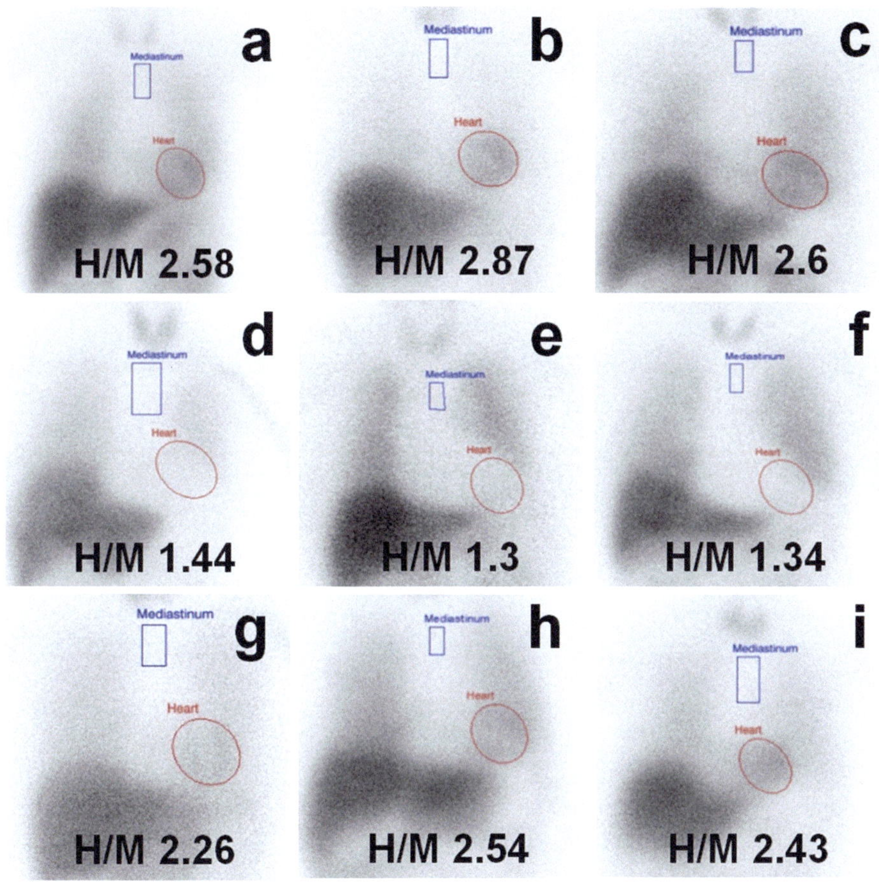

Fig. 15.1 ^{123}I-MIBG cardiac scintigraphy in representative cases of Parkinson's disease and related disorders. Cardiac MIBG uptake is lower than normal control in patients with Lewy body diseases such as PD, DLB, and PAF, but not in patients with other related disorders. (**a**) control, (**b**) vascular parkinsonism, (**c**) essential tremor, (**d**) Parkinson's disease, (**e**) dementia with Lewy bodies, (**f**) pure autonomic failure, (**g**) multiple system atrophy-parkinsonism type, (**h**) progressive supra nuclear palsy, (**i**) corticobasal degeneration

the results of postmortem examination of a Japanese SCA2 patient with parkinsonism [100]. In addition to the classic SCA2 neuropathological changes, Lewy bodies and Lewy neurites were identified in brainstem nuclei. Furthermore, genetic analysis demonstrated the presence of shorter abnormal expansion of CAG repeats (less than 39). In comparison, the authors did not find Lewy body pathology in two SCA2 cases free of parkinsonism. The study provided neuropathological evidence for a correlation between Lewy body pathology and parkinsonism of SCA2.

Since ^{123}I-MIBG cardiac scintigraphy demonstrated impairment of cardiac sympathetic function in SCA2, with and without parkinsonism, Lewy body pathology does not seem to explain all the findings of ^{123}I-MIBG cardiac scintigraphy.

Machado-Joseph Disease (MJD)

MJD or SCA3 is an autosomal-dominant neurodegenerative disorder associated with expanded CAG trinucleotide repeat in ATXN3 gene. It is characterized by ataxia, ophthalmoplegia, peripheral neuropathy, pyramidal dysfunction, and movement disorders. Only a few MJD patients develop parkinsonism. Although MJD is a relatively rare disease, it is the most frequent spinocerebellar ataxia with a worldwide distribution.

In a study of 19 patients with MJD who underwent ^{123}I-MIBG cardiac scintigraphy, the delayed H/M ratio was significantly lower in the patients than the control subjects, whereas the early H/M ratio was comparable between the two groups [101]. Six of the 19 patients showed abnormal sympathetic skin responses (SSR). The delayed H/M ratio was significantly lower in the latter group than in patients with normal SSR. These results suggest the presence of cardiac sympathetic dysfunction in MJD, as detected by ^{123}I-MIBG cardiac scintigraphy, which appears to correlate with sudomotor sympathetic dysfunction.

15.5 ^{123}I-MIBG Cardiac Scintigraphy Findings in Comorbidities

Finally, we describe the comorbidities associated with reduced cardiac MIBG uptake for careful interpretation of ^{123}I-MIBG cardiac scintigraphy findings.

15.5.1 Congestive Heart Failure (CHF), Cardiomyopathy, and Ischemic Heart Disease

Kline et al. [102] were the first investigators to use ^{123}I-MIBG cardiac scintigraphy and explore its utility in quantitative measurement of myocardial catecholamine content. Activation of the sympathetic nervous system is one of the main pathophysiological abnormalities associated with heart failure. The H/M ratio correlates significantly with myocardial noradrenaline concentration and with left ventricular ejection fraction in patients with idiopathic dilated cardiomyopathy [103]. Reduced cardiac MIBG uptake and high washout, especially reduced delayed H/M ratio, have been described in CHF [104], and the extent of such reduction correlated with the severity of CHF and response of treatment, prognosis, and mortality.

The failing heart requires sympathetic stimulation to increase cardiac performance, but paradoxically shows depletion of noradrenaline [105]. Recent studies have suggested that it is due to decreased noradrenaline reuptake [104, 106] and synthesis [107]. In this regard, Kanazawa et al. [108] demonstrated neurotransmitter switching from predominantly catecholaminergic to cholinergic or cholinergic transdifferentiation of the cardiac sympathetic nervous system in patients with CHF, as an adaptive response. These results may explain the reduced cardiac MIBG uptake in patients with CHF.

With regard to ischemic heart disease, cardiac MIBG washout is globally increased after myocardial infarction within 14 days of early reperfusion therapy, even in patients with preserved left ventricular function [109]. Podio et al. [110] reported no changes in cardiac MIBG uptake from 1 week after infarction to after 30 months of follow-up. Enhanced washout may reflect increased sympathetic nerve tone and represent increased catecholamine turnover or impaired reuptake in the subacute phase of myocardial infarction [109].

15.5.2 Diabetes Mellitus (DM)

Evaluation of cardiac MIBG uptake in patients with DM began in 1988. The first report of use of ^{123}I-MIBG cardiac scintigraphy was in patients with diabetic autonomic neuropathy with sudden cardiac death presumably due to QTc interval prolongation [111], though the usefulness of imaging of the sympathetic nervous system in DM is unclear [112]. One study that followed 144 patients with DM and no other cardiac disease for 7 years after ^{123}I-MIBG cardiac scintigraphy showed that reduced cardiac MIBG uptake was associated with increased risk of cardiac mortality [113]. Coronary arterial and arteriolar narrowing is common in patients with diabetic autonomic neuropathy; therefore, reduced cardiac MIBG uptake may reflect a combination of denervation and decreased delivery of the trace to the sympathetic nerves due to coronary hypoperfusion [114]. However, reduced cardiac MIBG uptake can also occur in patients with DM without any evidence of coronary heart disease or autonomic neuropathy [115]. In addition, enhanced washout rate is independently associated with the incidence of major adverse cardiac and cerebrovascular events in type 2 diabetic patients free of structural heart disease [116].

Experimental studies in a rat model of DM showed a high cardiac MIBG washout rate, but unlike patients with heart failure. This was not due to systemic sympathetic hyperactivity. Rather, it was likely due to dysfunction of the reuptake and/or pooling mechanism since the plasma and myocardial noradrenaline concentrations in diabetic rats were significantly lower than those in nondiabetic rats [117].

Cardiac MIBG uptake is comparatively preserved in diabetic patients without heart failure [118] or even improved when blood sugar level is under control [119]. Thus, the presence of DM alone does not always cause reduced cardiac MIBG uptake. Recently Slaets et al. [120] described no significant difference in H/M ratio among DLB patients with DM, arterial hypertension, hyperlipidemia, ischemic heart disease, heart failure, and pharmacological treatment and those without clinically observable conditions. Furthermore, Otsuka et al. [121] examined differences in H/M ratio in patients with Alzheimer disease (AD) or amnestic MCI (aMCI) with or without DM. In their study, ^{123}I-MIBG cardiac scintigraphy was performed in both the AD or aMCI without DM (AD/DM(−), $n = 248$) and AD or aMCI with DM (AD/DM(+), $n = 46$) and in age-matched control subjects (C, $n = 28$). The early/delayed H/M ratios in AD/DM(−), AD/DM(+), and C were $2.39 \pm 0.38/2.37 \pm 0.44$, $2.37 \pm 0.3/2.31 \pm 0.34$, and $2.43 \pm 0.24/2.44 \pm 0.26$, respectively, with no significant difference among the three groups. Five of 46 patients of the (AD/DM(+)) group showed slightly low H/M ratios but none of these patients showed no accumulation of MIBG in the planar image as typically seen in PD patients. These results suggest that DM does not always have a significant effect on cardiac MIBG uptake, particularly DM without diabetic autonomic neuropathy. Further studies are needed to confirm the cardiac MIBG uptake in DM patients with or without autonomic neuropathy.

References

1. Hakusui S, Yasuda T, Yanagi T, Tohyama J, Hasegawa Y, Koike Y, et al. A radiological analysis of heart sympathetic functions with meta-[123I]iodobenzylguanidine in neurological patients with autonomic failure. J Auton Nerv Syst. 1994;49(1):81–4.
2. Shin DH, Lee PH, Bang OY, Joo IS, Huh K. Clinical implications of cardiac-MIBG SPECT in the differentiation of Parkinsonian Syndromes. J Clin Neurol. 2006;2(1):51–7. doi:10.3988/jcn.2006.2.1.51.
3. Nagayama H, Hamamoto M, Ueda M, Nagashima J, Katayama Y. Reliability of MIBG myocardial scintigraphy in the diagnosis of Parkinson's disease. J Neurol Neurosurg Psychiatry. 2005;76(2):249–51. doi:10.1136/jnnp.2004.037028.
4. Orimo S, Suzuki M, Inaba A, Mizusawa H. 123I-MIBG myocardial scintigraphy for differentiating Parkinson's disease from other neurodegenerative parkinsonism: a systematic review and meta-analysis. Parkinsonism Relat Disord. 2012;18(5):494–500. doi:10.1016/j.parkreldis.2012.01.009.
5. Orimo S, Ozawa E, Oka T, Nakade S, Tsuchiya K, Yoshimoto M, et al. Different histopathology accounting for a decrease in myocardial MIBG uptake in PD and MSA. Neurology. 2001;57(6):1140–1.
6. Orimo S, Amino T, Itoh Y, Takahashi A, Kojo T, Uchihara T, et al. Cardiac sympathetic denervation precedes neuronal loss in the sympathetic ganglia in Lewy body disease. Acta Neuropathol. 2005;109(6):583–8. doi:10.1007/s00401-005-0995-7.
7. Amino T, Orimo S, Itoh Y, Takahashi A, Uchihara T, Mizusawa H. Profound cardiac sympathetic denervation occurs in Parkinson disease. Brain Pathol. 2005;15(1):29–34.
8. Orimo S, Ozawa E, Nakade S, Hattori H, Tsuchiya K, Taki K, et al. [123I] meta-iodobenzylguanidine myocardial scintigraphy differentiates corticobasal degeneration from Parkinson's disease. Intern Med. 2003;42(1):127–8.
9. Hamada K, Hirayama M, Watanabe H, Kobayashi R, Ito H, Ieda T, et al. Onset age and severity of motor impairment are associated with reduction of myocardial 123I-MIBG uptake in Parkinson's disease. J Neurol Neurosurg Psychiatry. 2003;74(4):423–6.
10. Saiki S, Hirose G, Sakai K, Kataoka S, Hori A, Saiki M, et al. Cardiac 123I-MIBG scintigraphy can assess the disease severity and phenotype of PD. J Neurol Sci. 2004;220(1–2):105–11. doi:10.1016/j.jns.2004.02.018.
11. Suzuki M, Urashima M, Oka H, Hashimoto M, Taira K. Cardiac sympathetic denervation in bradykinesia-dominant Parkinson's disease. Neuroreport. 2007;18(17):1867–70. doi:10.1097/WNR.0b013e3282f1ab33.

12. Satoh A, Serita T, Seto M, Tomita I, Satoh H, Iwanaga K, et al. Loss of 123I-MIBG uptake by the heart in Parkinson's disease: assessment of cardiac sympathetic denervation and diagnostic value. J Nucl Med: Off Publ Soc Nucl Med. 1999;40(3):371–5.
13. Takatsu H, Nishida H, Matsuo H, Watanabe S, Nagashima K, Wada H, et al. Cardiac sympathetic denervation from the early stage of Parkinson's disease: clinical and experimental studies with radiolabeled MIBG. J Nucl Med: Off Publ Soc Nucl Med. 2000;41(1):71–7.
14. Matsui H, Nishinaka K, Oda M, Komatsu K, Kubori T, Udaka F. Does cardiac metaiodobenzylguanidine (MIBG) uptake in Parkinson's disease correlate with major autonomic symptoms? Parkinsonism Relat Disord. 2006;12(5):284–8. doi:10.1016/j.parkreldis.2005.12.008.
15. Chiaravalloti A, Stefani A, Tavolozza M, Pierantozzi M, Di Biagio D, Olivola E, et al. Different patterns of cardiac sympathetic denervation in tremor-type compared to akinetic-rigid-type Parkinson's disease: molecular imaging with (1)(2)(3)I-MIBG. Mol Med Rep. 2012;6(6):1337–42. doi:10.3892/mmr.2012.1104.
16. Spiegel J, Hellwig D, Farmakis G, Jost WH, Samnick S, Fassbender K, et al. Myocardial sympathetic degeneration correlates with clinical phenotype of Parkinson's disease. Mov Dis: Off J Mov Dis Soc. 2007;22(7):1004–8. doi:10.1002/mds.21499.
17. Chiaravalloti A, Stefani A, Di Biagio D, Pierantozzi M, Tavolozza M, Di Pietro B, et al. Cardiac sympathetic denervation is not related to nigrostriatal degeneration in Parkinson's disease. Ann Nucl Med. 2013;27(5):444–51. doi:10.1007/s12149-013-0702-z.
18. Dorschner J, Farmakis G, Behnke S, Hellwig D, Schneider S, Fassbender K, et al. Myocardial MIBG scintigraphy may predict the course of motor symptoms in Parkinson's disease. Parkinsonism Relat Disord. 2011;17(5):372–5. doi:10.1016/j.parkreldis.2011.03.001.
19. Braune S, Reinhardt M, Schnitzer R, Riedel A, Lucking CH. Cardiac uptake of [123I]MIBG separates Parkinson's disease from multiple system atrophy. Neurology. 1999;53(5):1020–5.
20. Reinhardt MJ, Jungling FD, Krause TM, Braune S. Scintigraphic differentiation between two forms of primary dysautonomia early after onset of autonomic dysfunction: value of cardiac and pulmonary iodine-123 MIBG uptake. Eur J Nucl Med. 2000;27(5):595–600.
21. Orimo S, Ozawa E, Nakade S, Sugimoto T, Mizusawa H. (123)I-metaiodobenzylguanidine myocardial scintigraphy in Parkinson's disease. J Neurol Neurosurg Psychiatry. 1999;67(2):189–94.
22. Adler CH. Premotor symptoms and early diagnosis of Parkinson's disease. Int J Neurosci. 2011;121 Suppl 2:3–8. doi:10.3109/00207454.2011.620192.
23. Ross GW, Abbott RD, Petrovitch H, Tanner CM, Davis DG, Nelson J, et al. Association of olfactory dysfunction with incidental Lewy bodies. Mov Disord: Off J Mov Disord Soc. 2006;21(12):2062–7. doi:10.1002/mds.21076.
24. Jennings D, Siderowf A, Stern M, Seibyl J, Eberly S, Oakes D, et al. Imaging prodromal Parkinson disease: the Parkinson Associated Risk Syndrome Study. Neurology. 2014;83(19):1739–46. doi:10.1212/WNL.0000000000000960.
25. Mizutani Y, Nakamura T, Okada A, Suzuki J, Watanabe H, Hirayama M, et al. Hyposmia and cardiovascular dysautonomia correlatively appear in early-stage Parkinson's disease. Parkinsonism Relat Disord. 2014;20(5):520–4. doi:10.1016/j.parkreldis.2014.02.010.
26. Poewe W. Psychosis in Parkinson's disease. Mov Disorder: Off J Mov Disord Soc. 2003;18 Suppl 6:S80–7. doi:10.1002/mds.10567.
27. Williams DR, Lees AJ. Visual hallucinations in the diagnosis of idiopathic Parkinson's disease: a retrospective autopsy study. Lancet Neurol. 2005;4(10):605–10. doi:10.1016/S1474-4422(05)70146-0.
28. Uchiyama Y, Momose M, Kondo C, Kusakabe K, Uchiyama S. Comparison of parameters of (123)I-metaiodobenzylguanidine scintigraphy for differential diagnosis in patients with parkinsonism: correlation with clinical features. Ann Nucl Med. 2011;25(7):478–85. doi:10.1007/s12149-011-0490-2.

29. Kitayama M, Wada-Isoe K, Irizawa Y, Nakashima K. Association of visual hallucinations with reduction of MIBG cardiac uptake in Parkinson's disease. J Neurol Sci. 2008;264 (1–2):22–6. doi:10.1016/j.jns.2007.07.017.
30. Spiegel J, Hellwig D, Jost WH, Farmakis G, Samnick S, Fassbender K, et al. Cerebral and extracranial neurodegeneration are strongly coupled in Parkinson's Disease. Open Neurol J. 2007;1:1–4. doi:10.2174/1874205X00701010001.
31. Spiegel J, Mollers MO, Jost WH, Fuss G, Samnick S, Dillmann U, et al. FP-CIT and MIBG scintigraphy in early Parkinson's disease. Mov Disord: Off J Mov Disord Soc. 2005;20 (5):552–61. doi:10.1002/mds.20369.
32. Treglia G, Cason E, Cortelli P, Gabellini A, Liguori R, Bagnato A, et al. Iodine-123 metaiodobenzylguanidine scintigraphy and iodine-123 ioflupane single photon emission computed tomography in Lewy body diseases: complementary or alternative techniques? J Neuroimag: Off J Am Soc Neuroimag. 2014;24(2):149–54. doi:10.1111/j.1552-6569.2012. 00774.x.
33. Matsui H, Udaka F, Miyoshi T, Hara N, Tamura A, Oda M, et al. Brain perfusion differences between Parkinson's disease and multiple system atrophy with predominant parkinsonian features. Parkinsonism Relat Disord. 2005;11(4):227–32. doi:10.1016/j.parkreldis.2005.01. 001.
34. Van Laere K, Santens P, Bosman T, De Reuck J, Mortelmans L, Dierckx R. Statistical parametric mapping of (99m)Tc-ECD SPECT in idiopathic Parkinson's disease and multiple system atrophy with predominant parkinsonian features: correlation with clinical parameters. J Nucl Med: Off Publ Soc Nucl Med. 2004;45(6):933–42.
35. Nagamachi S, Wakamatsu H, Kiyohara S, Fujita S, Futami S, Tamura S, et al. Usefulness of rCBF analysis in diagnosing Parkinson's disease: supplemental role with MIBG myocardial scintigraphy. Ann Nucl Med. 2008;22(7):557–64. doi:10.1007/s12149-008-0156-x.
36. Becker G, Seufert J, Bogdahn U, Reichmann H, Reiners K. Degeneration of substantia nigra in chronic Parkinson's disease visualized by transcranial color-coded real-time sonography. Neurology. 1995;45(1):182–4.
37. Berg D, Grote C, Rausch WD, Maurer M, Wesemann W, Riederer P, et al. Iron accumulation in the substantia nigra in rats visualized by ultrasound. Ultrasound Med Biol. 1999;25 (6):901–4.
38. Berg D, Roggendorf W, Schroder U, Klein R, Tatschner T, Benz P, et al. Echogenicity of the substantia nigra: association with increased iron content and marker for susceptibility to nigrostriatal injury. Arch Neurol. 2002;59(6):999–1005.
39. Zecca L, Berg D, Arzberger T, Ruprecht P, Rausch WD, Musicco M, et al. In vivo detection of iron and neuromelanin by transcranial sonography: a new approach for early detection of substantia nigra damage. Mov Disord: Off J Mov Disord Soc. 2005;20(10):1278–85. doi:10. 1002/mds.20550.
40. Berg D, Godau J, Riederer P, Gerlach M, Arzberger T. Microglia activation is related to substantia nigra echogenicity. J Neural Transm. 2010;117(11):1287–92. doi:10.1007/s00702-010-0504-6.
41. Kajimoto Y, Miwa H, Okawa-Izawa M, Hironishi M, Kondo T. Transcranial sonography of the substantia nigra and MIBG myocardial scintigraphy: complementary role in the diagnosis of Parkinson's disease. Parkinsonism Relat Disord. 2009;15(4):270–2. doi:10.1016/j. parkreldis.2008.06.004.
42. Behnke S, Hellwig D, Burmann J, Runkel A, Farmakis G, Kirsch CM, et al. Evaluation of transcranial sonographic findings and MIBG cardiac scintigraphy in the diagnosis of idiopathic Parkinson's disease. Parkinsonism Relat Disord. 2013;19(11):995–9. doi:10.1016/j. parkreldis.2013.06.019.
43. Shindo K, Kaneko E, Watanabe H, Sugimoto T, Ohta E, Ohashi K, et al. Analysis of the relationship between muscle sympathetic nerve activity and cardiac 123I-metaiodobenzylguanidine uptake in patients with Parkinson's disease. Mov Disord: Off J Mov Disord Soc. 2005;20(11):1419–24. doi:10.1002/mds.20580.

44. Hori N, Takamori M, Hirayama M, Watanabe H, Nakamura T, Yamashita F, et al. Pupillary supersensitivity and visual disturbance in Parkinson's disease. Clin Auton Res: Off J Clin Auton Res Soc. 2008;18(1):20–7. doi:10.1007/s10286-008-0453-4.
45. Yamashita F, Hirayama M, Nakamura T, Takamori M, Hori N, Uchida K, et al. Pupillary autonomic dysfunction in multiple system atrophy and Parkinson's disease: an assessment by eye-drop tests. Clin Auton Res: Off J Clin Auton Res Soc. 2010;20(3):191–7. doi:10.1007/s10286-009-0051-0.
46. Kollensperger M, Seppi K, Liener C, Boesch S, Heute D, Mair KJ, et al. Diffusion weighted imaging best discriminates PD from MSA-P: a comparison with tilt table testing and heart MIBG scintigraphy. Mov Disord: Off J Mov Disord Soc. 2007;22(12):1771–6. doi:10.1002/mds.21614.
47. Braak H, Del Tredici K, Rub U, de Vos RA, Jansen Steur EN, Braak E. Staging of brain pathology related to sporadic Parkinson's disease. Neurobiol Aging. 2003;24(2):197–211.
48. Sakakibara R, Tateno F, Kishi M, Tsuyusaki Y, Terada H, Inaoka T. MIBG myocardial scintigraphy in pre-motor Parkinson's disease: a review. Parkinsonism Relat Disord. 2014;20(3):267–73. doi:10.1016/j.parkreldis.2013.11.001.
49. Adler CH, Connor DJ, Hentz JG, Sabbagh MN, Caviness JN, Shill HA, et al. Incidental Lewy body disease: clinical comparison to a control cohort. Mov Disord: Off J Mov Disord Soc. 2010;25(5):642–6. doi:10.1002/mds.22971.
50. Beach TG, White 3rd CL, Hladik CL, Sabbagh MN, Connor DJ, Shill HA, et al. Olfactory bulb alpha-synucleinopathy has high specificity and sensitivity for Lewy body disorders. Acta Neuropathol. 2009;117(2):169–74. doi:10.1007/s00401-008-0450-7.
51. Orimo S, Uchihara T, Nakamura A, Mori F, Kakita A, Wakabayashi K, et al. Axonal alpha-synuclein aggregates herald centripetal degeneration of cardiac sympathetic nerve in Parkinson's disease. Brain: J Neurol. 2008;131(Pt 3):642–50. doi:10.1093/brain/awm302.
52. Nagayama H, Yamazaki M, Ueda M, Nishiyama Y, Hamamoto M, Katayama Y, et al. Low myocardial MIBG uptake in multiple system atrophy with incidental Lewy body pathology: an autopsy case report. Mov Disord: Off J Mov Disord Soc. 2008;23(7):1055–7. doi:10.1002/mds.22031.
53. Mori K, Iwasaki Y, Ito M, Mimuro M, Yoshida M. Decreased myocardial uptake of meta-iodobenzylguanidine in an autopsy-confirmed case of corticobasal degeneration with Lewy bodies restricted to the sympathetic ganglia. Rinsho shinkeigaku = Clin Neurol. 2012;52(6):405–10.
54. Zarranz JJ, Alegre J, Gomez-Esteban JC, Lezcano E, Ros R, Ampuero I, et al. The new mutation, E46K, of alpha-synuclein causes Parkinson and Lewy body dementia. Ann Neurol. 2004;55(2):164–73. doi:10.1002/ana.10795.
55. Tijero B, Gomez-Esteban JC, Lezcano E, Fernandez-Gonzalez C, Somme J, Llorens V, et al. Cardiac sympathetic denervation in symptomatic and asymptomatic carriers of the E46K mutation in the alpha synuclein gene. Parkinsonism Relat Disord. 2013;19(1):95–100. doi:10.1016/j.parkreldis.2012.08.001.
56. Tijero B, Gomez-Esteban JC, Llorens V, Lezcano E, Gonzalez-Fernandez MC, de Pancorbo MM, et al. Cardiac sympathetic denervation precedes nigrostriatal loss in the E46K mutation of the alpha-synuclein gene (SNCA). Clin Auton Res: Off J Clin Auton Res Soc. 2010;20(4):267–9. doi:10.1007/s10286-010-0068-4.
57. Ibanez P, Bonnet AM, Debarges B, Lohmann E, Tison F, Pollak P, et al. Causal relation between alpha-synuclein gene duplication and familial Parkinson's disease. Lancet. 2004;364(9440):1169–71. doi:10.1016/S0140-6736(04)17104-3.
58. Nishioka K, Hayashi S, Farrer MJ, Singleton AB, Yoshino H, Imai H, et al. Clinical heterogeneity of alpha-synuclein gene duplication in Parkinson's disease. Ann Neurol. 2006;59(2):298–309. doi:10.1002/ana.20753.
59. Orimo S, Uchihara T, Nakamura A, Mori F, Ikeuchi T, Onodera O, et al. Cardiac sympathetic denervation in Parkinson's disease linked to SNCA duplication. Acta Neuropathol. 2008;116(5):575–7. doi:10.1007/s00401-008-0428-5.

60. Singleton AB, Farrer M, Johnson J, Singleton A, Hague S, Kachergus J, et al. alpha-Synuclein locus triplication causes Parkinson's disease. Science. 2003;302(5646):841. doi:10.1126/science.1090278.
61. Singleton A, Gwinn-Hardy K, Sharabi Y, Li ST, Holmes C, Dendi R, et al. Association between cardiac denervation and parkinsonism caused by alpha-synuclein gene triplication. Brain: J Neurol. 2004;127(Pt 4):768–72. doi:10.1093/brain/awh081.
62. Paisan-Ruiz C, Jain S, Evans EW, Gilks WP, Simon J, van der Brug M, et al. Cloning of the gene containing mutations that cause PARK8-linked Parkinson's disease. Neuron. 2004;44 (4):595–600. doi:10.1016/j.neuron.2004.10.023.
63. Zimprich A, Biskup S, Leitner P, Lichtner P, Farrer M, Lincoln S, et al. Mutations in LRRK2 cause autosomal-dominant parkinsonism with pleomorphic pathology. Neuron. 2004;44 (4):601–7. doi:10.1016/j.neuron.2004.11.005.
64. Berg D, Schweitzer KJ, Leitner P, Zimprich A, Lichtner P, Belcredi P, et al. Type and frequency of mutations in the LRRK2 gene in familial and sporadic Parkinson's disease*. Brain: J Neurol. 2005;128(Pt 12):3000–11. doi:10.1093/brain/awh666.
65. Gilks WP, Abou-Sleiman PM, Gandhi S, Jain S, Singleton A, Lees AJ, et al. A common LRRK2 mutation in idiopathic Parkinson's disease. Lancet. 2005;365(9457):415–6. doi:10. 1016/S0140-6736(05)17830-1.
66. Valldeoriola F, Gaig C, Muxi A, Navales I, Paredes P, Lomena F, et al. 123I-MIBG cardiac uptake and smell identification in parkinsonian patients with LRRK2 mutations. J Neurol. 2011;258(6):1126–32. doi:10.1007/s00415-010-5896-6.
67. Quattrone A, Bagnato A, Annesi G, Novellino F, Morgante L, Savettieri G, et al. Myocardial 123metaiodobenzylguanidine uptake in genetic Parkinson's disease. Mov Disord: Off J Mov Disord Soc. 2008;23(1):21–7. doi:10.1002/mds.21701.
68. Tomiyama H, Li Y, Funayama M, Hasegawa K, Yoshino H, Kubo S, et al. Clinicogenetic study of mutations in LRRK2 exon 41 in Parkinson's disease patients from 18 countries. Mov Disord: Off J Mov Disord Soc. 2006;21(8):1102–8. doi:10.1002/mds.20886.
69. Kitada T, Asakawa S, Hattori N, Matsumine H, Yamamura Y, Minoshima S, et al. Mutations in the parkin gene cause autosomal recessive juvenile parkinsonism. Nature. 1998;392 (6676):605–8. doi:10.1038/33416.
70. Suzuki M, Hattori N, Orimo S, Fukumitsu N, Abo M, Kono Y, et al. Preserved myocardial [123I]metaiodobenzylguanidine uptake in autosomal recessive juvenile parkinsonism: first case report. Mov Disord: Off J Mov Disord Soc. 2005;20(5):634–6. doi:10.1002/mds.20384.
71. Orimo S, Amino T, Yokochi M, Kojo T, Uchihara T, Takahashi A, et al. Preserved cardiac sympathetic nerve accounts for normal cardiac uptake of MIBG in PARK2. Mov Disord: Off J Mov Disord Soc. 2005;20(10):1350–3. doi:10.1002/mds.20594.
72. Kim YD, Song IU, Kim JS, Chung SW, Lee KS. Cardiac (123)I-metaiodobenzylguanidine scintigraphy in a patient with familial Parkinsonism with Parkin gene mutation. J Mov Disord. 2010;3(2):42–4. doi:10.14802/jmd.10011.
73. Valente EM, Abou-Sleiman PM, Caputo V, Muqit MM, Harvey K, Gispert S, et al. Hereditary early-onset Parkinson's disease caused by mutations in PINK1. Science. 2004;304(5674):1158–60. doi:10.1126/science.1096284.
74. Bonifati V, Rizzu P, van Baren MJ, Schaap O, Breedveld GJ, Krieger E, et al. Mutations in the DJ-1 gene associated with autosomal recessive early-onset parkinsonism. Science. 2003;299(5604):256–9. doi:10.1126/science.1077209.
75. Bonifati V, Rohe CF, Breedveld GJ, Fabrizio E, De Mari M, Tassorelli C, et al. Early-onset parkinsonism associated with PINK1 mutations: frequency, genotypes, and phenotypes. Neurology. 2005;65(1):87–95. doi:10.1212/01.wnl.0000167546.39375.82.
76. Djarmati A, Hedrich K, Svetel M, Schafer N, Juric V, Vukosavic S, et al. Detection of Parkin (PARK2) and DJ1 (PARK7) mutations in early-onset Parkinson disease: Parkin mutation frequency depends on ethnic origin of patients. Hum Mutat. 2004;23(5):525. doi:10.1002/ humu.9240.

77. Kumazawa R, Tomiyama H, Li Y, Imamichi Y, Funayama M, Yoshino H, et al. Mutation analysis of the PINK1 gene in 391 patients with Parkinson disease. Arch Neurol. 2008;65 (6):802–8. doi:10.1001/archneur.65.6.802.
78. Itokawa K, Tamura N, Kawai N, Shimazu K, Ishii K. Parkinsonism in type I Gaucher's disease. Intern Med. 2006;45(20):1165–7.
79. Lebouvier T, Clairembault T, Devos D, Pallardy A, Coron E, Neunlist M, et al. Peripheral autonomic nervous system involvement in Gaucher-related parkinsonism. J Park Dis. 2014;4 (1):29–32. doi:10.3233/JPD-130333.
80. Li Y, Sekine T, Funayama M, Li L, Yoshino H, Nishioka K, et al. Clinicogenetic study of GBA mutations in patients with familial Parkinson's disease. Neurobiol Aging. 2014;35 (4):935 e3-8. doi:10.1016/j.neurobiolaging.2013.09.019.
81. Takano H, Yoshimura N. Markedly decreased cardiac uptake with 123I-MIBG scintigraphy in a case of pure progressive autonomic failure. Rinsho shinkeigaku = Clin Neurol. 1993;33 (7):784–6.
82. Hirayama M, Hakusui S, Koike Y, Ito K, Kato T, Ikeda M, et al. A scintigraphical qualitative analysis of peripheral vascular sympathetic function with meta-[123I]iodobenzylguanidine in neurological patients with autonomic failure. J Auton Nerv Syst. 1995;53(2–3):230–4.
83. Miyamoto T, Miyamoto M, Suzuki K, Nishibayashi M, Iwanami M, Hirata K. 123I-MIBG cardiac scintigraphy provides clues to the underlying neurodegenerative disorder in idiopathic REM sleep behavior disorder. Sleep. 2008;31(5):717–23.
84. Miyamoto T, Miyamoto M, Iwanami M, Hirata K. Idiopathic REM sleep behavior disorder: implications for the pathogenesis of Lewy body diseases. Park Dis. 2011;2011:941268. doi:10.4061/2011/941268.
85. Koyama S, Tachibana N, Masaoka Y, Homma I, Kawamura M. Decreased myocardial (123) I-MIBG uptake and impaired facial expression recognition in a patient with REM sleep behavior disorder. Mov Disord: Off J Mov Disord Soc. 2007;22(5):746–7. doi:10.1002/mds. 21450.
86. Yoshita M. Differentiation of idiopathic Parkinson's disease from striatonigral degeneration and progressive supranuclear palsy using iodine-123 meta-iodobenzylguanidine myocardial scintigraphy. J Neurol Sci. 1998;155(1):60–7.
87. Nagayama H, Ueda M, Yamazaki M, Nishiyama Y, Hamamoto M, Katayama Y. Abnormal cardiac [(123)I]-meta-iodobenzylguanidine uptake in multiple system atrophy. Mov Dis: Off J Mov Disord Soc. 2010;25(11):1744–7. doi:10.1002/mds.23338.
88. Cohen J, Low P, Fealey R, Sheps S, Jiang NS. Somatic and autonomic function in progressive autonomic failure and multiple system atrophy. Ann Neurol. 1987;22(6):692–9. doi:10.1002/ ana.410220604.
89. Kashihara K, Ohno M, Kawada S, Okumura Y. Reduced cardiac uptake and enhanced washout of 123I-MIBG in pure autonomic failure occurs conjointly with Parkinson's disease and dementia with Lewy bodies. J Nucl Med: Off Publ Soc Nucl Med. 2006;47(7):1099–101.
90. Taki J, Nakajima K, Hwang EH, Matsunari I, Komai K, Yoshita M, et al. Peripheral sympathetic dysfunction in patients with Parkinson's disease without autonomic failure is heart selective and disease specific. taki@med.kanazawa-u.ac.jp. Eur J Nucl Med. 2000;27 (5):566–73.
91. Kim JS, Lee PH, Lee KS, Park JW, Kim YI, Chung YA, et al. Cardiac [123I] metaiodobenzylguanidine scintigraphy for vascular Parkinsonism. Mov Disord: Off J Mov Disord Soc. 2006;21(11):1990–4. doi:10.1002/mds.21112.
92. Lee PH, Kim JW, Bang OY, Joo IS, Yoon SN, Huh K. Cardiac 123I-MIBG scintigraphy in patients with essential tremor. Mov Disord: Off J Mov Disord Soc. 2006;21(8):1235–8. doi:10.1002/mds.20908.
93. Novellino F, Arabia G, Bagnato A, Cascini GL, Salsone M, Nicoletti G, et al. Combined use of DAT-SPECT and cardiac MIBG scintigraphy in mixed tremors. Mov Disord: Off J Mov Disord Soc. 2009;24(15):2242–8. doi:10.1002/mds.22771.

94. Lee PH, Kim JS, Shin DH, Yoon SN, Huh K. Cardiac 123I-MIBG scintigraphy in patients with drug induced parkinsonism. J Neurol Neurosurg Psychiatry. 2006;77(3):372–4. doi:10.1136/jnnp.2005.073999.
95. Lee PH, Yeo SH, Yong SW, Kim YJ. Odour identification test and its relation to cardiac 123I-metaiodobenzylguanidine in patients with drug induced parkinsonism. J Neurol Neurosurg Psychiatry. 2007;78(11):1250–2. doi:10.1136/jnnp.2007.121285.
96. Kim JS, Oh YS, Kim YI, Yang DW, Chung YA, You Ie R, et al. Combined use of (1)(2)(3)I-metaiodobenzylguanidine (MIBG) scintigraphy and dopamine transporter (DAT) positron emission tomography (PET) predicts prognosis in drug-induced Parkinsonism (DIP): a 2-year follow-up study. Arch Gerontol Geriatr. 2013;56(1):124–8. doi:10.1016/j.archger.2012.05.001.
97. Koyano S, Baba Y, Kuroiwa Y. Spinocerebellar ataxia type 2 with parkinsonism. JMDD. 2007;17:43–6.
98. Yomono HS, Kurisaki H, Hebisawa A, Sakiyama Y, Saito Y, Murayama S. Autopsy case of SCA2 with Parkinsonian phenotype. Rinsho Shinkeigaku = Clin Neurol. 2010;50(3):156–62.
99. De Rosa A, Pappata S, Pellegrino T, De Leva MF, Maddaluno G, Fiumara G, et al. Reduced cardiac 123I-metaiodobenzylguanidine uptake in patients with spinocerebellar ataxia type 2: a comparative study with Parkinson's disease. Eur J Nucl Med Mol Imaging. 2013;40(12):1914–21. doi:10.1007/s00259-013-2524-6.
100. Takao M, Aoyama M, Ishikawa K, Sakiyama Y, Yomono H, Saito Y, et al. Spinocerebellar ataxia type 2 is associated with Parkinsonism and Lewy body pathology. BMJ Case Rep. 2011;2011. doi:10.1136/bcr.01.2011.3685.
101. Kazuta T, Hayashi M, Shimizu T, Iwasaki A, Nakamura S, Hirai S. Autonomic dysfunction in Machado-Joseph disease assessed by iodine123-labeled metaiodobenzylguanidine myocardial scintigraphy. Clin Auton Res: Off J Clin Auton Res Soc. 2000;10(3):111–5.
102. Kline RC, Swanson DP, Wieland DM, Thrall JH, Gross MD, Pitt B, et al. Myocardial imaging in man with I-123 meta-iodobenzylguanidine. J Nucl Med: Off Publ Soc Nucl Med. 1981;22(2):129–32.
103. Schofer J, Spielmann R, Schuchert A, Weber K, Schluter M. Iodine-123 meta-iodobenzylguanidine scintigraphy: a noninvasive method to demonstrate myocardial adrenergic nervous system disintegrity in patients with idiopathic dilated cardiomyopathy. J Am Coll Cardiol. 1988;12(5):1252–8.
104. Henderson EB, Kahn JK, Corbett JR, Jansen DE, Pippin JJ, Kulkarni P, et al. Abnormal I-123 metaiodobenzylguanidine myocardial washout and distribution may reflect myocardial adrenergic derangement in patients with congestive cardiomyopathy. Circulation. 1988;78(5 Pt 1):1192–9.
105. Chidsey CA, Kaiser GA, Sonnenblick EH, Spann JF, Braunwald E. Cardiac norepinephrine stores in experimental heart failure in the dog. J Clin Invest. 1964;43:2386–93. doi:10.1172/JCI105113.
106. Merlet P, Valette H, Dubois-Rande JL, Moyse D, Duboc D, Dove P, et al. Prognostic value of cardiac metaiodobenzylguanidine imaging in patients with heart failure. J Nucl Med: Off Publ Soc Nucl Med. 1992;33(4):471–7.
107. Kimura K, Ieda M, Kanazawa H, Yagi T, Tsunoda M, Ninomiya S, et al. Cardiac sympathetic rejuvenation: a link between nerve function and cardiac hypertrophy. Circ Res. 2007;100(12):1755–64. doi:10.1161/01.RES.0000269828.62250.ab.
108. Kanazawa H, Ieda M, Kimura K, Arai T, Kawaguchi-Manabe H, Matsuhashi T, et al. Heart failure causes cholinergic transdifferentiation of cardiac sympathetic nerves via gp130-signaling cytokines in rodents. J Clin Invest. 2010;120(2):408–21. doi:10.1172/JCI39778.
109. Bengel FM, Barthel P, Matsunari I, Schmidt G, Schwaiger M. Kinetics of 123I-MIBG after acute myocardial infarction and reperfusion therapy. J Nucl Med: Off Publ Soc Nucl Med. 1999;40(6):904–10.

110. Podio V, Spinnler MT, Spandonari T, Moretti C, Castellano G, Bessone M, et al. Regional sympathetic denervation after myocardial infarction: a follow-up study using [123I]MIBG. Q J Nucl Med: Off Publ Ital Assoc Nucl Med. 1995;39(4 Suppl 1):40–3.
111. Kahn JK, Sisson JC, Vinik AI. Prediction of sudden cardiac death in diabetic autonomic neuropathy. J Nucl Med: Off Publ Soc Nucl Med. 1988;29(9):1605–6.
112. Link JM, Caldwell JH. Diagnostic and prognostic imaging of the cardiac sympathetic nervous system. Nat Clin Pract Cardiovasc Med. 2008;5 Suppl 2:S79–86. doi:10.1038/ncpcardio1150.
113. Nagamachi S, Fujita S, Nishii R, Futami S, Tamura S, Mizuta M, et al. Prognostic value of cardiac I-123 metaiodobenzylguanidine imaging in patients with non-insulin-dependent diabetes mellitus. J Nucl Cardiol: Off Publ Am Soc Nucl Cardiol. 2006;13(1):34–42. doi:10.1016/j.nuclcard.2005.11.009.
114. Goldstein DS. Sympathetic neuroimaging. Handbook of clinical neurology. Amsterdam: Elsevier; 2013.
115. Mantysaari M, Kuikka J, Mustonen J, Tahvanainen K, Vanninen E, Lansimies E, et al. Noninvasive detection of cardiac sympathetic nervous dysfunction in diabetic patients using [123I]metaiodobenzylguanidine. Diabetes. 1992;41(9):1069–75.
116. Yufu K, Takahashi N, Okada N, Shinohara T, Nakagawa M, Hara M, et al. Cardiac iodine-123 metaiodobenzylguanidine (123I-MIBG) scintigraphy parameter predicts cardiac and cerebrovascular events in type 2 diabetic patients without structural heart disease. Circ J. 2012;76(2):399–404.
117. Togane Y. Evaluation of the cardiac autonomic nervous system in spontaneously non-insulin-dependent diabetic rats by 123I-metaiodobenzylguanidine imaging. Ann Nucl Med. 1999;13 (1):19–26.
118. Paolillo S, Rengo G, Pagano G, Pellegrino T, Savarese G, Femminella GD, et al. Impact of diabetes on cardiac sympathetic innervation in patients with heart failure: a 123I meta-iodobenzylguanidine (123I MIBG) scintigraphic study. Diabetes Care. 2013;36 (8):2395–401. doi:10.2337/dc12-2147.
119. Nagamachi S, Jinnouchi S, Kurose T, Nishii R, Kawai K, Futami S, et al. Serial change in 123I-MIBG myocardial scintigraphy in non-insulin-dependent diabetes mellitus. Ann Nucl Med. 2002;16(1):33–8.
120. Slaets S, Van Acker F, Versijpt J, Hauth L, Goeman J, Martin JJ, et al. Diagnostic value of MIBG cardiac scintigraphy for differential dementia diagnosis. Int J Geriatr Psychiatry. 2015;30(8):864–9. doi:10.1002/gps.4229.
121. Otsuka J, Sakamoto M, Azuma L, Takahashi M, Inaba A, Mizuno Y, et al. A study of an effect of diabetes mellitus on cardiac MIBG uptake – investigation in the differential diagnosis of dementia. Clin Neurol. 2013;53(12):1551.

Chapter 16
Findings of MIBG Cardiac Scintigraphy: Dementia with Lewy Bodies and Related Dementia

Hirohisa Watanabe and Gen Sobue

Abstract Dementia with Lewy bodies (DLB) is the second most common neurodegenerative form of dementia. The core features of DLB include progressive cognitive decline, parkinsonism, autonomic failure, and sleep disorders. Because the symptoms are similar between DLB and Alzheimer's disease (AD), it is often difficult to make a definitive diagnosis, in particular, during the early course of the illness. Cardiac ^{123}I-metaiodobenzylguanidine (MIBG) scintigraphy is one of the most useful tools for differentiating DLB from AD. DLB patients generally show markedly reduced cardiac MIBG uptake, regardless of the presence of parkinsonism. This finding is consistent with pathological findings of cardiac sympathetic nerve involvement in DLB. However, patients with AD, corticobasal degeneration (CBD), progressive supranuclear palsy (PSP), and vascular dementia (VaD) generally show normal cardiac MIBG uptake. Therefore, MIBG cardiac scintigraphy provides a powerful method for early differentiation of DLB from other forms of dementia, with equal or superior diagnostic capabilities when compared to dopamine transporter imaging or perfusion/metabolic imaging. Recent studies have demonstrated that most subjects with rapid eye movement (REM) sleep behavior disorder (RBD) progressing to DLB or Parkinson's disease (PD) during long follow-up periods also have abnormal cardiac MIBG uptake. Thus, MIBG cardiac scintigraphy may have the potential to detect prodromal DLB before obvious cognitive and functional changes are observed. This procedure could more accurately identify early dementia and could enhance the development and use of disease-modifying therapies for DLB.

Keywords Dementia with Lewy bodies • Alzheimer's disease • Diagnosis • Prodromal

H. Watanabe (✉)
Brain and Mind Research Center, Nagoya University Graduate School of Medicine, Nagoya University, 65 Tsurumai-cho, Showa-ku, Nagoya, Aichi 466-8550, Japan
e-mail: nabe@med.nagoya-u.ac.jp

G. Sobue
Nagoya University Graduate School of Medicine, 65 Tsurumai-cho, Showa-ku, Nagoya, Aichi 466-8550, Japan

S. Iwase et al. (eds.), *Clinical Assessment of the Autonomic Nervous System*,
DOI 10.1007/978-4-431-56012-8_16

16.1 Introduction

Dementia with Lewy bodies (DLB) is the second most common neurodegenerative dementia encompassing progressive cognitive decline, "fluctuations" in alertness and attention, visual hallucinations, parkinsonian motor symptoms, including rigidity, slowness of movement, difficulty walking, or falling, autonomic failure such as orthostatic hypotension and urinary disturbance, and sleep problems. DLB patients show widespread and abundant Lewy bodies mainly composed of α-synuclein in the brain, brain stem, spinal cord, and peripheral autonomic systems. Thus, DLB can be classified as a synucleinopathies similar to Parkinson's disease (PD) and multiple system atrophy (MSA).

Visual hallucination and cognitive fluctuation are characteristic psychological features of DLB compared to Alzheimer's disease (AD). However, these symptoms are not observed in all DLB patients particularly during the early course of the disease. Furthermore, some DLB patients present only with memory loss similar to AD. Mild parkinsonism is also observed not only in AD patients but also in healthy older subjects. Thus, without supportive imaging findings, DLB patients are often misdiagnosed as AD.

In clinical practice, differentiation of DLB from AD is very important because (1) progression and prognosis are different between DLB and AD, (2) DLB patients can show better responses to cholinesterase inhibitors than AD patients, and (3) supersensitivity to neuroleptic treatment is common in DLB patients. Moreover, diagnosis during the prodromal stage, using specific biomarkers to provide clues about the underlying pathological process is becoming increasingly important for the development of disease-modifying therapies.

Cardiac ^{123}I-metaiodobenzylguanidine (MIBG) scintigraphy can assess the pathophysiology of postganglionic presynaptic cardiac sympathetic endings. Cardiac scintigraphy shows significantly reduced cardiac MIBG uptake in PD but not in multiple system atrophy (MSA), progressive supranuclear palsy (PSP), corticobasal degeneration (CBD), or vascular parkinsonism. It has been widely used for differentiating PD from other parkinsonian syndromes.

As for dementia, several studies independently identified reduced cardiac MIBG uptake in DLB but not in AD in 2001 [1, 2]. These studies have demonstrated that ^{123}I-MIBG cardiac scintigraphy is a useful tool for differentiating DLB from AD. Additional studies have also verified the findings and extended these findings [3–12].

In this chapter, we would like to discuss the cardiac MIBG uptake in dementia focusing on DLB.

16.2 Cardiac MIBG Findings in DLB

16.2.1 MIBG in DLB: Results from a Single-Center Study

Figure 16.1 demonstrated the cardiac MIBG findings in patients with DLB and AD. Most of DLB patients showed markedly reduced cardiac MIBG uptake compared to age-matched healthy controls irrespective of the presence of parkinsonism [8]. Physiologically, cardiac MIBG uptake is significantly related to the blood pressure overshoot in phase IV of the Valsalva maneuver suggesting a relationship between cardiac MIBG uptake and cardiac sympathetic dysfunction in DLB [13]. Orimo and colleagues clearly and pathologically demonstrated the almost entirely disappearance of tyrosine hydroxylase-immunoreactive nerve fibers in the hearts of DLB patients [14]. These pathophysiological studies distinctly support the view that reduced cardiac MIBG uptake in DLB would reflect the cardiac sympathetic denervation (as discussed in Chap. 19).

On the contrary, most of AD patients show normal cardiac MIBG uptake, consistent with normal autonomic function test result and lack of pathological cardiac sympathetic nerve involvements. Patients with frontotemporal dementia (FTD), CBD, PSP, and vascular dementia (VaD) also show normal cardiac MIBG uptake. Thus, it has been reported that cardiac MIBG scintigraphy can be a useful tool for differentiating DLB from AD, FTD, CBD, PSP, and VaD [5, 6, 12].

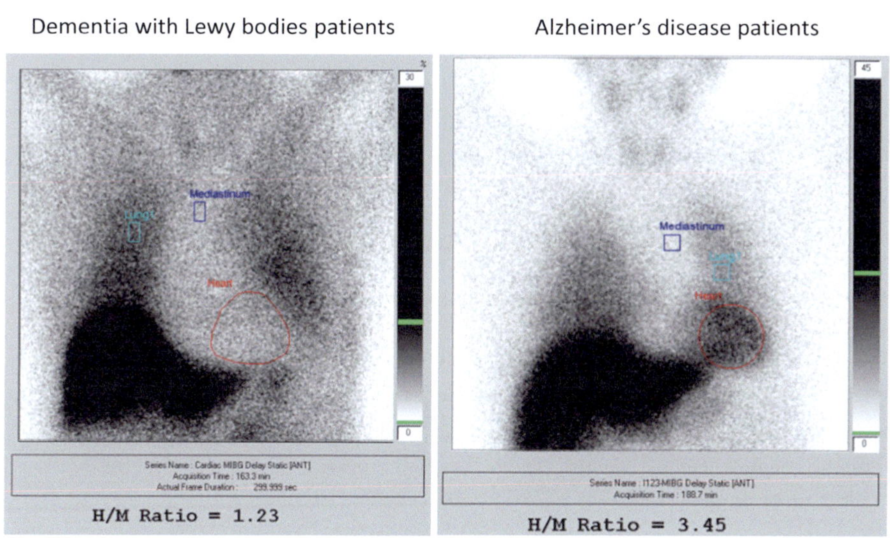

Fig. 16.1 Representative cardiac ^{123}I-metaiodobenzylguanidine (MIBG) scintigraphy findings in patients with dementia with Lewy bodies (DLB) and Alzheimer's disease (AD). Delayed cardiac MIBG scintigraphy showed significant decreased accumulation of MIBG in DLB patient but not in AD

Slaets et al. investigated the potential diagnostic value of MIBG cardiac scintigraphy in 20 patients to determine whether the diagnosis of DLB or AD was ambiguous at baseline [15]. The authors classified the patients into two groups with a cutoff of heart-to-mediastinum (H/M) ratio at 1.68 and followed them for more than 6 months ($n = 19$) or confirmed the classification with an autopsy ($n = 1$). The mean follow-up time was 17 ± 14 months (range: 6–57 months). Ninety-five percent (19/20) of the patients were correctly classified as compared with clinical or definite diagnosis at follow-up, with sensitivity and specificity values for diagnosing DLB of 100 % (16/16) and 75 % (3/4), respectively. This study suggested that MIBG cardiac scintigraphy could be valuable for earlier differential diagnosis of DLB and AD.

Although several promising results have been published that suggest the usefulness of MIBG cardiac scintigraphy for differentiation of DLB and AD, most of these studies were conducted under very strict exclusion criteria because abnormal cardiac MIBG uptake has been reported in patients with diabetes mellitus, thyroid diseases, heart failure, and ischemic heart disease. In clinical practices, older dementia patients could present with these highly confounding factors. Thus, if only applied to selected patients without any disease conditions, MIBG would not be an effective tool for estimation of dementia. Interestingly, Sayagues and colleagues demonstrated that prior cardiac or vascular involvement did not influence the results of MIBG cardiac scintigraphy in a cohort of patients with reduced MIBG cardiac uptake and DLB diagnosis [16]. More recently, Slaets et al. described no significant difference of H/M ratio among DLB patients with diabetes mellitus, arterial hypertension, hyperlipidemia, ischemic heart disease, heart failure, or pharmacological treatments and in those without in clinically observable conditions [15]. Further prospective cohort studies are needed to validate the application of MIBG for differentiation of DLB and AD even including not only dementia patients without heart and vascular conditions but also those patients without these conditions.

16.2.1.1 MIBG in DLB: Results from a Meta-Analysis

Treglia et al. performed a comprehensive computer literature search in PubMed/MEDLINE and in the Embase databases of studies published through May 2010 regarding MIBG cardiac scintigraphy performed for differential diagnosis between DLB and other dementias [17]. According to eight studies comprising a total of 346 patients with dementia (152 patients with DLB and 194 patients with other dementias), the pooled sensitivity of MIBG cardiac scintigraphy for detection of DLB was 98 % [95 % confidence interval (CI), 94–100 %] and the pooled specificity of MIBG cardiac scintigraphy for differential diagnosis between DLB and other dementias was 94 % (95 % CI, 90–97 %). The area under the receiver operating characteristic (ROC) curve was 99. The authors concluded that MIBG cardiac scintigraphy was an accurate test for differential diagnosis between DLB and other dementias.

16.2.1.2 MIBG in DLB: Results from Multicenter Study

According to a multicenter study achieved at ten Japanese sites, the heart-to-mediastinum ratio calculated with the automated system showed that the sensitivity was 68.9 % and the specificity was 89.1 % for differentiating probable DLB from probable AD using both early and delayed images [18, 19]. By visual assessment, the sensitivity and specificity were similar. Even in patients with mild dementia [mini-mental state examination (MMSE) $\geq$22], the delayed heart-to-mediastinum ratio showed has a sensitivity of 77.4 % and a specificity of 93.8 %. The reasons for the differences in the sensitivity and specificity between mild and moderate to severe dementia groups are unknown. They showed that AD patients with extrapyramidal signs and hallucinations in the DLB group and patients with comorbidity of DLB and AD pathology in the AD group may have affected the accuracy of diagnosis in moderate to severe dementia cases. Moreover, comparing to the results of previous reports, both sensitivity and specificity were relatively low, even in mild dementia cases. This result indicates that patterns of disease evolution are more variable than previously suggested in DLB.

16.3 MIBG in DLB: Comparison with FP-CIT

Dopamine transporter imaging using single-photon emission computed tomography (SPECT) or positron emission tomography (PET) imaging has enabled the diagnosis of DLB and is included as suggestive features in the Consensus Criteria. According to a recent systematic review of clinical trials using ^{123}I-fluoropropyl (FP) carbomethoxy-3 beta-(4-iodophenyltropane) (CIT), ^{123}I-FP-CIT, the sensitivity was 78.5 % and specificity was 90.1 % for differentiating probable DLB from probable AD [19]. However, the sensitivity decreased to 38.2 % in possible DLB [20]. Figure 16.2 demonstrated the ^{123}I-FP-CIT findings in patients with DLB and AD.

More recently, the usefulness of ^{123}I-FP-CIT for differentiating possible DLB from AD was reported. In this study, with 49 of 114 (43 %) individuals had an abnormal scans [21]. When comparing the diagnostic accuracy of both ^{123}I-MIBG cardiac scintigraphy and ^{123}I-FP-CIT SPECT in 20 patients with DLB and 11 with other dementias (6 AD, 3 FTD, and 1 VaD), the sensitivity (90 %), specificity (91 %), positive predictive value (95 %), and negative predictive value (83 %) were not exactly the same but were close in values [22].

Camacho et al. reported that 23 of 28 DLB patients who had reduced cardiac MIBG uptake also showed reduced ^{123}I-FP-CIT binding in basal ganglia with a positive correlation between the H/M ratio and the specific binding ratio of the striatum ($P < 0.01$) [23].

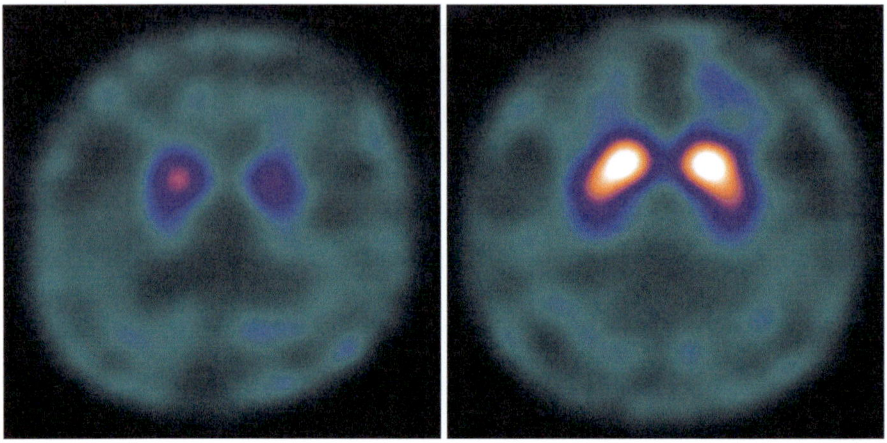

Fig. 16.2 Representative ^{123}I-fluoropropyl (FP) carbomethoxy-3 beta-(4-iodophenyltropane) (CIT) findings in patients with dementia with Lewy bodies (DLB) and Alzheimer's disease (AD). FP-CIT SPECT showed significant decreased accumulation of the putamen in patients DLB patient but not in AD

16.4 MIBG in DLB: Comparison with Brain Perfusion Imaging

Both occipital hypoperfusion and reduced cardiac MIBG uptake are supportive features in the diagnostic criteria for DLB. Comparisons between technetium-99 methyl cysteinate (99mTc-ethyl cysteinate) dimer brain perfusion SPECT and MIBG in 25 DLB patients showed that occipital hypoperfusion was observed in 68 %, and reduced cardiac MIBG uptake was observed in 96 % of the patients' ethylcysteinate dimer brain perfusion SPECT and MIBG in 25 DLB patients; occipital hypoperfusion was observed in 68 % and reduced cardiac MIBG uptake was in 96 % [24].

Inui et al. investigated N-isopropyl-p-^{123}I-iodoamphetamine (^{123}I-IMP) brain perfusion SPECT and MIBG cardiac scintigraphy findings in 12 patients with probable DLB and 9 patients with possible DLB [9]. This study showed that 83.3 % of patients with probable DLB and 11.1 % with possible DLB had severe reductions in the bilateral occipital lobe in ^{123}I-IMP and also showed low ^{123}I-MIBG uptake. In 1 of 12 patients with probable DLB and 2 of 9 patients with possible DLB, no occipital hypoperfusion was observed but reduced cardiac MIBG uptake was noted. In this same study, 8.3 % with probable DLB and 66.7 % with possible DLB did not show any occipital hypoperfusion and reduced cardiac MIBG uptake.

MIBG cardiac scintigraphy is superior to brain perfusion imaging for differentiating DLB from AD. In particular, cardiac MIBG may provide a powerful differential diagnostic tool when it is difficult to distinguish cases of DLB from AD using brain perfusion SPECT [4, 25].

Oda et al. compared the reliability of brain perfusion SPECT and MIBG in predicting the conversion of possible DLB to probable DLB [26]. In this study, 33 of 94 patients met the criteria for probable DLB after 1 year of follow-up. The areas under the ROC curves for SPECT for predicting the conversion to probable DLB from possible DLB based on the occipital/cerebellum and occipital/striatum cortex ratios of blood flow counts were 0.591 and 0.585, respectively. In contrast, the areas under the ROC curves for ^{123}I-MIBG based on the early H/M ratio, delayed H/M ratios, and washout rate were 0.935, 0.936, and 0.884, respectively. These results indicate that ^{123}I-MIBG cardiac scintigraphy is a good predictor of the future conversion of possible DLB to probable DLB.

16.5 Prodromal DLB

The term "prodrome" is derived from the Greek word prodromos, meaning the forerunner of an event [27]. In clinical medicine, a prodrome refers to the early symptoms and signs of an illness that precede the characteristic manifestations of the fully developed illness [28].

It is becoming increasingly apparent that a prodromal stage exists prior to the time when DLB can be diagnosed by current diagnostic criteria. The identification of prodromal DLB is therefore clinically important, because this can promote understanding of the initiation and progression mechanisms during the pathological changes to DLB. Knowledge of these mechanisms may provide further information as to the etiology and pathophysiology of DLB, including the mechanisms of cell vulnerability and pathological selectivity. Furthermore, this information will aid in selecting earlier and more appropriate supplementary treatments, may provide for better rehabilitation and patient education, and will provide additional time for the patient to plan for inheritance and other estate issues. Lastly, prodromal DLB patients are an important group for future clinical trials to develop disease-modifying therapies. The initiation of a disease-modifying therapy in the prodromal stage may be associated with better improvement of the progression and prognosis for patients with DLB.

Figure 16.3 showed a patient with prodromal DLB patients. He complained visual hallucination, but parkinsonism and cognitive decline were not observed. Brain SPECT imaging showed occipital hypoperfusion. Accumulation of cardiac MIBG scintigraphy was significantly decreased. However, FP-CIT finding was normal. After 2 years, he showed cognitive decline and fulfilled diagnostic criteria of probable DLB.

Fujishiro et al. retrospectively investigated the clinical courses of 90 patients with probable DLB, including olfactory dysfunction, dysautonomia, depression, and rapid eye movement sleep behavioral disorder (RBD), and reported that these DLB-related symptoms were observed in 87.8 % of the patients, preceding the onset

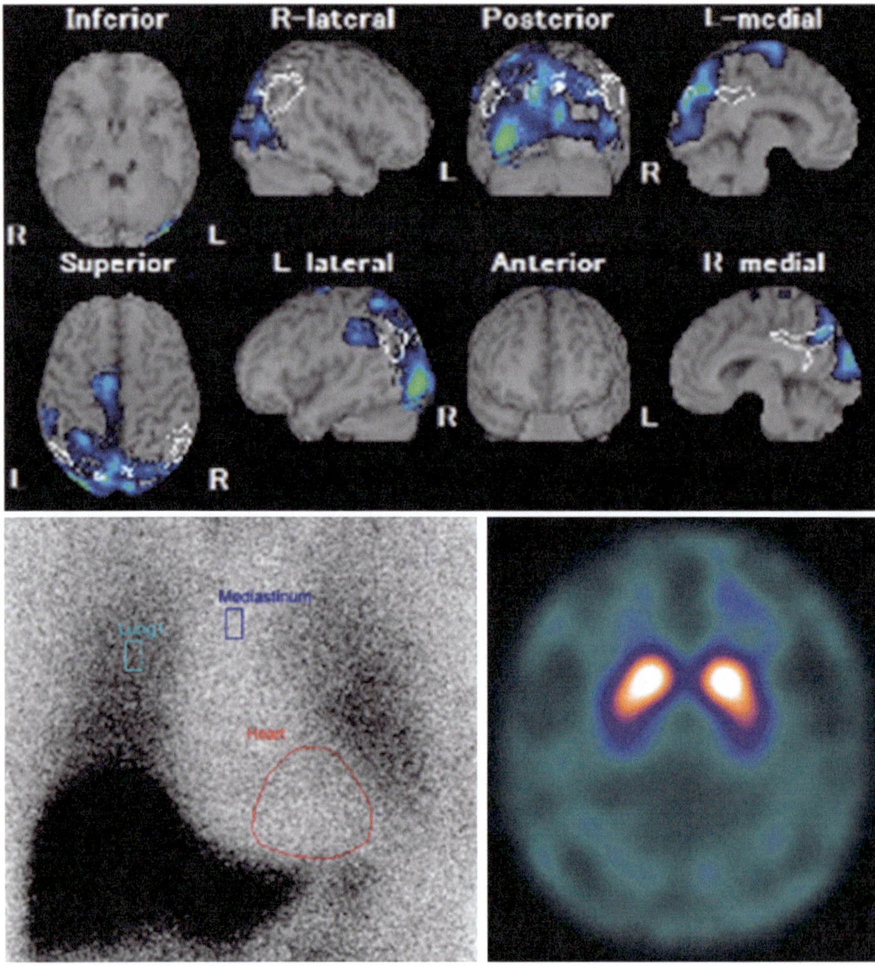

Fig. 16.3 Figure 16.3 showed 80 patients with prodromal DLB patient. He complained visual hallucination, but parkinsonism and cognitive decline were not observed. Brain SPECT imaging showed occipital hypoperfusion. Accumulation of cardiac MIBG scintigraphy was significantly decreased. However, FP-CIT finding was normal. After 2 years, he showed cognitive decline and fulfilled diagnostic criteria of probable DLB

of memory loss by between 1.2 and 9.3 years [29]. Particularly, recent studies demonstrated that RBD is frequently the first indication of forthcoming α-synuclein-associated disorders, including DLB, PD, or MSA. Approximately 82 % of RBD patients develop a neurodegenerative disorder and 32 % convert to DLB during long follow-up period [30].

As for the initiation of parkinsonism in RBD patients, it was estimated that a prodromal interval was ~4.5 years on the Unified Parkinson's Disease Rating Scale [31]. The estimated onset of autonomic dysfunction ranged from 11 to 20 years;

systolic blood pressure drop (prior to 20.4 years) and constipation (15.3 years) had the earliest estimates. Systolic blood pressure drop, erectile dysfunction, and constipation have a potential to discern disease up to 5 years before diagnosis of α-synuclein-associated disorders with sensitivity ranging from 50 % to 90 % [32]. Interestingly, pathological findings of RBD patient who did not present parkinsonism when still alive showed widespread peripheral autonomic nervous system pathology including cardiac plexus [33].

16.5.1 MIBG in REM Sleep Behavior Disorder

As previously discussed, RBD is one of the most important prodromal syndrome symptoms of α-synuclein-associated disorders and may show early cardiac sympathetic nerve involvement. In 2006, Miyamoto et al. reported the cardiac MIBG findings in 34 RBD patients. This study showed that markedly reduced cardiac MIBG uptake was observed in all RBD patients. The degrees of H/M ratio reduction in RBD were consistent with those in PD [34]. The study also reported that reduced cardiac MIBG uptake, which is independent of parkinsonism, may be more closely associated with idiopathic RBD than olfactory impairment [35].

According to follow-up PET studies in cases of idiopathic RBD, compared with MIBG cardiac scintigraphy and dopamine transporter PET images using ^{11}C-carbomethoxy fluorophenyl tropane, reduced cardiac MIBG uptake might precede reduced presynaptic dopaminergic function [36]. In PD, there were no statistical relationships between early and delayed cardiac MIBG uptake and ^{123}I FP-CIT striatal uptake in contralateral caudate and in contralateral putamen to the side mainly affected [37]. These results indicate that the cardiac sympathetic system and nigrostriatal system are differentially affected in RBD. Further prospective MIBG and FP-CIT studies will be needed to elucidate whether the cardiac sympathetic system is affected earlier than the nigrostriatal system in RBD patients.

16.5.1.1 MIBG and Brain Perfusion/Metabolic Imaging in Prodromal DLB

Sakakibara et al. reported cases of amnestic mild cognitive impairment (MCI) without the core clinical features of DLB with reduced cardiac MIBG uptake [38]. During a 3-year period at a university clinic, 44 patients fulfilled the criteria of amnestic MCI and 13 of them (30 %) showed reduced cardiac MIBG uptake. These patients had low frontal function, mild visual hallucination (5/13), nocturia (7/13), constipation (2/13), postural hypotension (1/13), and RBD (3/13). Unfortunately, there were no prospective follow-up data in this study, but the presence of low frontal dysfunction, mild visual hallucination, and autonomic and sleep disorders supported the idea that these patients may be classified as prodromal DLB and eventually develop DLB during the course of their illness. Notably, 5 of 12 (42 %)

patients showed occipital hypoperfusion by SPECT. Remarkable occipital hypoperfusion or hypometabolism, especially in the primary visual cortex, can support a diagnosis of DLB and differentiation of DLB from AD [39]. More recently, highly reproducible disease-related metabolic brain network data assessed by fluorodeoxyglucose positron emission tomography (FDG-PET) or perfusion SPECT, known as the PD-related covariance pattern in subjects with RBD, are associated with a greater likelihood of subsequent phenotypic conversion to a progressive alpha-synuclein associated disorders [40]. Prospective studies in combination with MIBG cardiac scintigraphy and brain perfusion/metabolic imaging in RBD and MCI subjects are therefore needed in the future.

References

1. Watanabe H, Ieda T, Katayama T, Takeda A, Aiba I, Doyu M, et al. Cardiac (123)I-metaiodobenzylguanidine (MIBG) uptake in dementia with Lewy bodies: comparison with Alzheimer's disease. J Neurol Neurosurg Psychiatry. 2001;70(6):781–3.
2. Yoshita M, Taki J, Yamada M. A clinical role for [(123)I]MIBG myocardial scintigraphy in the distinction between dementia of the Alzheimer's-type and dementia with Lewy bodies. J Neurol Neurosurg Psychiatry. 2001;71(5):583–8.
3. Oide T, Tokuda T, Momose M, Oguchi K, Nakamura A, Ohara S, et al. Usefulness of [123I] metaiodobenzylguanidine ([123I]MIBG) myocardial scintigraphy in differentiating between Alzheimer's disease and dementia with Lewy bodies. Intern Med. 2003;42(8):686–90.
4. Hanyu H, Shimizu S, Hirao K, Kanetaka H, Iwamoto T, Chikamori T, et al. Comparative value of brain perfusion SPECT and [(123)I]MIBG myocardial scintigraphy in distinguishing between dementia with Lewy bodies and Alzheimer's disease. Eur J Nucl Med Mol Imaging. 2006;33(3):248–53.
5. Hanyu H, Shimizu S, Hirao K, Sakurai H, Iwamoto T, Chikamori T, et al. The role of 123I-metaiodobenzylguanidine myocardial scintigraphy in the diagnosis of Lewy body disease in patients with dementia in a memory clinic. Dement Geriatr Cogn Disord. 2006;22(5–6):379–84.
6. Kashihara K, Ohno M, Kawada S, Okumura Y. Reduced cardiac uptake and enhanced washout of 123I-MIBG in pure autonomic failure occurs conjointly with Parkinson's disease and dementia with Lewy bodies. J Nucl Med. 2006;47(7):1099–101.
7. Suzuki M, Kurita A, Hashimoto M, Fukumitsu N, Abo M, Ito Y, et al. Impaired myocardial 123I-metaiodobenzylguanidine uptake in Lewy body disease: comparison between dementia with Lewy bodies and Parkinson's disease. J Neurol Sci. 2006;240(1–2):15–9.
8. Yoshita M, Taki J, Yokoyama K, Noguchi-Shinohara M, Matsumoto Y, Nakajima K, et al. Value of 123I-MIBG radioactivity in the differential diagnosis of DLB from AD. Neurology. 2006;66(12):1850–4.
9. Inui Y, Toyama H, Manabe Y, Sato T, Sarai M, Kosaka K, et al. Evaluation of probable or possible dementia with lewy bodies using 123I-IMP brain perfusion SPECT, 123I-MIBG, and 99mTc-MIBI myocardial SPECT. J Nucl Med. 2007;48(10):1641–50.
10. Wada-Isoe K, Kitayama M, Nakaso K, Nakashima K. Diagnostic markers for diagnosing dementia with Lewy bodies: CSF and MIBG cardiac scintigraphy study. J Neurol Sci. 2007;260(1–2):33–7.
11. Estorch M, Camacho V, Paredes P, Rivera E, Rodriguez-Revuelto A, Flotats A, et al. Cardiac (123)I-metaiodobenzylguanidine imaging allows early identification of dementia with Lewy bodies during life. Eur J Nucl Med Mol Imaging. 2008;35(9):1636–41.

12. Novellino F, Bagnato A, Salsone M, Cascini GL, Nicoletti G, Arabia G, et al. Myocardial (123)I-MIBG scintigraphy for differentiation of Lewy bodies disease from FTD. Neurobiol Aging. 2010;31(11):1903–11.
13. Oka H, Yoshioka M, Morita M, Onouchi K, Suzuki M, Ito Y, et al. Reduced cardiac 123I-MIBG uptake reflects cardiac sympathetic dysfunction in Lewy body disease. Neurology. 2007;69(14):1460–5.
14. Orimo S, Amino T, Itoh Y, Takahashi A, Kojo T, Uchihara T, et al. Cardiac sympathetic denervation precedes neuronal loss in the sympathetic ganglia in Lewy body disease. Acta Neuropathol. 2005;109(6):583–8.
15. Slaets S, Van Acker F, Versijpt J, Hauth L, Goeman J, Martin JJ, et al. Diagnostic value of MIBG cardiac scintigraphy for differential dementia diagnosis. Int J Geriatr Psychiatry. 2014;42:1239–50.
16. Marquie Sayagues M, Da Silva AL, Molina-Porcel L, Alcolea Rodriguez D, Sala Matavera I, Sanchez-Saudinos MB, et al. (123)I-MIBG myocardial scintigraphy in the diagnosis of Lewy body dementia. Neurologia. 2010;25(7):414–21.
17. Treglia G, Cason E. Diagnostic performance of myocardial innervation imaging using MIBG scintigraphy in differential diagnosis between dementia with lewy bodies and other dementias: a systematic review and a meta-analysis. J Neuroimaging. 2012;22(2):111–7.
18. Yoshita M, Arai H, Arai H, Arai T, Asada T, Fujishiro H, et al. Diagnostic accuracy of 123I-meta-iodobenzylguanidine myocardial scintigraphy in dementia with Lewy bodies: a multicenter study. PLoS ONE. 2015;10(3):e0120540.
19. O'Brien JT, Oertel WH, McKeith IG, Grosset DG, Walker Z, Tatsch K, et al. Is ioflupane I123 injection diagnostically effective in patients with movement disorders and dementia? Pooled analysis of four clinical trials. BMJ Open. 2014;4(7):e005122.
20. McKeith I, O'Brien J, Walker Z, Tatsch K, Booij J, Darcourt J, et al. Sensitivity and specificity of dopamine transporter imaging with 123I-FP-CIT SPECT in dementia with Lewy bodies: a phase III, multicentre study. Lancet Neurol. 2007;6(4):305–13.
21. Walker Z, Moreno E, Thomas A, Inglis F, Tabet N, Rainer M, et al. Clinical usefulness of dopamine transporter SPECT imaging with 123I-FP-CIT in patients with possible dementia with Lewy bodies: randomised study. Br J Psychiatry. 2015;206(2):145–52.
22. Treglia G, Cason E, Cortelli P, Gabellini A, Liguori R, Bagnato A, et al. Iodine-123 metaiodobenzylguanidine scintigraphy and iodine-123 ioflupane single photon emission computed tomography in Lewy body diseases: complementary or alternative techniques? J Neuroimaging. 2014;24(2):149–54.
23. Camacho V, Marquie M, Lleo A, Alves L, Artigas C, Flotats A, et al. Cardiac sympathetic impairment parallels nigrostriatal degeneration in probable dementia with Lewy bodies. Q J Nucl Med Mol Imaging. 2011;55(4):476–83.
24. Tateno M, Kobayashi S, Shirasaka T, Furukawa Y, Fujii K, Morii H, et al. Comparison of the usefulness of brain perfusion SPECT and MIBG myocardial scintigraphy for the diagnosis of dementia with Lewy bodies. Dement Geriatr Cogn Disord. 2008;26(5):453–7.
25. Inui Y, Toyama H, Manabe Y, Sarai M, Iwata N. Comparison of (123)I-MIBG myocardial scintigraphy, brain perfusion SPECT, and voxel-based MRI morphometry for distinguishing between dementia with Lewy bodies and Alzheimer's disease. Ann Nucl Med. 2014;28 (8):796–804.
26. Oda H, Ishii K, Terashima A, Shimada K, Yamane Y, Kawasaki R, et al. Myocardial scintigraphy may predict the conversion to probable dementia with Lewy bodies. Neurology. 2013;81 (20):1741–5.
27. Fava GA, Kellner R. Prodromal symptoms in affective disorders. Am J Psychiatry. 1991;148 (7):823–30.
28. Yung AR, McGorry PD. The prodromal phase of first-episode psychosis: past and current conceptualizations. Schizophr Bull. 1996;22(2):353–70.
29. Fujishiro H, Iseki E, Nakamura S, Kasanuki K, Chiba Y, Ota K, et al. Dementia with Lewy bodies: early diagnostic challenges. Psychogeriatrics. 2013;13(2):128–38.

30. Iranzo A, Tolosa E, Gelpi E, Molinuevo JL, Valldeoriola F, Serradell M, et al. Neurodegenerative disease status and post-mortem pathology in idiopathic rapid-eye-movement sleep behaviour disorder: an observational cohort study. Lancet Neurol. 2013;12 (5):443–53.
31. Postuma RB, Lang AE, Gagnon JF, Pelletier A, Montplaisir JY. How does parkinsonism start? Prodromal parkinsonism motor changes in idiopathic REM sleep behaviour disorder. Brain J Neurol. 2012;135(Pt 6):1860–70.
32. Postuma RB, Gagnon JF, Pelletier A, Montplaisir J. Prodromal autonomic symptoms and signs in Parkinson's disease and dementia with Lewy bodies. Mov Disord Off J Mov Disord Soc. 2013;28(5):597–604.
33. Iranzo A, Gelpi E, Tolosa E, Molinuevo JL, Serradell M, Gaig C, et al. Neuropathology of prodromal Lewy body disease. Mov Disord Off J Mov Disord Soc. 2014;29(3):410–5.
34. Miyamoto T, Miyamoto M, Inoue Y, Usui Y, Suzuki K, Hirata K. Reduced cardiac 123I-MIBG scintigraphy in idiopathic REM sleep behavior disorder. Neurology. 2006;67 (12):2236–8.
35. Miyamoto T, Miyamoto M, Iwanami M, Hirata K. Idiopathic REM sleep behavior disorder: implications for the pathogenesis of Lewy body diseases. Park Dis. 2011;2011:941268.
36. Miyamoto T, Orimo S, Miyamoto M, Hirata K, Adachi T, Hattori R, et al. Follow-up PET studies in case of idiopathic REM sleep behavior disorder. Sleep Med. 2010;11(1):100–1.
37. Chiaravalloti A, Stefani A, Di Biagio D, Pierantozzi M, Tavolozza M, Di Pietro B, et al. Cardiac sympathetic denervation is not related to nigrostriatal degeneration in Parkinson's disease. Ann Nucl Med. 2013;27(5):444–51.
38. Sakakibara R, Ogata T, Haruta M, Kishi M, Tsuyusaki Y, Tateno A, et al. Amnestic mild cognitive impairment with low myocardial metaiodobenzylguanidine uptake. Am J Neurodegener Dis. 2012;1(2):146–51.
39. Minoshima S, Foster NL, Sima AA, Frey KA, Albin RL, Kuhl DE. Alzheimer's disease versus dementia with Lewy bodies: cerebral metabolic distinction with autopsy confirmation. Ann Neurol. 2001;50(3):358–65.
40. Holtbernd F, Gagnon JF, Postuma RB, Ma Y, Tang CC, Feigin A, et al. Abnormal metabolic network activity in REM sleep behavior disorder. Neurology. 2014;82(7):620–7.

Chapter 17
Physiological Background of Reduced Cardiac ^{123}I-Meta-Iodobenzylguanidine Uptake

Tomohiko Nakamura and Gen Sobue

Abstract Many diseases have been reported to show reduced cardiac ^{123}I-metaiodobenzylguanidine (MIBG) uptake, and the reasons vary among the diseases. The reduced uptake from the early phase reflects cardiac sympathetic denervation, and patients with heart transplant, diabetes, and Parkinson's disease are well known. Patients with pheochromocytoma who shows high plasma level of noradrenaline may demonstrate reduced cardiac MIBG uptake from the early phase because of the competition between circulating MIBG and oversecreted noradrenaline in the sympathetic nerve terminals. Normal or slight reduced uptake in early phase but reduced uptake in the delayed phase, which means increase of washout rate, is well observed in patients with heart failure, and increase of noradrenaline spillover based on systemic sympathetic hyperactivity is considered. Also note that extremely low accumulation of cardiac MIBG can be observed in about 5 % of the subjects without any definite diseases, and aging alone can be the cause. Reduced cardiac MIBG uptake is not peculiar to the diseases, but a variety of drugs can be the cause. It is based on the functional mechanism of the drugs that interferes the various stages of MIBG uptake and storage. To understand the meaning of the results of MIBG study and avoid false positive, it is important to understand the physiological background of reduced cardiac MIBG uptake.

Keywords MIBG • Physiological background • Reduced uptake • Supersensitivity • Sympathetic denervation

---

T. Nakamura (✉)
Department of Neurology, Nagoya University Graduate School of Medicine, 65 Tsurumai-cho, Showa-ku, Nagoya, Aichi 466-8550, Japan
e-mail: tomohiko@med.nagoya-u.ac.jp

G. Sobue
Nagoya University Graduate School of Medicine, 65 Tsurumai-cho, Showa-ku, Nagoya, Aichi 466-8550, Japan

S. Iwase et al. (eds.), *Clinical Assessment of the Autonomic Nervous System*,
DOI 10.1007/978-4-431-56012-8_17

17.1 Introduction

Autonomic control of the heart plays an important role in the regulation of myocardial contraction, heart rate, and myocardial metabolism. Thus, autonomic nervous system impairments can lead to a variety of disorders. However, even during complete denervation of cardiac sympathetic nerves such as that in a heart transplant, humans do not experience orthostatic hypotension and can even exercise. Thus, disturbances of the cardiac sympathetic nerves are not visible. However, the advent of cardiac ^{123}I-metaiodobenzylguanidine (MIBG) scintigraphy has made it possible to detect local abnormalities and evaluate the functional activity of cardiac sympathetic nerves based on its accumulation in the heart in variety of diseases. This has led to clarification of the pathophysiology of cardiac sympathetic dysfunction. To avoid false-positive results, it is important to understand why and how reduced MIBG uptake occurs under various circumstances. This chapter provides an overview of the physiological background and its associated clinical features in a variety of diseases or conditions that demonstrate abnormal cardiac MIBG uptake.

17.2 Pharmacological Detection of Cardiac Sympathetic Denervation and Its Characteristics

The clinical features of reduced cardiac sympathetic denervation are obscure. In patients with reduced cardiac MIBG uptake with diabetes mellitus or Parkinson's disease (PD), the left ventricular ejection fraction at rest does not show any abnormalities [1, 2]. However, there is a classical method to evaluate sympathetic denervation, which is called denervation supersensitivity. To evaluate the existence of denervation of cardiac beta-1 receptor, dobutamine was used [3]. Low doses of dobutamine resulted in a hyperdynamic cardiac response and an excessive pressor response in patients with PD compared to that in controls (Fig. 17.1), and these hyperdynamic responses correlated negatively with the heart to mediastinal ratio (Fig. 17.2), and this result demonstrated that adrenal beta-1 denervation exists within the heart and is related to reduced cardiac MIBG uptake in patients with PD [3]. In addition, injection of the indirectly acting sympathomimetic amine, tyramine, demonstrated increased cardiac contractility in normal subjects, whereas cardiac contractility remained unchanged in PD patients with cardiac sympathetic denervation [4]. Tyramine acts as a catecholamine-releasing agent in peripheral sympathetic nerves. Therefore, cardiac sympathetic denervation attenuates increases in contractility responses to stimuli that normally increase inotropy via endogenous noradrenaline release, whereas contractility responses to directly acting adrenoceptor agonists remain intact [5].

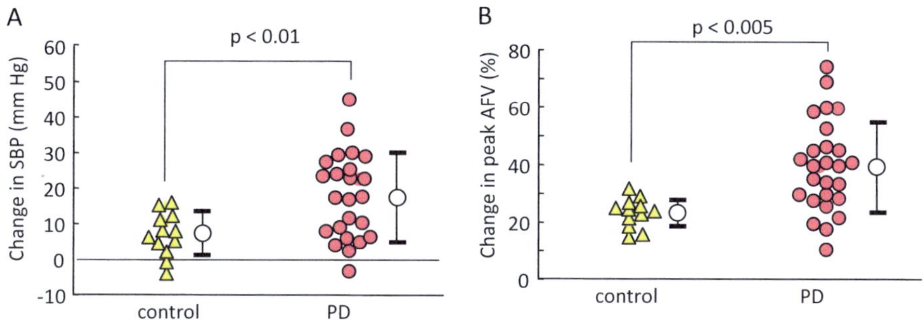

Fig. 17.1 Changes in SBP and peak AFV after infusion of 4 μg/kg/min dobutamine controls and PD patients. (**a**) The mean change in SBP is significantly greater in the PD group than in the control group. (**b**) The mean percentage change in AFV, an indicator of cardiac contractility, was significantly larger in the PD group than in the control group (Reprinted and adapted from Ref. [3] with permission from Elsevier). *SBP* systolic blood pressure, *AFV* aortic flow velocity, *PD* Parkinson's disease

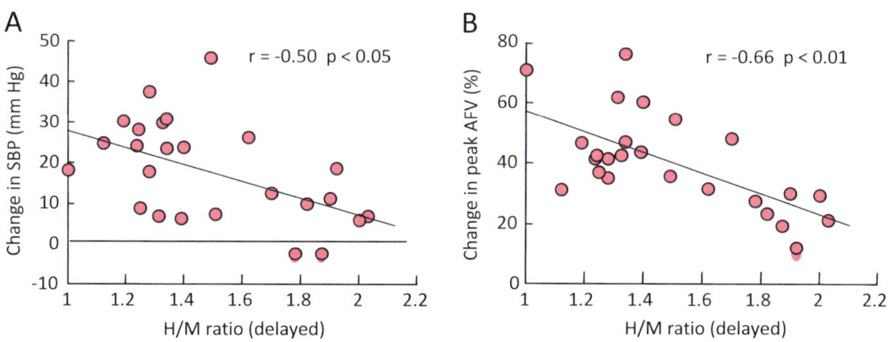

Fig. 17.2 Correlation between changes in SBP and peak AFV during infusion of 4 μg/kg/min dobutamine and the H/M ratios in patients with PD. Significant correlations between the delayed H/M ratio and (**a**) SBP changes and (**b**) changes in peak AFV were observed (Reprinted and adapted from Ref. [3] with permission from Elsevier). *SBP* systolic blood pressure, *AFV* aortic flow velocity, *PD* Parkinson's disease, *H/M* heart to mediastinum

17.3 Reduced MIBG Uptake in the Diseases

17.3.1 Pattern and Mechanism of Reduced Cardiac MIBG Uptake

In normal healthy subjects, MIBG accumulates in the early phase and is visible on the delayed imaging. MIBG uptake in the extraneural tissue is very limited as evidenced by patients receiving heart transplants showing an absence of MIBG uptake on early and delayed images [6, 7]; thus, early imaging reflects inhibition of sodium-dependent uptake (uptake-1), i.e., reflects presynaptic integrity and sympathetic nerve distribution, while delayed imaging reflects cardiac sympathetic functional activity. Therefore, there are two patterns of abnormal findings in the cardiac MIBG study: reduced MIBG uptake from the early phase and a normal or slightly reduced uptake in the early phase but reduced uptake in the delayed phase, which means an increased washout rate. The former is considered an organic denervated condition of cardiac sympathetic nerves caused by cardiac sympathetic failure, noradrenaline depletion, and dysfunction of the uptake-1 mechanism. In addition, in the condition of excess of catecholamine such as pheochromocytoma [8], increased noradrenaline can inhibit MIBG uptake, leading to low accumulation. In one study, extremely low accumulation of cardiac MIBG was observed in 24 of 511 subjects without any definite diseases, and aging alone was shown to cause such abnormalities [9], but these normal subjects may suffer from incidental Lewy body disease [10]. The latter is commonly observed in patients with heart failure and thought to be caused by increased noradrenaline spillover based on systemic sympathetic hyperactivity. In addition, in patients with severe cardiac ischemia, an increased washout rate is observed because of the non-exocytotic local metabolic release of noradrenaline [11]. The details of the typical diseases are discussed below.

17.3.2 Physiology of Disease-Induced Cardiac MIBG Uptake Reductions

17.3.2.1 Heart Failure

The use of MIBG scintigraphy in heart failure was first reported in 1988 by Schofer, who showed that the cardiac MIBG abnormality was significantly related to the left ventricular ejection fraction in patients with idiopathic dilated cardiomyopathy [12]. After that, other studies showed that delayed imaging of cardiac MIBG is a powerful predictor of overall cardiac death in patients with heart failure [13, 14]. In addition, both delayed imaging of cardiac MIBG and washout rate were powerful predictors of sudden cardiac death in patients with heart failure [15–18]. Myocardial adrenergic nerve activity is accelerated in proportion to heart failure severity, while myocardial washout activity is related to heart failure severity and correlates well with New York Heart Association functional classification [19]. The usefulness of reduced cardiac MIBG uptake in delayed imaging and an excessive washout rate in patients with heart failure have been established.

In addition, in patients with dilated or hypertrophic cardiomyopathy, reduced cardiac MIBG uptake and excessive washout rate are associated with reduced contractile strength during a dobutamine stress test or exercise, and cardiac MIBG uptake degree can be used to evaluate the existence of left ventricular functional reserve in such diseases [20–23].

In patients with heart failure, cardiac function is depressed, and systemic sympathetic nerve hyperactivity is observed as a compensatory mechanism to maintain cardiac output. This results in a chronic increase in noradrenaline release from the nerve endings. The neuronal uptake capacity is eventually exceeded, leading to increased overspill of noradrenaline into the plasma and a net neuronal loss [24]. This likely accounts for the increased washout rate in patients with heart failure [17]. As the cardiac dysfunction progresses, neuron loss and Net1 downregulation occur, which diminish presynaptic function and likely account for the decreased cardiac MIBG uptake seen in advanced disease [25]. In addition, recent studies revealed that during congestive heart failure, sympathetic neural tone is upregulated, but there is a paradoxical reduction in noradrenaline synthesis and reuptake in the cardiac sympathetic nervous system. This is because the cholinergic differentiation cytokines, leukemia inhibitory factor, and cardiotrophin-1 were strongly upregulated in congestive heart failure, and congestive heart failure causes target-dependent cholinergic transdifferentiation of the cardiac sympathetic nervous system via these cytokines secreted from the failing myocardium [26].

17.3.2.2 Ischemic Heart Disease

In patients with ischemic heart disease such as myocardial infarction and unstable angina, cardiac MIBG imaging and perfusion imaging findings are consistent with the site and size of the damaged region. Cardiac MIBG uptake is reduced on early imaging because it does not accumulate in the nerve endings in the absence of blood flow. The sympathetic neuronal damage measured by MIBG is often larger than the infarct size (the so-called mismatch defect) [27–30]. This region of damaged sympathetic nerves and surviving myocardial cells indicates that sympathetic nerves are more sensitive to ischemia than myocardial cells. In patients with vasospastic angina, a marked reduction in cardiac MIBG uptake is often observed on delayed imaging due to excessive washout. This is the result of impaired noradrenaline retention and sympathetic nerve hyperexcitability. This finding demonstrates the concept of "memory imaging" of ischemia because the emergence and the region of the ischemic attack can be diagnosed retrospectively even after the normalization of electrocardiogram and echocardiogram abnormalities [31–33]. The mismatch defect is also observed in patients with old myocardial infarction, indicating the existence of "ischemic but viable tissue" [34, 35]. This mismatch defect is also observed in various myocardial disorders such as dilated cardiomyopathy [36, 37] and hypertrophic cardiomyopathy [38], but its diagnostic significance in these disorders has yet to be determined. This mismatch is also observed in patients with subarachnoid hemorrhage with left ventricular systolic dysfunction and is thought to be associated with an excessive release of noradrenaline from myocardial sympathetic nerves, which can damage both myocytes and nerve terminals [39].

17.3.2.3 Transplanted Heart

The transplanted heart is completely denervated. Because of surgical denervation, presynaptic nerve terminals disappear, and myocardial storage of the neurotransmitter noradrenaline is depleted [40]. Therefore, the denervated transplanted heart has to rely on circulating catecholamines to adapt the cardiac output to meet the increased demand. However, there are usually no associated symptoms in daily life. Their ability to maintain systemic blood pressure during postural stress is unaffected, and increased systemic vascular resistance and well-preserved systolic blood pressure without orthostatic hypotension were observed during the head-up tilt test [41].

However, under exercise stress, this adaptation is limited and cannot achieve a normal increase in heart rate and cardiac muscle contractility [42]. Under stress conditions, the restoration of innervation is important to achieve an adequate contractile response. One study showed that transplant recipients who achieve reinnervation have a greater capacity for exercise than those who do not achieve denervation [43]. The chronotropic response to exercise is reported to increase at 3–6 weeks after transplantation [44].

17.3.2.4 Pheochromocytoma and Neuroblastoma

In patients with pheochromocytoma, the cardiac MIBG uptake is decreased on early imaging. Since these patients show high levels of plasma noradrenaline, the excess circulating noradrenaline caused by pheochromocytoma was thought to damage the cardiac sympathetic nerve function [45]. However, since the abnormal scintigraphic findings improved after tumor resection [8, 46], the abnormal MIBG finding was presumed to be caused by the high plasma catecholamine concentration itself [8]. In addition, both patients with pheochromocytoma and those with high serum noradrenaline levels with neuroblastoma showed reduced cardiac MIBG uptake before treatment and significantly increased cardiac MIBG uptake after treatment. One of the possible explanations for the reduced cardiac MIBG uptake before treatment in patients with neuroadrenergic tumors is the competition between circulating MIBG and oversecreted noradrenaline and adrenaline in the sympathetic nerve terminals [47]. The downregulation of the uptake pathway and the rapid turnover of MIBG in the cardiac sympathetic neurons due to the effects of excess catecholamines are also postulated [47, 48]. However, not all patients show improved cardiac MIBG uptake. Some patients with pheochromocytoma show only slight improvements in cardiac MIBG uptake with persistent cardiac dysfunction. Thus, another possible mechanism underlying reduced MIBG uptake in pheochromocytoma is that regional excess of catecholamines on the myocardium directly induces focal myofibril degeneration with inflammatory cellular infiltration [49]. Myocardial cell damage can result from reduced coronary perfusion and hypoxia caused by a vasospasm that is mediated by an adrenergic receptor [50] and a change in the calcium ion permeability of the cell membrane [46]. Functionally, excess noradrenaline leads to reduced global myocardial pump function [51]. These forms of pathological damage result in a decreased MIBG uptake owing to the reduced number of sympathetic nerve endings, impaired neuronal uptake function, and reduced synthesis or rapid turnover of noradrenaline within the neurons. Catecholamine-induced cardiomyopathy may be reversible after primary tumor removal, but persistent dysfunction can occur due to the long-term accumulation of myocardial damage [47].

17.3.2.5 Diabetes

According to previous studies, the left ventricular ejection fraction at rest is normal in diabetic patients [2, 52]. However, diabetic patients with reduced cardiac MIBG uptake showed an impaired response to exercise, as indicated by a smaller increase in ejection fraction, and it is suggested that subclinical left ventricular dysfunction is related to derangement of adrenergic cardiac innervation [2, 52, 53]. In diabetic rats, the MIBG washout rate was high, but unlike heart failure patients, this is not likely due to the systemic sympathetic hyperactivity. Rather, it is likely due to reuptake and/or pooling mechanism dysfunction since the plasma and myocardial noradrenaline concentrations in diabetic rats were significantly lower than those in nondiabetic rats [54]. Other details of clinical findings in diabetic patients are covered elsewhere (Chap. 15).

17.3.2.6 Parkinson's Disease

The first cardiac MIBG study in patients with neurodegenerative disorders, including PD, was reported by Hakusui et al. in 1994. Reduced cardiac MIBG uptake was observed in both patients with autonomic failure, mainly orthostatic hypotension, and in those without orthostatic hypotension [55, 56]. The details of the transition and the clinical significance of MIBG scintigraphy in PD are saved for another chapter, but an overview of the physiological significance of reduced cardiac MIBG uptake in patients with PD is provided here.

Orimo et al. reported that tyrosine hydroxylase-immunoreactive nerve fibers in the heart were markedly decreased in patients with PD, which suggests that the involvement of postganglionic sympathetic nerves accounts for reduced cardiac MIBG uptake in patients with PD [57]. Recently, their group showed that the degree of cardiac MIBG uptake is correlated with that of cardiac sympathetic denervation in pathologically confirmed Lewy body disease [58].

But, patients with PD and reduced cardiac MIBG uptake show normal left ventricular function on echocardiography [1], and the clinical symptoms of autonomic disorders associated with cardiac denervation are difficult to recognize. Reduced cardiac MIBG is not observed in patients with multiple system atrophy, a disease with characteristics of orthostatic hypotension, but is commonly observed in patients with PD with or without orthostatic hypotension; thus, many studies reported that reduced cardiac MIBG uptake is not associated with orthostatic hypotension [59–61]. However, reduced cardiac MIBG uptake is associated with a reduced overshoot of phase IV on the Valsalva maneuver, which indicates that reduced cardiac MIBG uptake clinically reflects cardiac sympathetic dysfunction in patients with PD [62]. The uptake of MIBG and 6-[18F]-fluorodopamine, which can also detect cardiac sympathetic denervation, was reportedly much lower in patients with orthostatic hypotension than in those without [63, 64]; thus, cardiac denervation is likely to be associated with orthostatic hypotension in patients with PD. A recent report on PD showed that under orthostatic stress, cardiac sympathetic denervation with failure to increase total peripheral resistance leads to large reductions in systolic blood pressure; however, patients without cardiac denervation exhibited a positive inotropic response against vasodilatation, which may prevent orthostatic hypotension [65]. On the other hand, patients with sufficient peripheral adrenergic innervation to elicit an increase in total peripheral resistance through vasoconstriction, a potential cardiac denervation effect, would be masked, and a normal blood pressure response is maintained (Fig. 17.3). Thus, cardiac sympathetic nerves play an important role in regulating blood pressure, and impaired cardiac contractility is one of the important contributing factors to the development of orthostatic hypotension in patients with PD.

Fatigue and exercise intolerance in patients with PD may be also explained by the cardiac sympathetic dysfunction that results in the failure to increase contractility in response to the release of noradrenaline from sympathetic nerves [4]. Patients with PD often complain of fatigue, and cardiac MIBG uptake was indeed more reduced in the fatigued group than in the non-fatigued group [66]. Blood pressure and heart rate response to exercise differ among many studies.

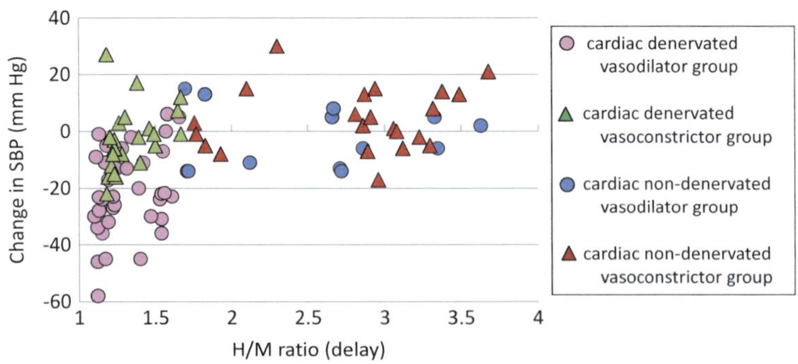

Fig. 17.3 Scatter plot of the association with change in SBP at fifth min of 60°tilt and cardiac MIBG. Association with cardiac MIBG H/M ratio. Patients with cardiac sympathetic denervation with failure to increase total peripheral resistance (cardiac denervated vasodilator group) leads to large reductions in systolic BP; however, patients without cardiac denervation exhibited a positive inotropic response against vasodilatation, which may prevent orthostatic hypotension (cardiac non-denervated vasodilator group) (Reprinted and adapted from Ref. [65] with permission from Elsevier). *SBP* systolic blood pressure, *H/M* heart to mediastinum

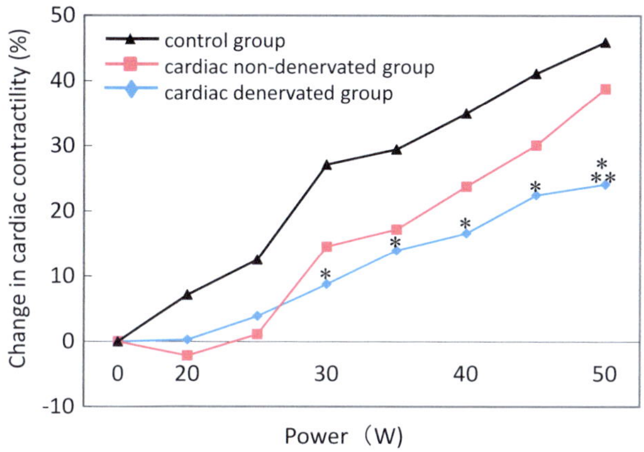

Fig. 17.4 Change in cardiac contractility. Change in cardiac contractility showing a significant difference at the first minute of the 30 W workload between the cardiac denervated and control groups. This difference continued till the 50 W workload. A similar tendency was observed between the patient groups, in which the cardiac denervated group had a significantly lower cardiac response than the cardiac non-denervated group at the first minute of the 50 W workload. *; $p < 0.05$ vs control, **; $p < 0.05$ vs cardiac non-denervated group (Reprinted and adapted from Ref. [71] with permission from John Wiley and Sons)

Some patients with PD reportedly had a statistically lower systolic blood pressure change upon exercising [67], whereas others reportedly had blood pressure changes that did not differ from those of control subjects [68]. Regarding the chronotropic response, some authors reported that the heart rate change is higher in patients with PD than in control subjects [69], whereas others reported that chronotropic insufficiency against cardiac stress testing was already observed in individuals who developed the motor features of PD several years after cardiac stress testing and that it may constitute an early sign of cardiac sympathetic dysfunction in PD [70]. The exercise study on cardiac sympathetic dysfunction showed that patients with PD having reduced cardiac MIBG uptake had lower cardiac contractility than cardiac non-denervated subjects and control subjects during exercise (Fig. 17.4), suggesting that this response represents impaired exercise capacity of patients with PD and cardiac sympathetic denervation [71]; this result was consistent with those of previous studies conducted in heart transplant and diabetic patients as described above.

17.3.2.7 Other Diseases

In other diseases, such as Brugada syndrome, reduced cardiac MIBG uptake is considered important finding in the pathophysiology and arrhythmogenesis of the disease, but whether this is a primary adrenergic dysfunction or an imbalance of sympathetic and parasympathetic innervation of the heart is unclear [72]. In addition, many other diseases such as postural tachycardia syndrome [73, 74], spinocerebellar ataxia type 2 [75], amyloidosis [76], amyotrophic lateral sclerosis [77], and temporal lobe epilepsy [78] demonstrate reduced cardiac MIBG uptake. The reduced uptake in these diseases may be due to neuropathy, sympathetic hyperactivity, or dysfunction of postganglionic cardiac sympathetic fibers. However, these reports are sporadic, so further studies reflecting the pathophysiology or those that can consider the physiological background are warranted.

17.4 Drug and Diet-Induced Cardiac MIBG Uptake Reductions

In addition to diseases, a variety of drugs and food can reduce cardiac MIBG uptake by interfering with the various stages of radiolabeled MIBG uptake and storage, which increases the likelihood of a false-negative study.

17.4.1 Classification Based on MIBG Kinetics

According to the physiological features of MIBG kinetics, the following mechanisms of reduction of cardiac MIBG uptake were identified [79].

17.4.1.1 Inhibition of Sodium-Dependent Uptake (Uptake-1) in the Synaptic Cleft

Uptake-1 in the synaptic cleft is the main mechanism underlying MIBG uptake, a process that is inhibited by cocaine. Cocaine interacts with the catecholamine transport protein involved in the neuronal uptake-1 system. Tricyclic antidepressant also affects MIBG uptake [80]. In dogs, pretreatment with desmethylimipramine (also known as desipramine) 1 h before MIBG administration led to a tenfold reduction in uptake of the radiopharmaceutical by the adrenal medullae [81]. A reduced uptake of MIBG caused by other tricyclic antidepressants, such as imipramine, amitriptyline, clomipramine, and doxepin, has also been well documented [82, 83]. Labetalol, an antihypertensive agent with combined alpha- and beta-blocking properties that has been used to manage patients with suspected pheochromocytoma, also reduces MIBG uptake [84]. The inhibitory effect of labetalol on MIBG uptake in the sympathomedullary tissues is likely to be a result of the drug's little-known additional properties of blocking uptake-1 and depleting the storage vesicle contents [84]. In contrast, conventional alpha- or beta-adrenergic receptor blockers do not have such an inhibitory effect.

17.4.1.2 Inhibited Uptake by Active Transport into the Vesicles from the Cytoplasm

Reserpine and tetrabenazine are known to inhibit active transport into the neurosecretory vesicles [79]. MIBG is transported across the neurosecretory vesicle membrane and can reduce MIBG uptake. Other *Rauwolfia* alkaloids as well as antidepressants such as viloxazine may also have similar action and interfere with MIBG uptake [79].

17.4.1.3 Competition for Transport into the Vesicles

MIBG translocation across the neurosecretory vesicle membrane also occurs via a monoamine transport system, which is shared by noradrenaline, serotonin, and MIBG. The translocation of these neurotransmitters depends upon the concentration of the relevant substrate in the cytoplasm. Adrenergic neuron blockers such as guanethidine share the same transport system and compete with MIBG for this pathway. Selegiline, a drug used to treat PD, works by inhibiting monoamine oxidase (MAO) to slow noradrenaline breakdown and may increase noradrenaline leading to the competition for transport into the vesicles with MIBG.

17.4.1.4 Depletion of the Storage Vesicle Contents

Many medicines are known to deplete the storage vesicle contents, most commonly sympathomimetic drugs such as reserpine, labetalol, guanethidine, and amphetamines. Although weak, selegiline has a known amphetamine effect, and its breakdown product methamphetamine may reduce MIBG uptake. Phenylpropanolamine and phenylephedrine may also reduce MIBG uptake. Even small amounts of the medications delivered intranasally may be sufficient to reduce MIBG uptake.

17.4.2 Classification Based on Medication Category [83]

From a clinical perspective, a recent review classified medication interference with cardiac MIBG uptake based on medication category (Table 17.1). Drugs are categorized using the following groupings. This classification is more directly clinically relevant than the classification based on the presumed kinetics. Recommendations of these drugs to be withheld prior to cardiac MIBG imaging are also provided in Table 17.1.

Table 17.1 Classification of the medication interference with cardiac MIBG uptake

Medication category	Subcategory	Most commonly tested medication(s)	Strength of MIBG inhibitory effect	Recommendation/comment
Drugs acting on adrenoreceptors	Beta-blockers	Labetalol, propranolol	Labetalol: strong	Labetalol should be withheld
			Others: none	Others do not require withholding
	Beta-agonists	Salbutamol	None	Based on pharmacologic mechanism, withdrawal not required
	Alpha-antagonists	Phenoxybenzamine	Uncertain[a]	Uncertain; withdrawal probably not necessary
	Alpha-agonists	Clonidine	Uncertain[a]	Uncertain; withdrawal probably not necessary
Drugs affecting noradrenaline transport, retention, or release	Tricyclic antidepressants	Desipramine, imipramine	Moderate to strong	Should be withheld if clinically feasible
	SSRIs and SNRIs	Fluvoxamine, milnacipran	Moderate to strong for some agents	Consider withholding agents with documented effect on noradrenaline transporter
	Other antidepressants	Trazodone	None	Evidence insufficient for a recommendation
	Noradrenaline depleters	Reserpine	Strong	If still in clinical use, should be withheld
	Sympathomimetics	Phenylpropanolamine, phenylephrine	Strong	Should be withheld if clinically feasible
	Monoamines	Cocaine	Moderate	In cases of suspected use, screen patients prior to MIBG dosing
Calcium channel blockers		Nifedipine, cilnidipine	None	Effect (if any) is on release, not uptake of MIBG. Not necessary to withhold
Miscellaneous	Neuroleptics	Haloperidol, levomepromazine	Uncertain[a]	Uncertain
	Anesthetics	Ketamine, xylazine, pentobarbital	Uncertain[a]	Uncertain
	Cardiac glycosides	Digoxin	Uncertain[a]	Uncertain; withdrawal probably not necessary
	Antiarrhythmics	Amiodarone	None	Effect (if any) is beneficial; not necessary to withhold

[a]Data judged insufficient for making this estimate (Reprinted and adapted from Ref [83] with permission from Springer)

17.4.2.1 Drugs Acting on Adrenoreceptors

As mentioned above, labetalol, a combined alpha- and beta-blocking properties, has a strong inhibitory effect on cellular uptake of MIBG but no effect or uncertain effect of all other drugs acting on adrenoreceptors [83].

17.4.2.2 Drugs Affecting Noradrenaline Transport, Retention, or Release

In addition to tricyclic antidepressants, other antidepressants have effect on cardiac MIBG uptake. The selective serotonin reuptake inhibitors (SSRI) have variable effects on noradrenaline transporter activity, and strong reduction of MIBG uptake by fluvoxamine was reported [85–87]. In addition, patients taking milnacipran, a serotonin/noradrenaline reuptake inhibitors (SNRI), also showed reduce cardiac MIBG uptake [88]. But there is limited literature on the effect of these drugs to judge for reducing cardiac MIBG uptake. Noradrenaline depleters, sympathomimetics, and monoamines are classified here but are already mentioned above.

17.4.2.3 Calcium Channel Blockers

Unlike other drugs, calcium channel blockers reduce excessive sympathetic drive and preserve postsynaptic function, resulting in retention of MIBG in the delayed image and decreased washout rate.

17.4.2.4 Miscellaneous

A number of other drugs such as neuroleptics, anesthetics, cardiac glycosides, and antiarrhythmics have been reported. But these are small studies and the significance of these findings is uncertain.

17.4.3 Diet-Induced Cardiac MIBG Uptake Reductions

Foods containing vanilla and catecholamine-like ingredients (such as chocolate and blue cheese) might influence the uptake of MIBG [89]. These diets contain high tyramine. Tyramine acts as a catecholamine-releasing agent and increases noradrenaline release resulting in competing with MIBG uptake.

References

1. Orimo S, Ozawa E, Nakade S, Sugimoto T, Mizusawa H. (123)I-metaiodobenzylguanidine myocardial scintigraphy in Parkinson's disease. J Neurol Neurosurg Psychiatry. 1999;67 (2):189–94.
2. Kreiner G, Wolzt M, Fasching P, Leitha T, Edlmayer A, Korn A, et al. Myocardial m-[123I] iodobenzylguanidine scintigraphy for the assessment of adrenergic cardiac innervation in patients with IDDM. Comparison with cardiovascular reflex tests and relationship to left ventricular function. Diabetes. 1995;44(5):543–9.
3. Nakamura T, Hirayama M, Ito H, Takamori M, Hamada K, Takeuchi S, et al. Dobutamine stress test unmasks cardiac sympathetic denervation in Parkinson's disease. J Neurol Sci. 2007;263(1–2):133–8. doi:10.1016/j.jns.2007.07.005.
4. Imrich R, Eldadah BA, Bentho O, Pechnik S, Sharabi Y, Holmes C, et al. Functional effects of cardiac sympathetic denervation in neurogenic orthostatic hypotension. Parkinsonism Relat Disord. 2009;15(2):122–7. doi:10.1016/j.parkreldis.2008.04.002.
5. Goldstein DS. Dysautonomia in Parkinson disease. Compr Physiol. 2014;4(2):805–26. doi:10.1002/cphy.c130026.
6. Dae MW, De Marco T, Botvinick EH, O'Connell JW, Hattner RS, Huberty JP, et al. Scintigraphic assessment of MIBG uptake in globally denervated human and canine hearts – implications for clinical studies. J Nucl Med Off Publ Soc Nucl Med. 1992;33 (8):1444–50.
7. Glowniak JV, Turner FE, Gray LL, Palac RT, Lagunas-Solar MC, Woodward WR. Iodine-123 metaiodobenzylguanidine imaging of the heart in idiopathic congestive cardiomyopathy and cardiac transplants. J Nucl Med Off Publ Soc Nucl Med. 1989;30(7):1182–91.
8. Matsumoto K, Sugihara H, Ito K, Terada K, Taniguchi Y, Ohtsuki K, et al. 123I-metaiodobenzylguanidine myocardial scintigraphy in a case of pheochromocytoma before and after resection of it. Kaku Igaky Japan J Nucl Med. 1995;32(9):1023–8.
9. Sakata K, Iida K, Motiduki N, Nakaya Y. Frequency and characteristics of extremely low accumulation of 123I-metaiodobenzylguanidine (MIBG) in patients with and without cardiovascular diseases. Ann Nucl Med. 2008;22(9):743–9. doi:10.1007/s12149-008-0169-5.
10. Orimo S, Kanazawa T, Nakamura A, Uchihara T, Mori F, Kakita A, et al. Degeneration of cardiac sympathetic nerve can occur in multiple system atrophy. Acta Neuropathol. 2007;113 (1):81–6. doi:10.1007/s00401-006-0160-y.
11. Schomig A. Catecholamines in myocardial ischemia. Systemic and cardiac release. Circulation. 1990;82(3 Suppl):II13–22.
12. Schofer J, Spielmann R, Schuchert A, Weber K, Schluter M. Iodine-123 metaiodobenzylguanidine scintigraphy: a noninvasive method to demonstrate myocardial adrenergic nervous system disintegrity in patients with idiopathic dilated cardiomyopathy. J Am Coll Cardiol. 1988;12(5):1252–8.
13. Merlet P, Valette H, Dubois-Rande JL, Moyse D, Duboc D, Dove P, et al. Prognostic value of cardiac metaiodobenzylguanidine imaging in patients with heart failure. J Nucl Med Off Publ Soc Nucl Med. 1992;33(4):471–7.
14. Nakata T, Miyamoto K, Doi A, Sasao H, Wakabayashi T, Kobayashi H, et al. Cardiac death prediction and impaired cardiac sympathetic innervation assessed by MIBG in patients with failing and nonfailing hearts. J Nucl Cardiol Off Publ Am Soc Nucl Cardiol. 1998;5 (6):579–90.
15. Tamaki S, Yamada T, Okuyama Y, Morita T, Sanada S, Tsukamoto Y, et al. Cardiac iodine-123 metaiodobenzylguanidine imaging predicts sudden cardiac death independently of left ventricular ejection fraction in patients with chronic heart failure and left ventricular systolic dysfunction: results from a comparative study with signal-averaged electrocardiogram, heart rate variability, and QT dispersion. J Am Coll Cardiol. 2009;53(5):426–35. doi:10.1016/j.jacc.2008.10.025.

16. Kasama S, Toyama T, Sumino H, Nakazawa M, Matsumoto N, Sato Y, et al. Prognostic value of serial cardiac 123I-MIBG imaging in patients with stabilized chronic heart failure and reduced left ventricular ejection fraction. J Nucl Med Off Publ Soc Nucl Med. 2008;49 (6):907–14. doi:10.2967/jnumed.107.047548.
17. Travin MI. Cardiac autonomic imaging with SPECT tracers. J Nucl Cardiol Off Publ Am Soc Nucl Cardiol. 2013;20(1):128–43. doi:10.1007/s12350-012-9655-1. quiz 46.
18. Ogita H, Shimonagata T, Fukunami M, Kumagai K, Yamada T, Asano Y, et al. Prognostic significance of cardiac (123)I metaiodobenzylguanidine imaging for mortality and morbidity in patients with chronic heart failure: a prospective study. Heart. 2001;86(6):656–60.
19. Imamura Y, Ando H, Mitsuoka W, Egashira S, Masaki H, Ashihara T, et al. Iodine-123 metaiodobenzylguanidine images reflect intense myocardial adrenergic nervous activity in congestive heart failure independent of underlying cause. J Am Coll Cardiol. 1995;26 (7):1594–9. doi:10.1016/0735-1097(95)00374-6.
20. Isobe S, Izawa H, Iwase M, Nanasato M, Nonokawa M, Ando A, et al. Cardiac 123I-MIBG reflects left ventricular functional reserve in patients with nonobstructive hypertrophic cardiomyopathy. J Nucl Med Off Publ Soc Nucl Med. 2005;46(6):909–16.
21. Ohshima S, Isobe S, Izawa H, Nanasato M, Ando A, Yamada A, et al. Cardiac sympathetic dysfunction correlates with abnormal myocardial contractile reserve in dilated cardiomyopathy patients. J Am Coll Cardiol. 2005;46(11):2061–8. doi:10.1016/j.jacc.2005.08.046.
22. Kobayashi M, Izawa H, Cheng XW, Asano H, Hirashiki A, Unno K, et al. Dobutamine stress testing as a diagnostic tool for evaluation of myocardial contractile reserve in asymptomatic or mildly symptomatic patients with dilated cardiomyopathy. J Am Coll Cardiol Img. 2008;1 (6):718–26. doi:10.1016/j.jcmg.2008.04.012.
23. Ohshima S, Isobe S, Hayashi D, Abe S, Kato K, Murohara T. Myocardial 123I-MIBG scintigraphy predicts an impairment in myocardial functional reserve during dobutamine stress in patients with idiopathic dilated cardiomyopathy. Eur J Nucl Med Mol Imaging. 2013;40 (2):262–70. doi:10.1007/s00259-012-2256-z.
24. Link JM, Caldwell JH. Diagnostic and prognostic imaging of the cardiac sympathetic nervous system. Nat Clin Pract Cardiovasc Med. 2008;5 Suppl 2:S79–86. doi:10.1038/ncpcardio1150.
25. Chen GP, Tabibiazar R, Branch KR, Link JM, Caldwell JH. Cardiac receptor physiology and imaging: an update. J Nucl Cardiol Off Publ Am Soc Nucl Cardiol. 2005;12(6):714–30. doi:10.1016/j.nuclcard.2005.09.008.
26. Kanazawa H, Ieda M, Kimura K, Arai T, Kawaguchi-Manabe H, Matsuhashi T, et al. Heart failure causes cholinergic transdifferentiation of cardiac sympathetic nerves via gp130-signaling cytokines in rodents. J Clin Invest. 2010;120(2):408–21. doi:10.1172/JCI39778.
27. Matsunari I, Schricke U, Bengel FM, Haase HU, Barthel P, Schmidt G, et al. Extent of cardiac sympathetic neuronal damage is determined by the area of ischemia in patients with acute coronary syndromes. Circulation. 2000;101(22):2579–85.
28. McGhie AI, Corbett JR, Akers MS, Kulkarni P, Sills MN, Kremers M, et al. Regional cardiac adrenergic function using I–123 meta-iodobenzylguanidine tomographic imaging after acute myocardial infarction. Am J Cardiol. 1991;67(4):236–42.
29. Simula S, Lakka T, Laitinen T, Remes J, Kettunen R, Kuikka J, et al. Cardiac adrenergic denervation in patients with non-Q-wave versus Q-wave myocardial infarction. Eur J Nucl Med. 2000;27(7):816–21.
30. Simoes MV, Barthel P, Matsunari I, Nekolla SG, Schomig A, Schwaiger M, et al. Presence of sympathetically denervated but viable myocardium and its electrophysiologic correlates after early revascularised, acute myocardial infarction. Eur Heart J. 2004;25(7):551–7. doi:10.1016/j.ehj.2004.02.016.
31. Ha JW, Lee JD, Jang Y, Chung N, Kwan J, Rim SJ, et al. 123I-MIBG myocardial scintigraphy as a noninvasive screen for the diagnosis of coronary artery spasm. J Nucl Cardiol Off Publ Am Soc Nucl Cardiol. 1998;5(6):591–7.

32. Sakata K, Shirotani M, Yoshida H, Kurata C. Iodine-123 metaiodobenzylguanidine cardiac imaging to identify and localize vasospastic angina without significant coronary artery narrowing. J Am Coll Cardiol. 1997;30(2):370–6.
33. Taki J, Yasuhara S, Takamatsu T, Nakajima K, Tatami R, Ishise S, et al. Value of iodine-123 metaiodobenzylguanidine scintigraphy in patients with vasospastic angina. Eur J Nucl Med. 1998;25(3):229–34.
34. Dae MW, Herre JM, O'Connell JW, Botvinick EH, Newman D, Munoz L. Scintigraphic assessment of sympathetic innervation after transmural versus nontransmural myocardial infarction. J Am Coll Cardiol. 1991;17(6):1416–23.
35. Simula S, Lakka T, Kuikka J, Laitinen T, Remes J, Kettunen R, et al. Cardiac adrenergic innervation within the first 3 months after acute myocardial infarction. Clin Physiol. 2000;20 (5):366–73.
36. Yamazaki J, Muto H, Kabano T, Yamashina S, Nanjo S, Inoue A. Evaluation of beta-blocker therapy in patients with dilated cardiomyopathy – clinical meaning of iodine 123-metaiodobenzylguanidine myocardial single-photon emission computed tomography. Am Heart J. 2001;141(4):645–52. doi:10.1067/mhj.2001.112783.
37. Parthenakis FI, Prassopoulos VK, Koukouraki SI, Zacharis EA, Diakakis GF, Karkavitsas NK, et al. Segmental pattern of myocardial sympathetic denervation in idiopathic dilated cardiomyopathy: relationship to regional wall motion and myocardial perfusion abnormalities. J Nucl Cardiol Off Publ Am Soc Nucl Cardiol. 2002;9(1):15–22.
38. Shimizu M, Ino H, Yamaguchi M, Terai H, Hayashi K, Nakajima K, et al. Heterogeneity of cardiac sympathetic nerve activity and systolic dysfunction in patients with hypertrophic cardiomyopathy. J Nucl Med Off Publ Soc Nucl Med. 2002;43(1):15–20.
39. Banki NM, Kopelnik A, Dae MW, Miss J, Tung P, Lawton MT, et al. Acute neurocardiogenic injury after subarachnoid hemorrhage. Circulation. 2005;112(21):3314–9. doi:10.1161/CIRCULATIONAHA.105.558239.
40. Cooper T, Willman VL, Jellinek M, Hanlon CR. Heart autotransplantation: effect on myocardial catecholamine and histamine. Science. 1962;138(3536):40–1. doi:10.1126/science.138.3536.40.
41. Banner NR, Williams TD, Patel N, Chalmers J, Lightman SL, Yacoub MH. Altered cardiovascular and neurohumoral responses to head-up tilt after heart-lung transplantation. Circulation. 1990;82(3):863–71.
42. Bengel FM, Ueberfuhr P, Schiepel N, Nekolla SG, Reichart B, Schwaiger M. Effect of sympathetic reinnervation on cardiac performance after heart transplantation. N Engl J Med. 2001;345(10):731–8. doi:10.1056/NEJMoa010519.
43. Schwaiblmair M, von Scheidt W, Uberfuhr P, Ziegler S, Schwaiger M, Reichart B, et al. Functional significance of cardiac reinnervation in heart transplant recipients. J Heart Lung Transplant Off Publ Int Soc Heart Transplant. 1999;18(9):838–45.
44. Scott CD, Dark JH, McComb JM. Evolution of the chronotropic response to exercise after cardiac transplantation. Am J Cardiol. 1995;76(17):1292–6.
45. Izumi C, Himura Y, Konishi T. Abnormal cardiac sympathetic nerve function in a patient with pheochromocytoma. An analysis using 123I metaiodobenzylguanidine scintigraphy. Int J Cardiol. 1995;50(2):189–92.
46. Agostini D, Darlas Y, Filmont JE, Dronne F, Babatasi G, Grollier G, et al. The reversibility of cardiac neuronal function after removal of a pheochromocytoma: an I-123 MIBG scintigraphic study. Clin Nucl Med. 1999;24(7):514–8.
47. Suga K, Ogasawara N, Ariga M, Motoyama K, Hara A, Kume N, et al. Alteration of myocardial metaiodobenzylguanidine uptake after treatment of phaeochromocytoma and neuroblastoma. Eur J Nucl Med. 2000;27(5):574–82.
48. Nakajo M, Shimabukuro K, Miyaji N, Shimada J, Shirono K, Sakata H, et al. Rapid clearance of iodine-131 MIBG from the heart and liver of patients with adrenergic dysfunction and pheochromocytoma. J Nucl Med Off Publ Soc Nucl Med. 1985;26(4):357–65.

49. Sardesai SH, Mourant AJ, Sivathandon Y, Farrow R, Gibbons DO. Phaeochromocytoma and catecholamine induced cardiomyopathy presenting as heart failure. Br Heart J. 1990;63 (4):234–7.
50. Simons M, Downing SE. Coronary vasoconstriction and catecholamine cardiomyopathy. Am Heart J. 1985;109(2):297–304.
51. Fripp RR, Lee JC, Downing SE. Inotropic responsiveness of the heart in catecholamine cardiomyopathy. Am Heart J. 1981;101(1):17–21.
52. Scognamiglio R, Avogaro A, Casara D, Crepaldi C, Marin M, Palisi M, et al. Myocardial dysfunction and adrenergic cardiac innervation in patients with insulin-dependent diabetes mellitus. J Am Coll Cardiol. 1998;31(2):404–12.
53. Nagaoka H, Iizuka T, Kubota S, Kato N, Suzuki T, Inoue T, et al. Depressed contractile response to exercise in diabetic patients in the absence of cardiovascular disease: relationship to adrenergic cardiac dysinnervation. Nucl Med Commun. 1997;18(8):761–70.
54. Togane Y. Evaluation of the cardiac autonomic nervous system in spontaneously non-insulin-dependent diabetic rats by 123I-metaiodobenzylguanidine imaging. Ann Nucl Med. 1999;13 (1):19–26.
55. Hakusui S, Yasuda T, Yanagi T, Takahashi A, Hasegawa Y, Inoue M. 123I-MIBG myocardial scintigraphical analysis in patients with and without autonomic disorder. Rinsho Shinkeigaku Clin Neurol. 1994;34(4):402–4.
56. Hakusui S, Yasuda T, Yanagi T, Tohyama J, Hasegawa Y, Koike Y, et al. A radiological analysis of heart sympathetic functions with meta-[123I]iodobenzylguanidine in neurological patients with autonomic failure. J Auton Nerv Syst. 1994;49(1):81–4.
57. Orimo S, Ozawa E, Oka T, Nakade S, Tsuchiya K, Yoshimoto M, et al. Different histopathology accounting for a decrease in myocardial MIBG uptake in PD and MSA. Neurology. 2001;57(6):1140–1.
58. Takahashi M, Ikemura M, Oka T, Uchihara T, Wakabayashi K, Kakita A, et al. Quantitative correlation between cardiac MIBG uptake and remaining axons in the cardiac sympathetic nerve in Lewy body disease. J Neurol Neurosurg Psychiatry. 2015;86(9):939–44. doi:10.1136/jnnp-2015-310686.
59. Matsui H, Nishinaka K, Oda M, Komatsu K, Kubori T, Udaka F. Does cardiac metaiodobenzylguanidine (MIBG) uptake in Parkinson's disease correlate with major autonomic symptoms? Parkinsonism Relat Disord. 2006;12(5):284–8. doi:10.1016/j.parkreldis.2005.12.008.
60. Shibata M, Morita Y, Shimizu T, Takahashi K, Suzuki N. Cardiac parasympathetic dysfunction concurrent with cardiac sympathetic denervation in Parkinson's disease. J Neurol Sci. 2009;276(1–2):79–83. doi:10.1016/j.jns.2008.09.005.
61. Takatsu H, Nishida H, Matsuo H, Watanabe S, Nagashima K, Wada H, et al. Cardiac sympathetic denervation from the early stage of Parkinson's disease: clinical and experimental studies with radiolabeled MIBG. J Nucl Med Off Publ Soc Nucl Med. 2000;41(1):71–7.
62. Oka H, Toyoda C, Yogo M, Mochio S. Reduced cardiac 123I-MIBG uptake reflects cardiac sympathetic dysfunction in de novo Parkinson's disease. J Neural Transm. 2011;118 (9):1323–7. doi:10.1007/s00702-011-0598-5.
63. Spiegel J, Hellwig D, Farmakis G, Jost WH, Samnick S, Fassbender K, et al. Myocardial sympathetic degeneration correlates with clinical phenotype of Parkinson's disease. Mov Disord Off J Mov Disord Soc. 2007;22(7):1004–8. doi:10.1002/mds.21499.
64. Goldstein DS, Holmes CS, Dendi R, Bruce SR, Li ST. Orthostatic hypotension from sympathetic denervation in Parkinson's disease. Neurology. 2002;58(8):1247–55.
65. Nakamura T, Hirayama M, Hara T, Mizutani Y, Suzuki J, Watanabe H, et al. Role of cardiac sympathetic nerves in preventing orthostatic hypotension in Parkinson's disease. Parkinsonism Relat Disord. 2014;20(4):409–14. doi:10.1016/j.parkreldis.2014.01.003.
66. Nakamura T, Hirayama M, Hara T, Hama T, Watanabe H, Sobue G. Does cardiovascular autonomic dysfunction contribute to fatigue in Parkinson's disease? Mov Disord Off J Mov Disord Soc. 2011;26(10):1869–74. doi:10.1002/mds.23744.

67. Reuter I, Engelhardt M, Freiwaldt J, Baas H. Exercise test in Parkinson's disease. Clin Auton Res Off J Clin Auton Res Soc. 1999;9(3):129–34.
68. Werner WG, DiFrancisco-Donoghue J, Lamberg EM. Cardiovascular response to treadmill testing in Parkinson disease. J Neurol Phys Ther JNPT. 2006;30(2):68–73.
69. Protas EJ, Stanley RK, Jankovic J, MacNeill B. Cardiovascular and metabolic responses to upper- and lower-extremity exercise in men with idiopathic Parkinson's disease. Phys Ther. 1996;76(1):34–40.
70. Palma JA, Carmona-Abellan MM, Barriobero N, Trevino-Peinado C, Garcia-Lopez M, Fernandez-Jarne E, et al. Is cardiac function impaired in premotor Parkinson's disease? A retrospective cohort study. Mov Disord Off J Mov Disord Soc. 2013;28(5):591–6. doi:10.1002/mds.25431.
71. Nakamura T, Hirayama M, Yamashita F, Uchida K, Hama T, Watanabe H, et al. Lowered cardiac sympathetic nerve performance in response to exercise in Parkinson's disease. Mov Disord Off J Mov Disord Soc. 2010;25(9):1183–9. doi:10.1002/mds.23127.
72. Wichter T, Matheja P, Eckardt L, Kies P, Schafers K, Schulze-Bahr E, et al. Cardiac autonomic dysfunction in Brugada syndrome. Circulation. 2002;105(6):702–6.
73. Haensch CA, Lerch H, Schlemmer H, Jigalin A, Isenmann S. Cardiac neurotransmission imaging with 123I-meta-iodobenzylguanidine in postural tachycardia syndrome. J Neurol Neurosurg Psychiatry. 2010;81(3):339–43. doi:10.1136/jnnp.2008.168484.
74. Haensch CA, Tosch M, Katona I, Weis J, Isenmann S. Small-fiber neuropathy with cardiac denervation in postural tachycardia syndrome. Muscle Nerve. 2014;50(6):956–61. doi:10.1002/mus.24245.
75. De Rosa A, Pappata S, Pellegrino T, De Leva MF, Maddaluno G, Fiumara G, et al. Reduced cardiac 123I-metaiodobenzylguanidine uptake in patients with spinocerebellar ataxia type 2: a comparative study with Parkinson's disease. Eur J Nucl Med Mol Imaging. 2013;40 (12):1914–21. doi:10.1007/s00259-013-2524-6.
76. Hongo M, Urushibata K, Kai R, Takahashi W, Koizumi T, Uchikawa S, et al. Iodine-123 metaiodobenzylguanidine scintigraphic analysis of myocardial sympathetic innervation in patients with AL (primary) amyloidosis. Am Heart J. 2002;144(1):122–9.
77. Druschky A, Spitzer A, Platsch G, Claus D, Feistel H, Druschky K, et al. Cardiac sympathetic denervation in early stages of amyotrophic lateral sclerosis demonstrated by 123I-MIBG-SPECT. Acta Neurol Scand. 1999;99(5):308–14.
78. Druschky A, Hilz MJ, Hopp P, Platsch G, Radespiel-Troger M, Druschky K, et al. Interictal cardiac autonomic dysfunction in temporal lobe epilepsy demonstrated by [(123)I] metaiodobenzylguanidine-SPECT. Brain J Neurol. 2001;124(Pt 12):2372–82.
79. Solanki KK, Bomanji J, Moyes J, Mather SJ, Trainer PJ, Britton KE. A pharmacological guide to medicines which interfere with the biodistribution of radiolabelled meta-iodobenzylguanidine (MIBG). Nucl Med Commun. 1992;13(7):513–21.
80. Guilloteau D, Baulieu JL, Huguet F, Viel C, Chambon C, Valat C, et al. Meta-iodobenzylguanidine adrenal medulla localization: autoradiographic and pharmacologic studies. Eur J Nucl Med. 1984;9(6):278–81.
81. Tobes MC, Jaques Jr S, Wieland DM, Sisson JC. Effect of uptake-one inhibitors on the uptake of norepinephrine and metaiodobenzylguanidine. J Nucl Med Off Publ Soc Nucl Med. 1985;26 (8):897–907.
82. Nakajo M, Shapiro B, Sisson JC, Swanson DP, Beierwaltes WH. Salivary gland accumulation of meta-[131I]iodobenzylguanidine. J Nucl Med Off Publ Soc Nucl Med. 1984;25(1):2–6.
83. Jacobson AF, Travin MI. Impact of medications on mIBG uptake, with specific attention to the heart: comprehensive review of the literature. J Nucl Cardiol Off Publ Am Soc Nucl Cardiol. 2015;22:980–93. doi:10.1007/s12350-015-0170-z.
84. Khafagi FA, Shapiro B, Fig LM, Mallette S, Sisson JC. Labetalol reduces iodine-131 MIBG uptake by pheochromocytoma and normal tissues. J Nucl Med Off Publ Soc Nucl Med. 1989;30(4):481–9.

85. Guilloteau D, Huguet F, Chalon S, Baulieu JL, Gerard P, Narcisse G, et al. MIBG and catecholamine storage in the brain: an in-vitro study. Int J Radiat Appl Inst Part B Nucl Med Biol. 1988;15(6):723–4.
86. Rutgers M, Tytgat GA, Verwijs-Janssen M, Buitenhuis C, Voute PA, Smets LA. Uptake of the neuron-blocking agent meta-iodobenzylguanidine and serotonin by human platelets and neuro-adrenergic tumour cells. Int J Cancer J Int Cancer. 1993;54(2):290–5.
87. Muraoka T, Oku E, Sugataka K, Yamada S. A case of severe parkinsonism associated with short-term treatment with milnacipran. Clin Neuropharmacol. 2008;31(5):299–300. doi:10.1097/WNF.0b013e31815947c7.
88. Yokoyama K, Yamada T, Terachi S, Pu S, Ohta Y, Yamanashi T, et al. Milnacipran influences the indexes of I-metaiodobenzylguanidine scintigraphy in elderly depressed patients. Psychiatry Clin Neurosci. 2014;68(3):169–75.
89. Teresinska A. Metaiodobenzylguanidine scintigraphy of cardiac sympathetic innervation. Nucl Med Rev Central East Eur. 2012;15(1):61–70. doi:10.5603/nmr-18732.

Chapter 18
Pathological Background of Reduced Cardiac MIBG Uptake

Satoshi Orimo

Abstract Cardiac MIBG uptake is reduced in Lewy body diseases, including Parkinson's disease (PD), dementia with Lewy bodies, and pure autonomic failure. It is useful to differentiate Lewy body diseases from other related disorders. Postmortem studies have shown that tyrosine hydroxylase (TH)-immunoreactive axons in the heart are decreased, primarily due to degeneration of the cardiac sympathetic nerve in pathologically confirmed Lewy body disease but not in other related disorders. This supports the findings that reduced cardiac MIBG uptake is found in Lewy body disease. In incidental Lewy body disease (iLBD), TH-immunoreactive axons are relatively preserved, while α-synuclein aggregates accumulate in the heart in abundant numbers. In PD, α-synuclein aggregates are reduced in the heart but are increased in the mother neurons of the cardiac sympathetic nerve in the paravertebral sympathetic ganglia. This distal-dominant degeneration of the cardiac sympathetic nervous system may represent the pathological mechanism underlying the common degenerative process in PD. Furthermore, degeneration of the cardiac sympathetic nerve can occur in familial PD due to PARK1, PARK4, and PARK8 with Lewy bodies in the brain. Therefore, degeneration of the cardiac sympathetic nerve is closely related to the presence of Lewy bodies in a wide range of neurodegenerative processes.

Keywords Tyrosine hydroxylase • Neurofilament • α-Synuclein • Cardiac sympathetic nerve • Degeneration • Lewy bodies • Lewy neurites

18.1 The Pathological Progression in Parkinson's Disease

The histological hallmarks of Parkinson's disease (PD) and dementia with Lewy bodies (DLB) are neuronal α-synuclein aggregates called Lewy bodies and Lewy neurites (Lewy body pathology). There are many molecules that have been identified in Lewy body pathology, but α-synuclein is considered to be a major

S. Orimo (✉)
Division of Neurology, Kanto Central Hospital of the Mutual Aid Association of Public School Teachers, 6-25-1 Kami-Yoga, Setagaya-ku, Tokyo 158-8531, Japan
e-mail: sorimo0307@nifty.com

S. Iwase et al. (eds.), *Clinical Assessment of the Autonomic Nervous System*,
DOI 10.1007/978-4-431-56012-8_18

constituent of these inclusions [1]. Braak et al. examined incidental Lewy body disease (iLBD) and PD brains and proposed a pathological staging schema for PD [2]. Patients with iLBD have no apparent parkinsonism and dementia in life, but Lewy body pathology is incidentally found on postmortem examination. These are considered to be presymptomatic PD or DLB. In a proposed pathological staging scheme for PD, α-synuclein aggregates are detected in the dorsal vagal nucleus and in the olfactory bulb in the earliest stage of disease. This staging system characterizes a progression from the dorsal vagal nucleus (stage I), through the pontine tegmentum (stage II), into the midbrain and neostriatum (stage III), into the basal prosencephalon and mesocortex (stage IV), and finally though the neocortex (stages V and VI) [2].

In the peripheral nervous system, Lewy body pathology was found in the stomach and the distal part of the esophagus of all five patients with iLBD and PD [3], in the Meissner's plexus and the sympathetic trunk of 14/17 (82 %) patients with iLBD [4], and in the epicardial nerve fascicle of 18/20 (90 %) patients with iLBD [5]. Lewy bodies were found in the sinoatrial node of 33–50 % patients with Lewy body disease, which is presumably related to arrhythmia and ischemic heart disease [6]. Degeneration of the cardiac nerve fibers in the right atrium and conducting system of the heart were also observed [7]. Minguez-Castellanos examined surgical specimens from 100 patients without apparent neurological disorders (44–84 years of age) and reported that α-synuclein aggregates were found in the abdominopelvic autonomic plexuses in 9 % of the entire patient sample but were more common in vesicoprostatic (26.1 %) specimens than in digestive tract specimens (3.9 %) [8]. Navarro-Otano examined the surgical heart specimens from 91 patients without parkinsonism (31–84 years of age) and reported that α-synuclein aggregates were found in the epicardial fat tissues in 7.7 % of the entire patient sample [9]. Recently, Gelpi reported results from a postmortem histopathological study of the brain and peripheral tissues from 28 patients (10 with PD, 5 with DLB, and 13 with non-Lewy body disease). α-Synuclein aggregates were found in the peripheral autonomic nervous system of all 15 individuals with Lewy body disease in the stellate and the sympathetic ganglia (100 %), the vagus nerve (86.7 %), the gastrointestinal tract (86.7 %), the adrenal gland and/or surrounding fat (53.3 %), the heart (100 %), and the genitourinary tract (13.3 %) [10]. These findings demonstrated that the dorsal vagal nucleus and the olfactory bulb exhibit the first lesions in PD prior to the substantia nigra in the brain. Moreover, lesions in the peripheral autonomic nervous system begin early in the disease process.

18.2 Anatomy of the Cardiac Sympathetic Nerve

Sympathetic cardiopulmonary nerves arise from the stellate ganglia and the caudal halves of the cervical sympathetic trunks below the level of the cricoid cartilage. Parasympathetic cardiopulmonary nerves arise from the recurrent laryngeal nerves and the thoracic vagi just distal to them. These nerves connect with the sympathetic cardiopulmonary nerves anterior and posterior to the main pulmonary artery to form the ventral and dorsal cardiopulmonary plexuses. Three large cardiac nerves arise from these plexuses and are called the right and left coronary cardiac nerves and left lateral cardiac nerve [11]. Thus, in the cardiac nerves, there are different kinds of nerve fibers, including both efferent sympathetic and parasympathetic nerves and afferent nerves, which are divided into unmyelinated or myelinated axons [12]. The efferent cardiac sympathetic nerve fibers are basically unmyelinated axons because they are postganglionic axons. However, the precise components of these axons still need to be clarified.

18.3 Degeneration of the Cardiac Sympathetic Nerve in Parkinson's Disease

Lewy bodies and Lewy neurites are found not only in the central nervous system but also in the peripheral autonomic nervous system, such as in the sympathetic ganglia, the enteric nervous system of the alimentary tract, the cardiac plexus, the pelvic plexus, and the adrenal medulla, in patients with PD [13]. To clarify the pathological basis of reduced cardiac MIBG uptake in PD, Orimo et al. used immunohistochemistry to examine cardiac tissues from pathologically confirmed patients with PD, pure autonomic failure (PAF), and multiple system atrophy (MSA) by using an antibody against tyrosine hydroxylase (TH), the rate-limiting enzyme of catecholamine biosynthesis [14, 15]. The left ventricular anterior wall was selected in the study, because cardiac MIBG uptake is mainly observed in the left ventricle and TH-immunoreactive nerve fibers are more numerous in the anterior than in the posterior wall of the left ventricle [16]. These studies demonstrated a near-complete depletion of TH-immunoreactive nerve fibers in the epicardial nerve fascicles in PD and PAF, but not in MSA and control [14, 15]. These findings revealed the involvement of the cardiac sympathetic nerve in PD and PAF, which presumably accounts for the reduced cardiac MIBG uptake seen in these disorders. Moreover, Amino et al. observed that in addition to TH-immunoreactive nerve fibers, nerve fibers immunoreactive to neurofilament (NF. an axonal marker) and myelin basic protein (MBP, a myelin marker) were markedly decreased in four pathologically confirmed patients with PD (Hoehn–Yahr stage 4–5) (Fig. 18.1) [17]. However, S100 protein, a Schwann cell marker, was well preserved in all patients with PD [17] (Fig. 18.1). Degeneration of the cardiac sympathetic nerve in a patient with autopsy-confirmed PD was also reported by Mitsui [18]. These findings clearly demonstrate that the degeneration of the cardiac sympathetic nerve occurs in PD.

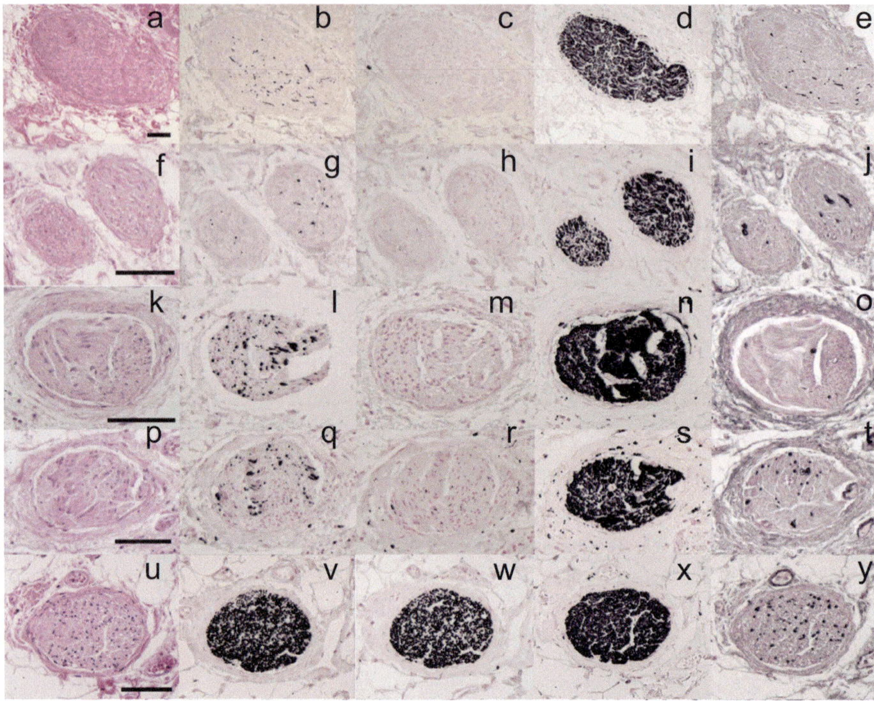

Fig. 18.1 Hematoxylin and eosin (H&E) and immunohistochemical staining of the epicardial nerve fascicle from controls and patients with PD. In all patients with PD (PD 1, 2, 3, 4), neurofilament (NF)-immunoreactive and tyrosine hydroxylase (TH)-immunoreactive nerve fibers were markedly decreased, compared to control. In contrast, S100 protein-immunoreactive structures were well preserved in all patients with PD as well as the control subject. A few myelin basic protein (MBP)-immunoreactive structures were seen in the epicardial nerve fascicles of all patients with PD; this number was smaller in patients with PD compared to the control subject. **a**, **b**, **c**, **d**, **e**, PD 1; **f**, **g**, **h**, **i**, **j**, PD 2; **k**, **l**, **m**, **n**, **o**, PD 3; **p**, **q**, **r**, **s**, **t**, PD 4; **u**, **v**, **w**, **x**, **y**, control; **a**, **f**, **k**, **p**, **u**, H&E; **b**, **g**, **l**, **q**, **v**, neurofilament; **c**, **h**, **m**, **r**, **w**, tyrosine hydroxylase; **d**, **i**, **n**, **s**, **x**, S100 protein; **e**, **j**, **o**, **t**, **y**, myelin basic protein. Scale bar = 100 μm (Adapted from Ref. [17] with permission)

18.4 Degeneration of the Cardiac Sympathetic Nerve Is Specific to Lewy Body Disease

Cardiac tissues from patients with PD and a variety of other neurodegenerative disorders were immunohistochemically examined using an antibody against TH to determine whether degeneration of the cardiac sympathetic nerve is specific to Lewy body disease [19]. TH-immunoreactive nerve fibers in the epicardial nerve fascicles were abundant in MSA, progressive supranuclear palsy (PSP), and Alzheimer disease (AD) as well as in controls, whereas no TH-immunoreactive nerve fibers were observed in PD and DLB. To assess the number of TH-immunoreactive nerve fibers in the epicardial nerve fascicles, a semiquantitative rating scale was used: absent or nearly absent, sparse, moderate, and numerous. The

TH-immunoreactive nerve fibers were absent or nearly absent in almost all patients with PD, DLB, and DLB/AD regardless of age, illness duration, and the presence of orthostatic hypotension. In contrast, the number of TH-immunoreactive nerve fibers was moderate to numerous in patients with MSA, PSP, CBD, and pure AD. The number of TH-immunoreactive nerve fibers was also moderate to numerous in all except one patient with MSA. In the myocardium, TH-immunoreactive nerve fibers were markedly decreased in PD, DLB, and DLB/AD, but were preserved in MSA, PSP, and pure AD. These findings demonstrate that degeneration of the cardiac sympathetic nerve is specific to Lewy body disease.

18.5 Degeneration of the Cardiac Sympathetic Nerve Begins in the Early Disease Stages of Parkinson's Disease

Cardiac MIBG uptake is frequently reduced even in the early stages of PD (Hoehn–Yahr stage 1 or 2), which suggests early involvement of the cardiac sympathetic nerve, even though routine autonomic function tests fail to detect abnormal autonomic functions [20, 21]. Braak et al. have reported detailed pathological stages for PD progression. Early pathological changes begin in the lower part of the brain stem, particularly in the dorsal vagal nucleus before nigral involvement [2]. However, when and how the degeneration of the cardiac sympathetic nerve begins in the disease process of PD remains to be clarified. The cardiac tissues, sympathetic ganglia, and medulla oblongata at the level of the dorsal vagal nucleus from 20 pathologically confirmed patients with iLBD and ten control subjects were immunohistochemically examined using antibodies against TH and NF to determine when degeneration of the cardiac sympathetic nerve begins [22]. TH- and NF-immunoreactive nerve fibers of the epicardial nerve fascicles were well preserved in ten of the 20 patients with iLBD as well as in control subjects (Fig. 18.2). In contrast, TH-immunoreactive nerve fibers had almost entirely disappeared in six patients with iLBD and were moderately decreased in four of the 20 patients with iLBD. Neuronal cell loss in the dorsal vagal nucleus and the sympathetic ganglia was not detectable in any of the examined iLBD patients (Fig. 18.2). These findings suggest that the degeneration of the cardiac sympathetic nerve begins early in the disease process of PD, prior to neuronal cell loss in the dorsal vagal nucleus, accounting for the reduced cardiac MIBG uptake in early PD stages [22]. Fujishiro reported that in PD, the degree of Lewy body pathology in the epicardial nerve fascicles correlated with PD duration, Hoehn–Yahr stage, and Braak' stage [23]. Miki reported on a 35-year-old man without parkinsonism who was later shown to have Lewy body pathology only in the cardiac sympathetic nerve and stellate ganglia during autopsy [24]. The pathological mechanism of this case cannot be explained by the "dual hit" theory [25].

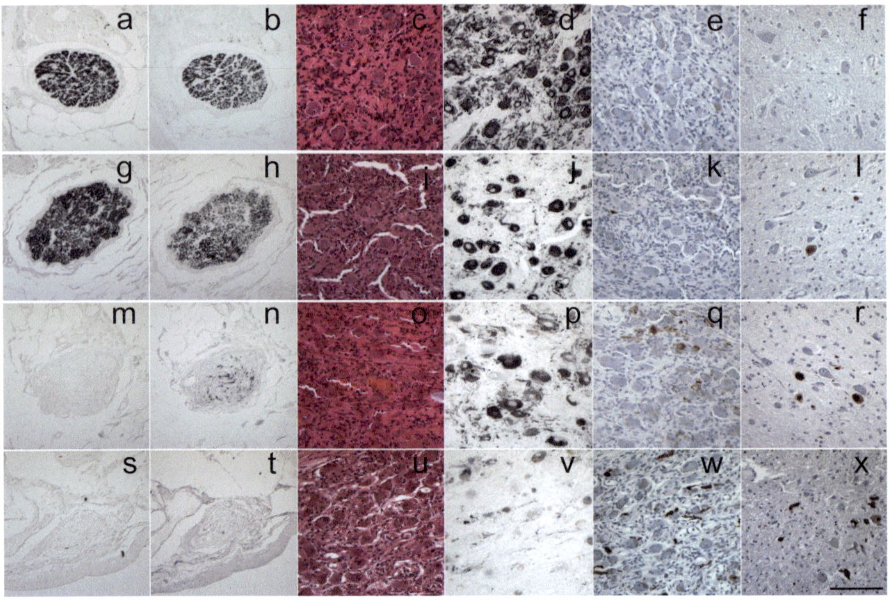

Fig. 18.2 Immunohistochemical staining of the epicardial nerve fascicle, sympathetic ganglia, and medulla oblongata at the level of the dorsal vagal nucleus in incidental Lewy body disease (iLBD) and control subjects. In the control, tyrosine hydroxylase (TH)- or neurofilament (NF)-immunoreactive nerve fibers of the epicardial nerve fascicle were well preserved (**a**, **b**). No neuronal cell loss (**c**), abundant TH-immunoreactive neurons (**d**), and no α-synuclein-immunoreactive Lewy neurites and Lewy bodies (**e**) were observed in the sympathetic ganglia. Neuronal cell loss and α-synuclein-immunoreactive Lewy neurites and Lewy bodies were not observed in the dorsal vagal nucleus (**f**). In iLBD, there were different degenerating stages of nerve fibers in the fascicles (**g**, **h**, **m**, **n**, **s**, **t**). No neuronal cell loss was noticed in the sympathetic ganglia (**i**, **o**, **u**). TH immunoreactivity in neurons was well preserved in 18 patients (**j**). The number of TH-immunonegative neurons was slightly increased in one patient (**p**) and moderately increased in one patient with iLBD (**v**). The severity of Lewy body pathology was slight (**k**), moderate (**q**), and severe (**w**) in the sympathetic ganglia and slight (l, **r**) and moderate (**x**) in the dorsal vagal nucleus. **a**, **b**, **c**, **d**, **e**, **f**, control; **g**, **h**, **i**, **j**, **k**, **l**, patient 3; **m**, **n**, **o**, **p**, **q**, **r**, patient 18; **s**, **t**, **u**, **v**, **w**, **x**, patient 20; **a**, **g**, **m**, **s**, cardiac tissue/TH; **b**, **h**, **n**, **t**, cardiac tissue/NF; **c**, **i**, **o**, **u**, sympathetic ganglia (SG)/hematoxylin and eosin; **d**, **j**, **p**, **v**, SG/TH; **e**, **k**, **q**, **w**, SG/phosphorylated α-synuclein; **f**, **l**, **r**, **x**, dorsal vagal nucleus/phosphorylated α-synuclein. Bar = 50 μm (Adapted from Ref. [22] with permission)

18.6 Degeneration of the Cardiac Sympathetic Nerve in Familial Parkinson's Disease

To date, there have been 20 reported familial PD mutations [26]. It has been reported that cardiac MIBG uptake varies depending on the type of mutation. In four different familial PD mutations, there have been reports regarding whether degeneration of the cardiac sympathetic nerve is pathologically involved.

18.6.1 *PARK1 (PD Associated with an* α-Synuclein *Mutation)*

The PARK1 clinical phenotype is characterized by severe parkinsonism with dementia and sleep disturbances [27]. Neuropathological examinations have revealed extensive Lewy bodies and Lewy neurites in the cortical and subcortical structures [28]. Tijero et al. previously reported that one asymptomatic carrier of the E41K mutation of *α-synuclein* showed markedly reduced cardiac MIBG uptake [29]. Subsequently, they performed cardiac MIBG scintigraphy in six patients with PARK1, four of whom were symptomatic (ages: 28, 46, 52, and 59 years old) and two who were asymptomatic carriers (ages: 29 and 52 years old). The four symptomatic carriers and the older asymptomatic carrier had markedly reduced cardiac MIBG uptake. There was a complete absence of TH-immunoreactive nerve fibers in the epicardial nerve fascicles from the two autopsied patients [30].

18.6.2 *PARK2 (Parkin Disease, PD Linked to* parkin *Mutation)*

Orimo et al. described the first three cases of pathologically confirmed familial PD with a *parkin* mutation [31]. Patient 1 was a 73-year-old man with a homozygous exon 4 deletion in the *parkin* gene. Cardiac MIBG uptake at the age of 70 (46 years after onset) was normal, both in the early and the delayed phase. Postmortem examination revealed the loss of pigmented neurons and gliosis in the substantia nigra pars compacta and locus coeruleus, with no Lewy body pathology. Patient 2 was a 66-year-old man with a homozygous exon 4 deletion in the *parkin* gene. Cardiac MIBG uptake at the age of 63 years (35 years after onset) was normal in the early phase and slightly reduced in the delayed phase. Postmortem examination revealed a loss of pigmented neurons and gliosis in the substantia nigra pars compacta and locus coeruleus, with no Lewy body pathology. The weight of the heart was 400 g and a histological examination was unremarkable. Patient 3 was a 70-year-old man with a homozygous exon 4 deletion in the *parkin* gene. He had been previously described clinically and neuropathologically [32]. Postmortem examination revealed a loss of pigmented neurons and gliosis in the substantia nigra pars compacta and locus coeruleus, with no Lewy pathology. Cardiac tissues from the three patients were immunohistochemically examined using an antibody against TH to determine whether the cardiac sympathetic nerve was pathologically involved. Postmortem examination revealed that TH-immunoreactive nerve fibers in the epicardial nerve fascicles were well preserved in all three patients. This study confirmed that the cardiac sympathetic nerve is well preserved in PARK2 with a homozygous exon deletion, which accounts for the normal cardiac MIBG uptake. Moreover, normal cardiac MIBG uptake may have potential diagnostic value to indicate the absence of Lewy body pathology, even in patients with levodopa-responsive parkinsonism, such as in parkin disease.

18.6.3 PARK4 (PD Linked to Multiplication of α-Synuclein*)*

Familial PD linked to the multiplication of *α-synuclein* is autosomal dominant and is characterized by an early onset age, autonomic disturbance, and rapidly progressive dopa-responsive parkinsonism, followed by dementia [33]. Reduced cardiac MIBG uptake or fluorodopamine uptake on positron emission tomography has been reported in patients with *α-synuclein* multiplication [34, 35]. Cardiac tissues from three patients with PD linked to *α-synuclein* duplication were immunohistochemically examined to determine whether the cardiac sympathetic nerve was pathologically involved [36]. The proband (patient 1) developed parkinsonism at the age of 61 and died at the age of 70. Her uncle (patient 2), who was also her husband, developed parkinsonism at the age of 71 and died at the age of 78. Their elder son (patient 3) developed parkinsonism at the age of 39 and died at the age of 54. The clinical and neuropathological findings of these patients have been reported elsewhere [37]. Quantitative PCR analysis demonstrated that all the patients were heterozygous for an *α-synuclein* duplication [38]. In each of the three patients, sections from the anterior wall of the left ventricle and the paravertebral sympathetic ganglia were immunostained with antibodies against TH, NF, or phosphorylated α-synuclein. TH-immunoreactive nerve fibers were markedly decreased and NF-immunoreactive nerve fibers were moderately decreased, with sparse α-synuclein aggregates, in the epicardial nerve fascicles from all three patients [36]. This study indicates that degeneration of the cardiac sympathetic nerve occurs in PD linked to *α-synuclein* duplication, accounting for the reduced cardiac MIBG or fluorodopamine uptake. Moreover, it further supports our previous conclusion that the degeneration of the cardiac sympathetic nerve is closely related to the presence of Lewy bodies.

18.6.4 PARK8 (PD Associated with LRRK2 *Mutations)*

PD associated with *LRRK2* mutations is an autosomal dominant form of familial PD and is characterized by middle to late onset as well as dopa-responsive parkinsonism similar to sporadic PD. Dementia is not common [39]. Neuropathological findings are mostly inconsistent, showing both Lewy pathology (and sometimes tau- and ubiquitin-containing inclusions) and pure nigral degeneration without Lewy bodies, with or without neurofibrillary tangles [40]. In patients with PD associated with a *LRRK2* mutation, such as the Sagamihara family [41] with I2020T *LRRK2* mutation, cardiac MIBG uptake was normal to slightly reduced in most patients in this family [42]. Cardiac fluorodopamine uptake was reduced in a patient with PD linked to G2019S *LRRK2* mutation [43]. Cardiac tissues from four patients with PD associated with *LRRK2* mutation were immunohistochemically examined to determine whether the cardiac sympathetic nerve was pathologically involved. The clinical and pathological findings of the three patients have been previously reported [40–42]. In one of the three patients, Lewy body pathology was found in the central nervous system but was absent in the other two patients. The fourth patient was a 68-year-old woman. At the age of 51, she developed a gait disturbance and was diagnosed with PD. These features responded well to levodopa. She developed "wearing-off" motor fluctuations since the age of 57 and visual hallucination at the age of 64. At the age of 68, she died of pneumonia. Cognitive impairments and autonomic dysfunctions were not found. The fourth patient was genetically determined to have an I2020T amino acid substitution in *LRRK2*. Cardiac MIBG uptake at the age of 66 (in Hoehn and Yahr stage 5) was normal in the early phase and slightly reduced in the delayed phase. Macroscopic examinations of the brain revealed a marked depigmentation in the substantia nigra, where marked neuronal loss and gliosis with extraneuronal melanin were observed. These changes were prominent in the substantia nigra pars reticulata. In the locus coeruleus, only mild gliosis was observed and the dorsal motor nucleus of the vagus was almost normal. No Lewy bodies and Lewy neurites were found in any brain lesions using hematoxylin and eosin and phosphorylated α-synuclein immunostaining. In three of four patients who had no Lewy bodies and Lewy neurites, the cardiac sympathetic nerves were well preserved. In one patient who had Lewy bodies in the brain, the cardiac sympathetic nerve was markedly reduced, and a small number of α-synuclein aggregates in the epicardial nerve fascicles were observed. The present study indicates that degeneration of the cardiac sympathetic nerve is closely related to the presence of Lewy bodies in the central nervous system in PD associated with *LRRK2* mutations in the Sagamihara family.

18.7 Degeneration of the Cardiac Sympathetic Nerve Can Occur in Multiple System Atrophy

In patients with MSA, cardiac MIBG uptake, which is a sensitive biological marker for the differential diagnosis between MSA and PD, is usually normal [20, 44–48]. Cardiac MIBG uptake was slightly reduced in some patients with MSA [20, 44–46]. However, the pathophysiological mechanism accounting for the slight reduction in cardiac MIBG uptake in MSA remains to be clarified. Subsequently, cardiac tissue and sympathetic ganglia from patients with MSA were immunohistochemically examined to determine whether the cardiac sympathetic nerve was pathologically involved [49]. Fifteen patients with pathologically confirmed MSA, along with ten control subjects, were pathologically examined using antibodies against TH and NF. TH-immunoreactive nerve fibers in the epicardial nerve fascicle were preserved in 8 of 15 patients with MSA, as well as in ten control subjects. The number of TH-immunoreactive, but not NF-immunoreactive, nerve fibers in the epicardial nerve fascicle was mildly or moderately decreased in six patients with MSA, four of whom showed a decrease of TH immunoreactivity in the neuronal somata in the sympathetic ganglia (Fig. 18.3). Moreover, TH- and NF-immunoreactive nerve fibers almost entirely disappeared in the heart of one patient with MSA, in whom Lewy neurites were present in the sympathetic ganglia (Fig. 18.3). These findings suggest that mild degeneration of the cardiac sympathetic nerve can occur in MSA, which is closely related to the pathological change in the sympathetic ganglia neurons, accounting for the slight reduction in cardiac MIBG uptake [49]. The intermediolateral nucleus of the spinal cord is more severely involved in MSA than in control [50, 51] and PD [52, 53] subjects. Loss of excitatory preganglionic inputs from the intermediolateral cell column could reduce firing rates of sympathetic ganglion cells and thus downregulate TH expression in an activity-dependent manner. Taken together, mild degeneration of the cardiac sympathetic nerve can

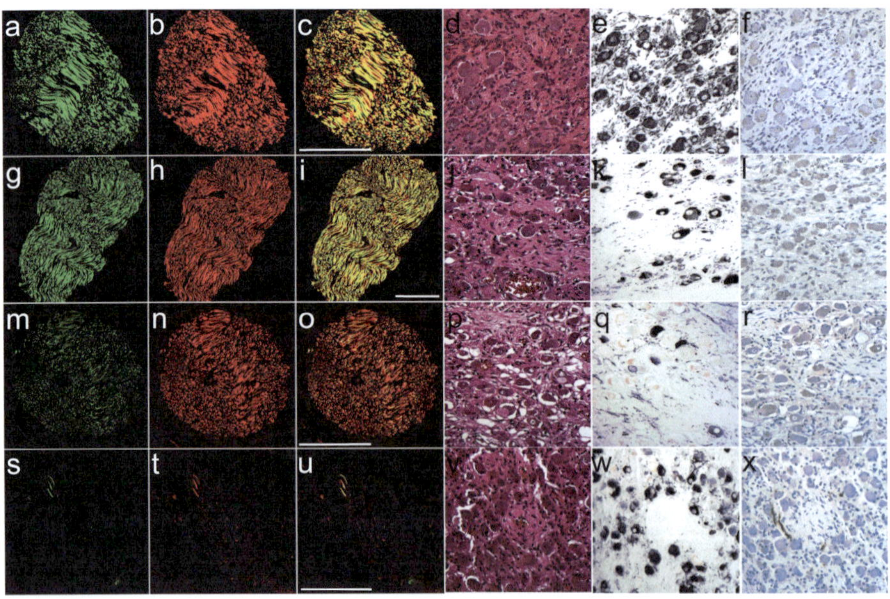

Fig. 18.3 Immunohistochemical staining of the epicardial nerve fascicle and the sympathetic ganglia in multiple system atrophy (MSA) and control subjects. In the control, tyrosine hydroxylase (TH)- or neurofilament (NF)-immunoreactive nerve fibers of the epicardial nerve fascicle were well preserved (**a**, **b**, **c**). No neuronal cell loss, abundant TH-immunoreactive neurons, and no α-synuclein-immunoreactive structures were observed in the sympathetic ganglia (**d**, **e**, **f**). In MSA, there were different patterns of degeneration of the nerve fibers in the epicardial nerve fascicle (**g**, **h**, **i**, **m**, **n**, **o**, **s**, **t**, **u**). No neuronal cell loss was observed in the sympathetic ganglia (**j**, **p**, **v**). TH-immunoreactive neurons were well preserved in patients 1 and 15 (**k**, **w**), while a number of TH-immunoreactive neurons were decreased in patient 12 (**q**). α-Synuclein-immunoreactive Lewy neurites were observed in patient 15 (**x**). **a**, **b**, **c**, **d**, **e**, **f**, control; **g**, **h**, **i**, **j**, **k**, **l**, patient 1; **m**, **n**, **o**, **p**, **q**, **r**, patient 12; **s**, **t**, **u**, **v**, **w**, **x**, patient 15; **a**, **g**, **m**, **s**, cardiac tissue/TH; **b**, **h**, **n**, **t**, cardiac tissue/NF; **c**, **i**, **o**, **u**, cardiac tissue/merge; **d**, **j**, **p**, **v**, SG/H&E; **e**, **k**, **q**, **w**, SG/TH; **f**, **l**, **r**, **x**, SG/phosphorylated α-synuclein. White bar (**a**, **b**, **c**, **g**, **h**, **i m**, **n**, **o**, **s**, **t**, **u**) = 100 μm, black bar (**d**, **e**, **f**, **j**, **k**, **l**, **p**, **q**, **r**, **v**, **w**, **x**) = 50 μm (Adapted from Ref. [49] with permission)

occur in MSA, and the mechanism might be through transsynaptic degeneration [54]. Moreover, concurrent Lewy body pathology in the sympathetic ganglia may accelerate the degeneration of the cardiac sympathetic nerve, even in MSA.

18.8 Axonal α-Synuclein Aggregates Herald the Centripetal Degeneration of Cardiac Sympathetic Nerve in Parkinson's Disease

Degeneration of the cardiac sympathetic nerve occurs in both PD and DLB and begins during early stages of PD, accounting for the reduced cardiac MIBG uptake, even in the early stages of Lewy body disease. Orimo et al. previously demonstrated that degeneration of the distal axons of the cardiac sympathetic nerve precedes loss of their mother neurons in the paravertebral sympathetic ganglia [19], suggesting a distal-dominant degeneration of the cardiac sympathetic nerve in PD. Because α-synuclein is one of the key molecules in PD pathogenesis, Orimo et al. further investigated how α-synuclein aggregates are involved in this distal-dominant degeneration in PD [5]. Both cardiac tissues and paravertebral sympathetic ganglia were obtained from 20 patients with iLBD, 10 with PD, 20 with MSA, and 10 control subjects for comparison. Immunohistochemical analysis was performed using antibodies against TH, NF, and phosphorylated α-synuclein. These results demonstrated that (1) α-synuclein aggregates in the epicardial nerve fascicles, namely, the distal axons of the cardiac sympathetic nerve, were much more abundant in iLBD; (2) α-synuclein aggregates in the epicardial nerve fascicles were closely associated with the disappearance of TH-immunoreactive nerve fibers; (3) in cases of iLBD with preserved TH-immunoreactive nerve fibers, α-synuclein aggregates were consistently more abundant in the epicardial nerve fascicles than in the paravertebral sympathetic ganglia; (4) this distal-dominant accumulation of α-synuclein aggregates was reversed in iLBD with decreased TH-immunoreactive nerve fibers and PD, both of which showed decreased or depleted TH-immunoreactive nerve fibers, but more abundant α-synuclein aggregates in the paravertebral sympathetic ganglia; and (5) MSA had a completely different profile compared to iLBD and PD, based on the preservation of TH-immunoreactive nerve fibers and the scarcity of α-synuclein aggregates in either cardiac tissues or paravertebral sympathetic ganglia. These findings indicate that accumulation of α-synuclein aggregates in the distal axons of the cardiac sympathetic nervous system precedes aggregate accumulation in neuronal somata or neurites in the paravertebral sympathetic ganglia. This heralds the centripetal degeneration of the cardiac sympathetic nerve in PD, which is in sharp contrast to the slight changes in MSA (Fig. 18.4). This chronological and dynamic relationship between α-synuclein aggregates and distal-dominant degeneration of the cardiac sympathetic nervous system may represent a pathological mechanism underlying a common degenerative process in PD [5].

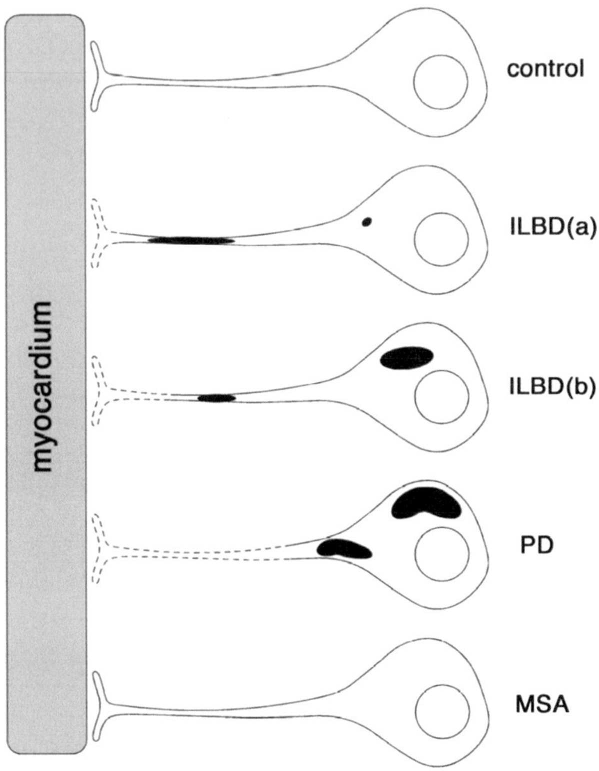

Fig. 18.4 Schematic illustration of the degenerative process of the cardiac sympathetic nervous system. In iLBD (**a**), α-synuclein aggregates abundantly accumulated in the distal axons in contrast to the sparse number of α-synuclein aggregates in the paravertebral sympathetic ganglia. In iLBD (**b**), α-synuclein aggregates in the distal axons were diminished but were increased in number and quantity in the paravertebral sympathetic ganglia. In PD, α-synuclein aggregates in the distal axons disappeared, along with a regression of TH-immunoreactive axons, whereas α-synuclein aggregates were much more abundant in the paravertebral sympathetic ganglia. In MSA, α-synuclein aggregates were generally not similar to controls, with a few exceptions. Black shading indicates α-synuclein aggregates. The line indicates the outline of TH-immunoreactive axons or their mother neurons. The *dotted line* indicates degeneration of TH-immunoreactive axons. iLBD (**a**), iLBD with preserved TH-immunoreactive axons; iLBD (**b**), iLBD with decreased TH-immunoreactive axons (Adapted from Ref. [5] with permission)

18.9 Unmyelinated Cardiac Nerve Axons Are More Vulnerable to Degeneration Than Myelinated Cardiac Nerve Axons in Parkinson's Disease

To determine whether there was a difference in the degenerative process between myelinated and unmyelinated cardiac nerve axons, cardiac tissue from four pathologically confirmed PD patients, nine patients with iLBD, and five control subjects was immunohistochemically examined using antibodies against NF, MBP, and phosphorylated α-synuclein [55]. First, the number of NF-immunoreactive unmyelinated (MBP negative) and myelinated (MBP positive) axons were counted. Next, the number of unmyelinated and myelinated axons with α-synuclein aggregates was counted. The percentage of unmyelinated axons in PD (77.5 ± 9.14 %) and iLBD (80.4 ± 9.54 %) was significantly lower compared to control subjects (91.6 ± 2.36 %). The ratio of unmyelinated axons with α-synuclein aggregates to total axons with α-synuclein aggregates ranged from 94.4 to 100 (98.2 ± 2.18 %). Among axons with α-synuclein aggregates, unmyelinated axons comprised the overwhelming majority (98.2 %). These findings suggest that in PD, unmyelinated cardiac nerve axons are more vulnerable to degeneration than myelinated cardiac nerve axons, because α-synuclein aggregates accumulated more abundantly in unmyelinated axons [55]. Figure 18.5 shows triple immunofluorescence labeling of the epicardial nerve fascicles with NF, MBP, and α-synuclein in iLBD and a control subject. In the control subject, unmyelinated and myelinated axons were seen (Fig. 18.5a), but there were no α-synuclein aggregates observed in the epicardial nerve fascicle (Fig. 18.5j). In iLBD, α-synuclein aggregates (Fig. 18.5k) were seen in unmyelinated axons but not in myelinated axons (Fig. 18.5b). In another iLBD patient, most α-synuclein aggregates co-localized with unmyelinated axons (Fig. 18.5c), whereas some aggregates were observed in myelinated axons (Fig. 18.5c, arrow head, inset). Sometimes, swollen axons with varicose-like accumulations of α-synuclein aggregates were observed only in unmyelinated axons (Fig. 18.5c). Braak et al. reported that unmyelinated axons and long or sparsely myelinated axons were preferentially seen in the PD brain [2, 56]. Their study strongly suggested that unmyelinated or sparsely myelinated axons are more vulnerable to degeneration compared to myelinated axons in the PD brain. Inversely, myelinated axons appeared to be more resistant to degeneration. This relative resistance of myelinated axons may be due to the well-developed myelin sheath. The velocity of axonal conduction increases with the growing thickness of the myelin sheath. A thick myelin sheath reduces the metabolic demands of neurons for impulse transmission [57]. In other words, neurons with unmyelinated or sparsely myelinated axons require such high metabolic demands that these neurons are chronically subjected to high-energy turnover, resulting in permanent exposure to oxidative stress [58]. Indeed, the loss of myelin would require an approximately 5000-fold increase in neuronal energy expenditure to maintain neurotransmission levels [57, 59].

18.10 The Ultrastructural Profile of the Cardiac Nerve in PD

Previously, Orimo et al. had clearly demonstrated the characteristic profile of unmyelinated and myelinated axons of epicardial nerve fascicles from the left cardiac ventricle in PD and control subjects, using electron microscopy [55]. There were many unmyelinated axons (Fig. 18.6a, open arrow head) intermingled

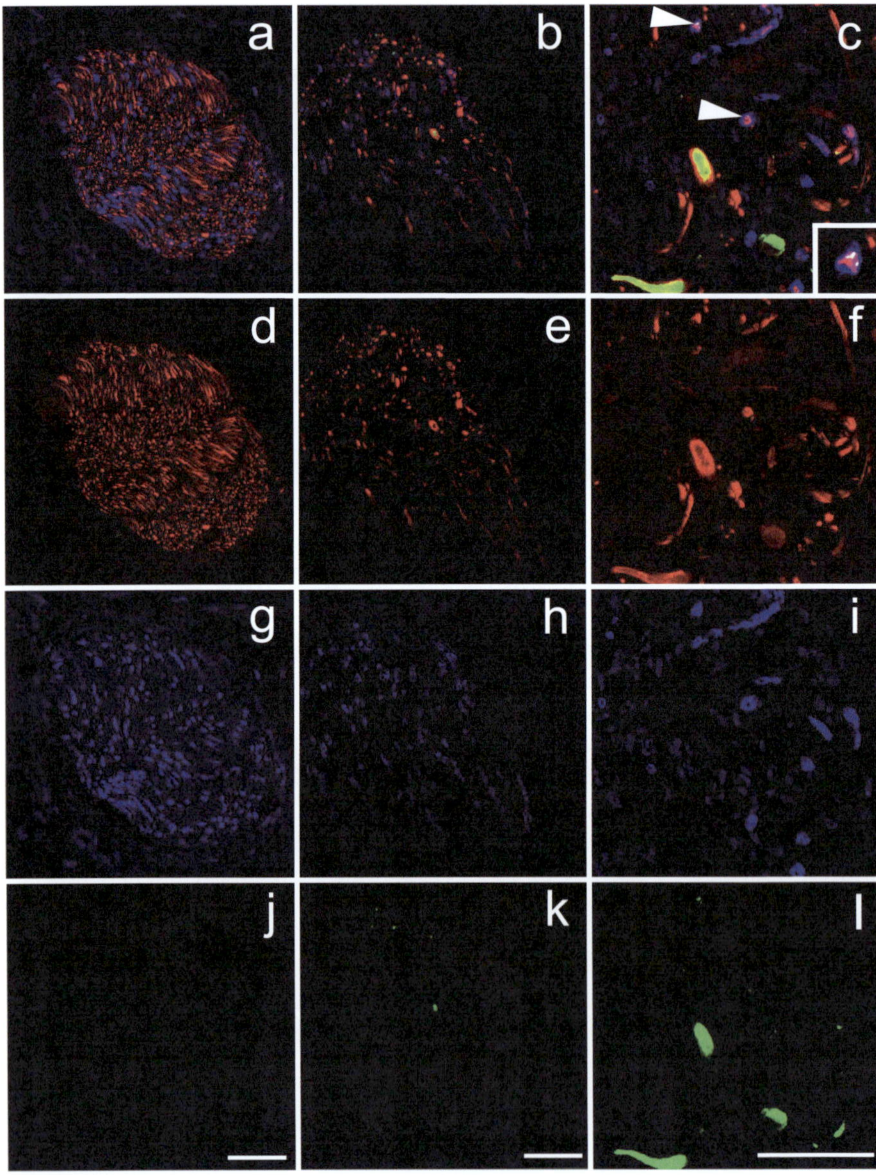

Fig. 18.5 Triple immunofluorescence labeling with neurofilament (NF), myelin basic protein (MBP), and α-synuclein of epicardial nerve fascicles in iLBD and a control subject. In the control subject, unmyelinated and myelinated axons were seen (**a**), but there were no α-synuclein aggregates observed in the epicardial nerve fascicle (**j**). In iLBD, α-synuclein aggregates (**k**) were observed in unmyelinated axons but not in myelinated axons (**b**). Most of the α-synuclein aggregates were co-localized with unmyelinated axons (**c**), whereas some were observed in myelinated axons (**c**; *arrow* head, inset). Some swollen axons with varicose-like accumulations of α-synuclein aggregates were observed only in unmyelinated axons (**c**). **a**, **b**, **c**, merge; **d**, **e**, **f**, NF; **g**, **h**, **i**, MBP; **j**, **k**, **l**, α-synuclein; **a**, **d**, **g**, **j**, control, **b**, **e**, **h**, **k**, patient 13; **c**, **f**, **i**, **l**, patient 7. Bar = 50 μm (Adapted from Ref. [55] with permission)

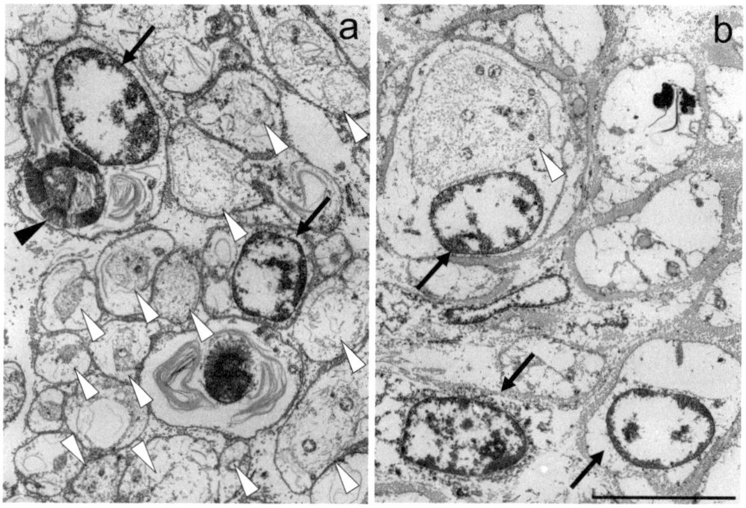

Fig. 18.6 Electron microscopic profiles of epicardial nerve fascicles in PD and a control subject. There were many unmyelinated axons (**a**, *open arrow head*) intermingled with Schwann cells (**a**, *arrow*) with or without a myelinated axon (**a**, *closed arrow head*) in the control subject. There were a markedly decreased number of unmyelinated axons (**b**, *open arrow head*) with preserved Schwann cells (**b**, *arrow*) in PD. Bar = 5 μm (Adapted from Ref. [55] with permission)

with Schwann cells (Fig. 18.6a, arrow), with or without a myelinated axon (Fig. 18.6a, closed arrow head) in the control subject. In contrast, there was a marked decrease in the number of unmyelinated axons (Fig. 18.6b, open arrow head) with preserved Schwann cells (Fig. 18.6b, arrow) in PD. In addition, Schwann cells seemed to be slightly hypertrophic.

18.11 Quantitative Correlation Between Cardiac MIBG Uptake and the Remaining Axons in the Cardiac Sympathetic Nerve in Lewy Body Disease

Takahashi et al. examined the relationship between the degree of degeneration of the cardiac sympathetic nerve and MIBG cardiac uptake across lifespan [60]. Twenty-three patients with pathologically confirmed Lewy body disease (17 men and 6 women; mean age at death: 77.5 ± 6.9 years) who underwent MIBG cardiac scintigraphy were enrolled in the study. One patient with MSA and one patient with AD served as controls. Sections of the left ventricular anterior wall were immunostained with antibodies against TH and NF. The immunoreactive areas of the residual cardiac nerve axons were quantified, and the relationship between the degree of cardiac nerve axons and H/M ratios on MIBG cardiac scintigraphy were examined. Cardiac MIBG uptake was reduced in 90.9 % and 95.7 % in the early and delayed phases of the subjects with Lewy body disease, respectively (Fig. 18.7). The area of TH-immunoreactive nerve fibers correlated with the degree of cardiac MIBG uptake both in early (correlation coefficient: $r = 0.57$, $p < 0.01$) and delayed ($r = 0.54$, $p < 0.01$) phases (Fig. 18.8). The area of NF-immunoreactive nerve fibers correlated with the degree of cardiac MIBG uptake both in the early ($r = 0.65$, $p < 0.01$) and delayed ($r = 0.58$, $p < 0.01$) phases, respectively (Fig. 18.8). These findings confirm that MIBG cardiac scintigraphy is a useful

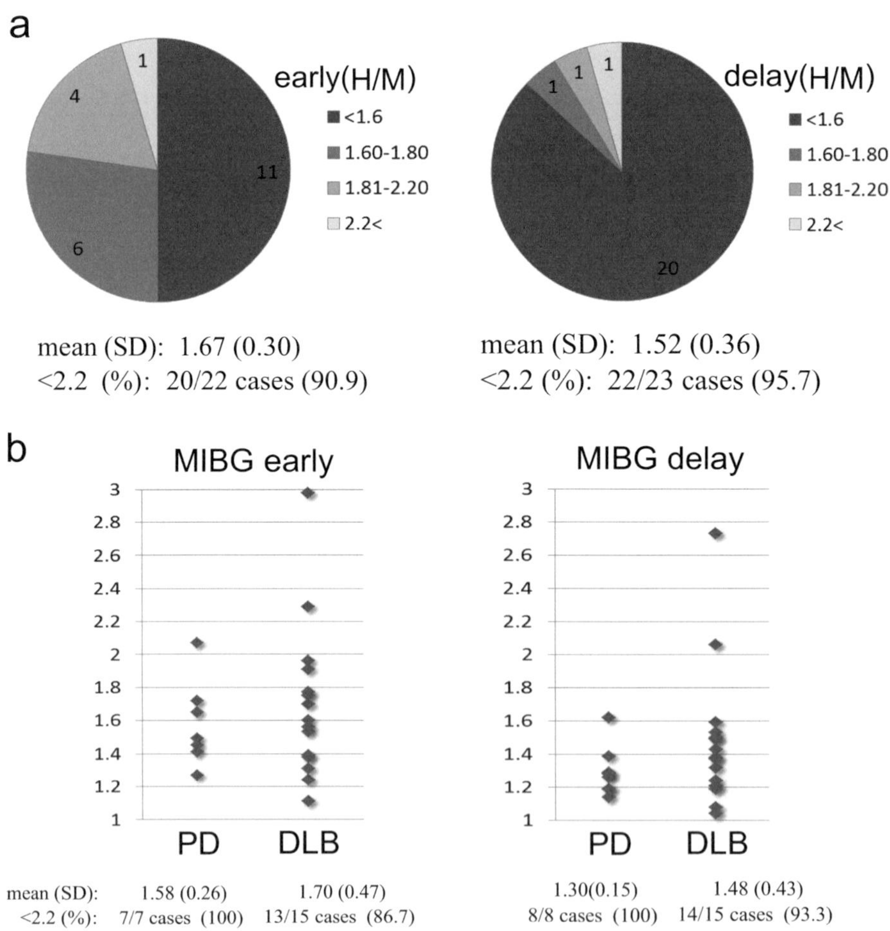

Fig. 18.7 Heart to mediastinum (H/M) ratio in Lewy body disease. (**a**) The proportion of standardized H/M ratios in the early and delayed phases of 23 patients with Lewy body disease confirmed by postmortem pathology. One patient did not undergo an early phase on MIBG cardiac scintigraphy. The H/M ratio was decreased in 90.9 % (20/22 patients) and 95.7 % (22/23 patients) of patients with Lewy body disease in the early and delayed phases, respectively. (**b**) The H/M ratio of 23 patients with Lewy body disease was confirmed by pathology (8 PD and 15 DLB patients). SD = standard deviation (Adapted from Ref. [60] with permission)

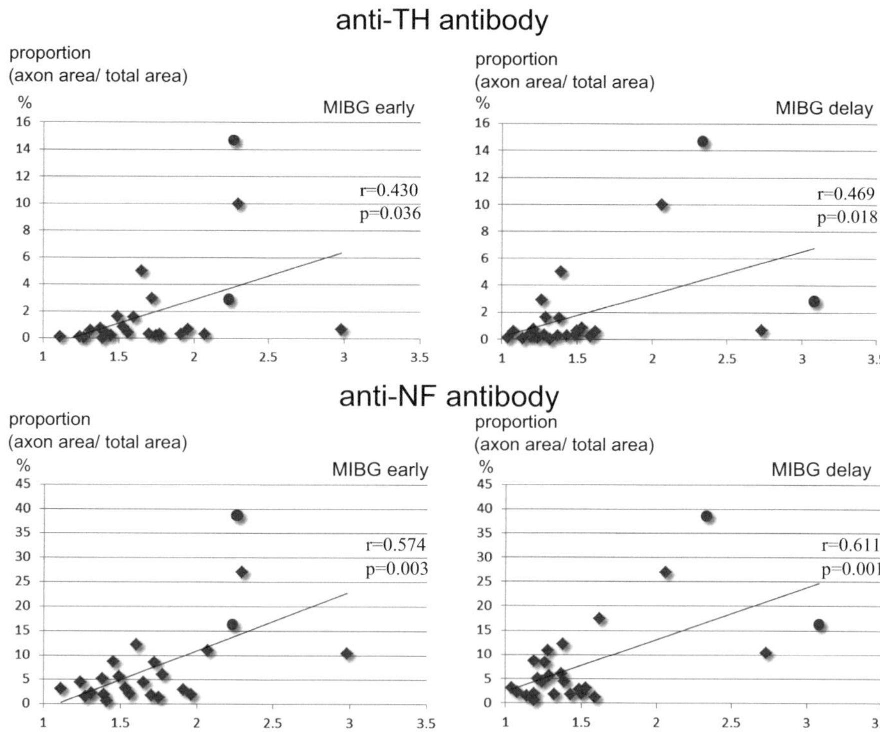

Fig. 18.8 Correlation between cardiac MIBG uptake and the proportion of residual sympathetic axons. The relationship between the proportion of residual sympathetic axons and cardiac MIBG uptake (Adapted from Ref. [60] with permission)

imaging tool for a clinical diagnosis of Lewy body disease and can assess the degree of degeneration of the cardiac sympathetic nerve.

18.12 Experimental Studies on Cardiac MIBG Uptake

The neurotoxic chemical agent 1-methyl-4-phenyl-1, 2, 3, 6-tetrahydroxypyridine (MPTP) has been used to produce animal models of PD [61, 62]. Short-term administration of MPTP to C57BL/6 mice resulted in an almost complete, permanent, and selective degeneration of nigrostriatal dopaminergic neurons [63, 64]. Takatsu et al. demonstrated that MIBG uptake was reduced in MPTP-treated cultured PC12 cells in a dose-dependent manner [65] and that the MPTP-treated mice had reduced cardiac MIBG uptake, as seen in patients with PD [66]. Subsequently, Fukumitsu et al. noted a reduction in noradrenaline transporter density in the hearts of MPTP-treated mice using a binding assay [67], which accounted for the reduced cardiac MIBG uptake. However, histopathological alterations of the cardiac sympathetic nerve in MPTP-treated mice have not yet been demonstrated. Amino et al. examined the hearts of MPTP-treated mice using immunohistochemistry and Western blot analyses [68]. MPTP-treated mice exhibited significant decreases in levels of cardiac noradrenaline and dopamine, suggesting sympathetic dysfunction. Synaptophysin-, TH-, or noradrenaline transporter-immunoreactive nerve fibers were abundant in the hearts of control mice and MPTP-treated mice, without apparent differences between the two groups. Western blot analyses also showed no differences in these protein levels. Cardiac nerve fibers were well preserved in MPTP-treated mice, even though an apparent cardiac sympathetic dysfunction was detected [68].

References

1. Fujiwara H, Hasegawa M, Dohmae N, Kawashima A, Masliah E, Goldberg MS, et al. - α-synuclein is phosphorylated in synucleinopathy lesions. Nat Cell Biol. 2002;4:160–4.
2. Braak H, Del Tredici K, Rüb U, de Vos RA, Jasen Steur EN, Braak E. Staging of brain pathology related to sporadic Parkinson's disease. Neurobiol Aging. 2003;24:197–211.
3. Braak H, de Vos RA, Bohl J, Del Tredici K. Gastric α-synuclein immunoreactive inclusions in Meissner's and Auerbach's plexuses in cases staged for Parkinson's disease-related brain pathology. Neurosci Lett. 2006;396:67–72.
4. Bloch A, Probst A, Bissig H, Adams H, Tolnay M. α-synuclein pathology of the spinal and peripheral autonomic nervous system in neurologically unimpaired elderly subjects. Neuropathol Appl Neurobiol. 2006;32:284–95.
5. Orimo S, Uchihara T, Nakamura A, Mori F, Kakita A, Wakabayashi K, et al. Axonal α-synuclein aggregates herald centripetal degeneration of cardiac sympathetic nerve in Parkinson disease. Brain. 2008;131:642–50.
6. Okada Y, Ito Y, Aida J, Yasuhara M, Ohkawa S, Hirokawa K. Lewy bodies in the sinoatrial nodal ganglion: clinicopathological studies. Pathol Int. 2004;54:682–7.
7. Ghebremedhin E, Del Tredici K, Langston JW, Braak H. Diminished tyrosine hydroxylase immunoreactivity in the cardiac conduction system and myocardium in Parkinson's disease: an anatomical study. Acta Neuropathol. 2009;118:777–84.
8. Minguez-Castellanos A, Chamorro CE, Escamilla-Sevilla F, Ortega-Moreno A, Rebollo AC, Gomez-Rio M, et al. Do α-synuclein aggregates in autonomic plexuses predate Lewy body disorders? A cohort study. Neurology. 2007;68:2012–8.
9. Navarro-Otano J, Gelpi E, Mestres CA, Quintana E, Rauek S, Ribalta T, et al. Alpha-synuclein aggregates in epicardial fat tissue in living subjects without parkinsonism. Parkinsonism Relat Disord. 2013;19:27–31.
10. Gelpi E, Navarro-Otano J, Tolosa E, Gaig C, Compta Y, Rey MJ, et al. Multiple organ involvement by alpha-synuclein pathology in Lewy body disorders. Mov Disord. 2014;29: 1010–8.
11. Janes RD, Brandys JC, Hopkins DA, Johnstone DE, Murphy DA, Armour JA. Anatomy of human extrinsic cardiac nerves and ganglia. Am J Cardiol. 1986;57:299–309.
12. Bannister R, Mathias CJ. Introduction and classification of autonomic disorders. In: Mathias CJ, Bannister R, editors. Autonomic failure. A textbook of clinical disorders of the autonomic nervous system. 4th ed. New York: Oxford University Press; 1999. p. xvii–xii.
13. Wakabayashi K, Takahashi H, Ohama E, Takeda S, Ikuta F. Lewy bodies in the visceral autonomic nervous system in Parkinson's disease. In: Narabayashi H, Nagatsu N, Yanagisawa N, Mizuno Y, editors. Advances in neurology. New York: Lippincott-Raven; 1993. p. 609–12.

14. Orimo S, Ozawa E, Oka T, Nakade S, Tsuchiya K, Yoshimoto M, et al. Different histopathology accounting for a decrease in myocardial MIBG uptake in PD and MSA. Neurology. 2001;57: 1140–1.
15. Orimo S, Oka T, Miura H, Tsuchiya K, Mori F, Wakabayashi K, et al. Sympathetic cardiac denervation in Parkinson's disease and pure autonomic failure but not in multiple system atrophy. J Neurol Neurosurg Psychiatry. 2002;73:776–7.
16. Kawano H, Okada R, Yano K. Histological study on the distribution of autonomic nerves in the human heart. Heart Vessel. 2003;18:32–9.
17. Amino T, Orimo S, Ihoh Y, Takahashi A, Uchihara H, Mizusawa H. Profound cardiac sympathetic denervation occurs in Parkinson disease. Brain Pathol. 2005;15:29–34.
18. Mitsui J, Saito Y, Momose T, Shimizu J, Arai N, Shibahara J, et al. Pathology of the sympathetic nervous system corresponding to the decreased cardiac uptake in ^{123}I-metaiodobenzylguanidine (MIBG) scintigraphy in a patient with Parkinson disease. J Neurol Sci. 2006; 15(243):101–4.
19. Orimo S, Amino T, Ito Y, Takahashi A, Kojo T, Uchihara T, et al. Cardiac sympathetic denervation precedes neuronal loss in the sympathetic ganglia in Lewy body disease. Acta Neuropathol. 2005;109:583–8.
20. Orimo S, Ozawa E, Nakade S, Sugimoto T, Mizusawa H. ^{123}I-metaiodobenzylguanidine myocardial scintigraphy in Parkinson's disease. J Neurol Neurosurg Psychiatry. 1999;67:189–94.
21. Courbon F, Brefel-Courbon C, Thalamas C, Alibelli MJ, Berry I, Montastruc JL, et al. Cardiac MIBG scintigraphy is a sensitive tool for detecting cardiac sympathetic denervation in Parkinson's disease. Mov Disord. 2003;18:890–7.
22. Orimo S, Takahashi A, Uchihara T, Mori F, Kakita A, Wakabayashi K, et al. Degeneration of cardiac sympathetic nerve begins in the early disease process of Parkinson's disease. Brain Pathol. 2007;17:24–30.
23. Fujishiro H, Frigerio R, Burnett M, Klos KJ, Josephs KA, Delledonne A, et al. Cardiac sympathetic denervation correlates with clinical and pathologic stages of Parkinson's disease. Mov Disord. 2008;23:1085–92.
24. Miki Y, Mori F, Wakabayashi K, Kuroda N, Orimo S. Incidental Lewy body disease restricted to the heart and stellate ganglia. Mov Disord. 2009;24:2299–301.
25. Hawkes CH, Del Tredici K, Braak H. Parkinson's disease: a dual-hit hypothesis. Neuropathol Appl Neurobiol. 2007;33:599–614.
26. Tomiyama H, Lesage S, Tan EK, Jeon BS. Familial Parkinson's disease/parkinsonism. Biomed Res Int. 2015;736915. 2 pages.
27. Zarranz JJ, Alegre J, Gomez-Esteban JC, Lezcano E, Ros R, Ampuero I, et al. The new mutation, E46K, of alpha-synuclein causes Parkinson and Lewy body dementia. Ann Neurol. 2004;55: 164–73.
28. Zarranz JJ, Fernandez-Bedoya A, Lambarri I, Gomez-Esteban JC, Lezcano E, Zamacona J, et al. Abnormal sleep architecture is an early feature in the E46K familial synucleinopathy. Mov Disord. 2005;20:1310–5.
29. Tijero B, Gomez-Esteban JC, Llorens V, Lezcano E, Gonzalez-Fernandez MC, de Pancorbo MM, et al. Cardiac sympathetic denervation precedes nigrostriatal loss in the E46K mutation of the alpha-synuclein gene (SNCA). Clin Auton Res. 2010;20:267–9.
30. Tijero B, Gómez-Esteban JC, Lezcano E, Fernández-González C, Somme J, Llorens V, et al. Cardiac sympathetic denervation in symptomatic and asymptomatic carriers of the E46K mutation in the α synuclein gene. Parkinsonism Relat Disord. 2013;19:95–100.
31. Orimo S, Amino T, Yokochi M, Kojo T, Uchihara T, Takahashi T, et al. Preserved cardiac sympathetic nerve accounts for normal cardiac uptake of MIBG in PARK2. Mov Disord. 2005; 20:1350–3.
32. Hayashi S, Wakabayashi K, Ishikawa A, Nagai H, Saito M, Maruyama M, et al. An autopsy case of autosomal-recessive juvenile parkinsonism with a homozygous exon 4 deletion in the parkin gene. Mov Disord. 2000;15:884–8.

33. Farrer M, Kachergus J, Forno L, Lincoln S, Wang DS, Hulihan M, et al. Comparison of kindreds with parkinsonism and α-synuclein genomic multiplications. Ann Neurol. 2004;55: 174–9.
34. Singleton A, Gwinn-Hardy K, Sharabi Y, Li ST, Holmes C, Dendi R, et al. Association between cardiac denervation and parkinsonism caused by alpha-synuclein gene triplication. Brain. 2004;127:768–72.
35. Nishioka K, Hayashi S, Farrer MJ, Singleton AB, Yoshino H, Imai H, et al. Clinical heterogeneity of α-synuclein gene duplication in Parkinson's disease. Ann Neurol. 2006;59:298–309.
36. Orimo S, Uchihara T, Nakamura A, Mori F, Ikeuchi T, Onodera O, et al. Cardiac sympathetic denervation in Parkinson's disease linked to SNCA duplication. Acta Neuropathol. 2008;116: 575–7.
37. Wakabayashi K, Hayashi S, Ishikawa A, Hayashi T, Okuizumi K, Tanaka H, et al. Autosomal dominant diffuse Lewy body disease. Acta Neuropathol. 1998;96:207–10.
38. Ikeuchi T, Kakita A, Shiga A, Kasuga K, Kaneko H, Tan CF, et al. Patients homozygous and heterozygous for SNCA duplication in a family with parkinsonism and dementia. Arch Neurol. 2008;65:514–9.
39. Khan NL, Jain S, Lynch JM, Pavese N, Abou-Sleiman P, Holton JL, et al. Mutations in the gene LRRK2 encoding dardarin (PARK8) cause familial Parkinson's disease: clinical, pathological, olfactory and functional imaging and genetic data. Brain. 2005;128:2786–96.
40. Giasson BI, Covy JP, Bonini NM, Hurtig HI, Farrer MJ, Farrer MJ, et al. Biochemical and pathological characterization of Lrrk2. Ann Neurol. 2006;59:315–22.
41. Nukada H, Kowa H, Saitoh T, Tazaki Y, Miura S. A big family of paralysis agitans. Rinsho Sinkeigaku. 1978;18:627–34 (in Japanese).
42. Hasegawa K, Stoessl AJ, Yokoyama T, Kowa H, Wszolek ZK, Yagishita S. Familial parkinsonism: study of original Sagamihara PARK8 (I2020T) kindred with variable clinicopathologic outcomes. Parkinsonism Relat Disord. 2009;15:300–6.
43. Goldstein DS, Imrich R, Peckham E, Holmes C, Lopex G, Crews C, et al. Neurocirculatory and nigrostriatal abnormalities in Parkinson disease from LRRK2 mutation. Neurology. 2007;69: 1580–4.
44. Braune S. The role of cardiac metaiodobenzylguanidine uptake in the differential diagnosis of parkinsonian syndromes. Clin Auton Res. 2001;11:351–5.
45. Braune S, Reinhardt M, Schnitzer R, Riedel A, Lücking CH. Cardiac uptake of [^{123}I]MIBG separates Parkinson's disease from multiple system atrophy. Neurology. 1999;53:1020–5.
46. Druschky A, Hilz MJ, Platsch G, Radespiel-Tröger M, Druschky K, Kuwert T, et al. Differentiation of Parkinson's disease and multiple system atrophy in early disease stages by means of I-123-MIBG-SPECT. J Neurol Sci. 2000;175:3–12.
47. Taki J, Nakajima K, Hwang E-H, Matsunari I, Komai K, Yoshita M, et al. Peripheral sympathetic dysfunction in patients with Parkinson's disease without autonomic failure is heart selective and disease specific. Eur J Nucl Med. 2000;27:566–73.
48. Yoshita M. Differentiation of idiopathic Parkinson's disease from striatonigral degeneration and progressive supranuclear palsy. Acta Neurol Scand. 1998;91:14–8.
49. Orimo S, Kanazawa T, Nakamura A, Uchihara T, Mori F, Kakita A, et al. Degeneration of cardiac sympathetic nerve can occur in multiple system atrophy. Acta Neuropathol. 2007;113: 81–6.
50. Gray F, Vincent D, Hauw JJ. Quantitative study of lateral horn cells in 15 cases of multiple system atrophy. Acta Neuropathol. 1988;75:513–8.
51. Kennedy PGE, Duchen LW. A quantitative study of intermediolateral column cells in motor neuron disease and the Shy-Drager syndrome. J Neurol Neurosurg Psychiatry. 1985; 48:1103–6.
52. Nakajima R, Takahashi K, Nakamura H, Otomo E, Kameyama M. A quantitative study on the intermediolateral cells of the thoracic cord in degenerative diseases of the nervous system. Clin Neurol. 1981;21:581–6.

53. Oppemheimer DR. Lateral horn cells in progressive autonomic failure. J Neurol Sci. 1980;46: 393–404.
54. Daniel SE. The neuropathology and neurochemistry of multiple system atrophy. In: Mathias CJ, Bannister SR, editors. Autonomic failure. Oxford: Oxford University Press; 1999. p. 321–8.
55. Orimo S, Uchihara T, Kanazawa T, Itoh Y, Wakabayashi K, Kakita A, et al. Unmyelinated axons are more vulnerable to degeneration than myelinated axons of the cardiac nerve in Parkinson's disease. J Neuropathol Appl Neurobiol. 2011;37:791–802.
56. Braak H, Del Tredici K. Poor protracted myelination as a contributory factor to neurodegenerative disorders. Neurobiol Aging. 2004;25:19–23.
57. Nieuwenhuys R. Structure and organization of fibre system. In: Nieuwenbuys R, Tendokelaar HJ, Nicholson C, editors. The central nervous system of vertebrates, vol. 1. Berlin: Springer; 1999. p. 113–57.
58. Bartzokis G, Cummings JL, Sultzer D. White matter structural integrity in healthy aging adults and patients with Alzheimer disease. Arch Neurol. 2003;60:393–8.
59. Hildebrand C, Remahl S, Persson H, Bjartmar C. Myelinated nerve fibers in the CNS. Prog Neurobiol. 1993;40:319–84.
60. Takahashi M, Ikemura M, Oka T, Uchihara T, Wakabayashi K, Kakita A, et al. Quantitative correlation between cardiac MIBG uptake and remaining axons in the cardiac sympathetic nerve in Lewy body disease. J Neurol Neurosurg Psychiatry. 2015;86:939–44.
61. Burns RS, Chiueh CC, Markey SP, Ebert MH, Jacobowitz DM, Kopin IJ. A primate model of parkinsonism: selective destruction of dopaminergic neurons in the pars compacta of the substantia nigra by N-methyl-4-phenyl-1, 2, 3, 6-tetrahydropyridine. Proc Natl Acad Sci U S A. 1983;80:4546–50.
62. Heikkila RE, Sonsalla PK. The MPTP-treated mouse as a model of parkinsonism: how good is it? Neurochem Int. 1992;20(Suppl):299S–303.
63. Sonsalla PK, Heikkila RE. The influence of dose and dosing interval on MPTP-induced dopaminergic neurotoxicity in mice. Eur J Pharmacol. 1986;129:339–5.
64. Sundström E, Strömberg I, Tsutsumi T, Olson L, Jonsson G. Studies on the effect of 1-methyl-4-phenyl-1, 2, 3, 6-tetrahydropyridine (MPTP) on central catecholamine neurons in C57BL/6 mice. Comparison with three other strains of mice. Brain Res. 1987;405:26–38.
65. Takatsu H, Wada H, Maekawa N, Takemura M, Saito K, Fujiwara H. Significant reduction of ^{125}I-meta-iodobenzylguanidine accumulation directly caused by 1-methyl-4-phenyl-1, 2, 3, 6-tetrahydroxypyridine, a toxic agent for inducing experimental Parkinson's disease. Nucl Med Commun. 2002;23:161–6.
66. Takatsu H, Nishida H, Matsuo H, Watanabe S, Nagashima K, Wada H, et al. Cardiac sympathetic denervation from the early stage of Parkinson's disease: clinical and experimental studies with radiolabeled MIBG. J Nucl Med. 2000;41:71–7.
67. Fukumitsu N, Suzuki M, Fukuda T, Kiyono Y, Kajiyama S, Saji H. Reduced ^{125}I-meta-iodobenzylguanidine uptake and norepinephrine transporter density in the hearts of mice with MPTP-induced parkinsonism. Nucl Med Biol. 2006;33:37–42.
68. Amino T, Uchihara T, Tsunekawa H, Takahata K, Shimazu S, Mizusawa H, et al. Myocardial nerve fibers are preserved in MPTP-treated mice, despite cardiac sympathetic dysfunction. Neurosci Res. 2008;60:314–8.

Chapter 19
Clinical Implications of Reduced Cardiac MIBG Uptake

Satoshi Orimo

Abstract Reduced cardiac MIBG uptake is caused by the degeneration of the cardiac sympathetic nerve. It is observed in Lewy body disease, Parkinson's disease (PD) associated with an *α-synuclein* mutation, PD associated with an *α-synuclein* multiplication, some cases of PD associated with *LRRK2* mutations with Lewy bodies in the brain, and various disorders with Lewy body disease or incidental Lewy body disease (iLBD), in which Lewy bodies are found in the brain. In contrast, cardiac MIBG uptake is usually normal in neurodegenerative parkinsonism and dementia, except for Lewy body disease in which Lewy bodies are not found in the brain. In neurodegenerative disorders and familial PD, degeneration of the cardiac sympathetic nerve is closely related to the presence of Lewy bodies. Therefore, we hypothesize that reduced cardiac MIBG uptake is a potential biomarker for the presence of Lewy bodies. MIBG cardiac scintigraphy allows us to estimate the presence of Lewy bodies.

Keywords MIBG • Biomarker • Differentiation • Parkinsonism • Dementia

19.1 Patients with Reduced Cardiac MIBG Uptake

As previously described, cardiac MIBG uptake was reduced in patients with Lewy body disease, including Parkinson's disease (PD), PD with dementia, dementia with Lewy bodies (DLB), and pure autonomic failure but not in the other related disorders [1–7]. Most patients with REM sleep behavior disorder (RBD) have reduced cardiac MIBG uptake, irrespective of an association with PD [8]. Patients with some familial PD, such as PD associated with *α-synuclein* mutation, PD associated with *α-synuclein* multiplication, and PD associated with *LRRK2* mutation, also had reduced cardiac MIBG uptake [9–11]. Moreover, pathologically confirmed subjects with multiple system atrophy (MSA) and corticobasal

S. Orimo (✉)
Division of Neurology, Kanto Central Hospital of the Mutual Aid Association of Public School Teachers, 6-25-1 kamiyoga, Setagaya-ku, Tokyo 158-8531, Japan
e-mail: sorimo0307@nifty.com

S. Iwase et al. (eds.), *Clinical Assessment of the Autonomic Nervous System*,
DOI 10.1007/978-4-431-56012-8_19

degeneration (CBD) with reduced cardiac MIBG also had PD or incidental Lewy body disease (iLBD) [12, 13].

19.2 Degeneration of the Cardiac Sympathetic Nerve Is Associated with Lewy Bodies

As previously described, the degeneration of the cardiac sympathetic nerve is closely associated with Lewy bodies in the nervous system [14–17]. Degeneration of the cardiac sympathetic nerve occurs in Lewy body disease and familial PD, including PD associated with *α-synuclein* mutation and PD associated with *LRRK2* mutation with Lewy bodies in the brain [9, 18]. In a pathology study, 10 out of 20 subjects with iLBD had various degrees of cardiac sympathetic nerve degeneration [19]. Moreover, pathologically confirmed subjects with MSA and CBD associated with PD or iLBD, who had reduced cardiac MIBG uptake, showed degeneration of the cardiac sympathetic nerve [12, 13]. Figure 19.1 shows immunohistochemical staining of the epicardial nerve fascicles from PD and various related disorders. Tyrosine hydroxylase (TH)- and neurofilament (NF)-immunoreactive axons were markedly decreased in PD, DLB, and PD associated with *α-synuclein* duplication and PD associated with *LRRK2* mutation with Lewy bodies in the brain and slightly to moderately decreased in two cases of iLBD. In contrast, TH- and NF-immunoreactive axons were well preserved in MSA, PSP, CBD, AD, FTD, parkin disease, and PD associated with *LRRK2* mutation with no Lewy bodies in the brain as well as the control subject. These findings confirmed that degeneration of the cardiac sympathetic nerve is closely associated with Lewy bodies in the nervous system.

19.3 Reduced Cardiac MIBG Uptake Is a Potential Biomarker for the Presence of Lewy Bodies

Reduced cardiac MIBG uptake is observed in Lewy body disease, PD associated with *α-synuclein* mutation, PD associated with *α-synuclein* multiplication, some cases of PD associated with *LRRK2* mutation with Lewy bodies in the brain, and various disorders with Lewy body disease or iLBD, as mentioned above. These disorders have Lewy bodies or Lewy neurites in the central nervous system and/or peripheral autonomic nervous system. In contrast, Lewy bodies and Lewy neurites are usually not found in MSA, progressive supranuclear palsy (PSP), CBD, Alzheimer disease (AD), frontotemporal dementia (FTD), parkin disease, and PD associated with *LRRK2* mutation without Lewy bodies in the brain. Figure 19.1 shows the TH- and NF-immunoreactive nerve fibers of epicardial nerve fascicles in

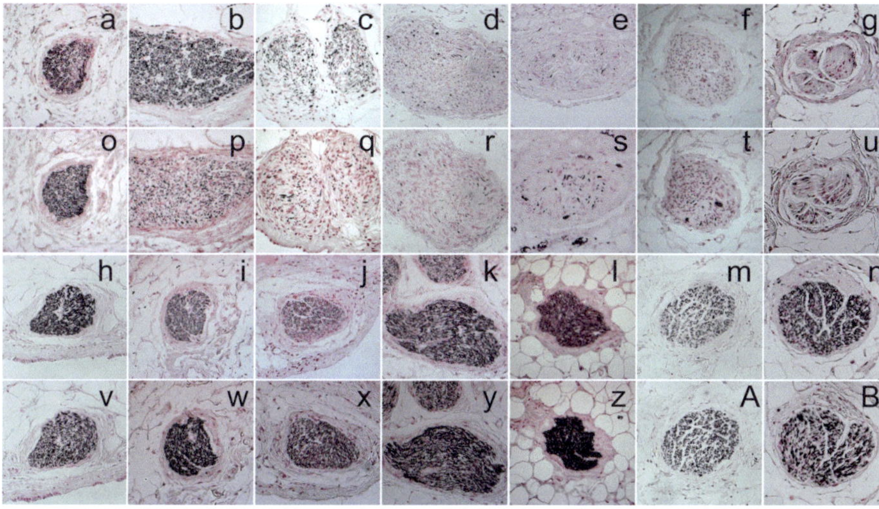

Fig. 19.1 Immunohistochemical staining of the epicardial nerve fascicles from Parkinson's disease and a wide range of neurodegenerative disorders. Tyrosine hydroxylase (TH)- and neurofilament (NF)-immunoreactive axons were markedly decreased in Parkinson's disease (PD), dementia with Lewy bodies (DLB), PD associated with *α-synuclein* duplication, and PD associated with *LRRK2* mutation with Lewy bodies in the brain and slightly to moderately decreased in two cases of incidental Lewy body disease (iLBD). In contrast, TH- and NF-immunoreactive axons were well preserved in multiple system atrophy (MSA), progressive supranuclear palsy (PSP), corticobasal degeneration (CBD), Alzheimer disease (AD), frontotemporal degeneration (FTD), parkin disease, and PD associated with *LRRK2* mutation with no Lewy bodies in the brain, as well as the control subject. **a o**, control; **b q**, iLBD 1; **c q**, iLBD 2; **d r**, PD; **e s**, DLB; **f t**, PD associated with *α-synuclein* duplication; **g u**, PD associated with *LRRK2* mutation with Lewy bodies in the brain; **h v**, MSA; **i w**, PSP; **j x**, CBD; **k y**, AD; **l z**, FTD; **m** *A*, parkin disease; **n** *B*, PD associated with *LRRK2* mutation with no Lewy bodies in the brain

a wide range of neurodegenerative disorders. In two cases of iLBD (b, c, I, j), PD (d, k), DLB (e, l), PD associated with *α-synuclein* duplication (f, m), PD associated with *LRRK2* mutation with Lewy bodies in the brain (g, n), and TH- and NF-immunoreactive nerve fibers were mildly to markedly decreased. In contrast, in MSA (o, v), PSP (p, w), CBD (q, x), AD (r, y), FTD (s, z), parkin disease (t, A), and PD linked to *LRRK2* mutation without Lewy bodies in the brain (u, B), TH- and NF-immunoreactive nerve fibers were well preserved. In neurodegenerative disorders and familial PD, degeneration of the cardiac sympathetic nerve was closely related to the presence of Lewy bodies. As previously mentioned, degeneration of the cardiac sympathetic nerve can cause a reduction in the cardiac sympathetic nerve. Therefore, reduced cardiac MIBG uptake can be a potential biomarker for the presence of Lewy bodies. MIBG cardiac scintigraphy can allow us to estimate the presence of Lewy bodies. However, attention needs to be paid to the interpretation of cardiac MIBG scintigraphy results, because some medications, peripheral autonomic neuropathies, and congestive heart failure can affect cardiac MIBG uptake. MIBG cardiac scintigraphy should be used as a supportive tool for differentiating parkinsonism or dementia.

References

1. Yoshida M, Fukumoto Y, Kuroda Y, Ohkoshi N. Sympathetic denervation of myocardium demonstrated by ^{123}I-MIBG scintigraphy in pure progressive autonomic failure. Eur Neurol. 1997;38:291–6.
2. Yoshita M. Differentiation of idiopathic Parkinson's disease from striatonigral degeneration and progressive supranuclear palsy using iodine-123 meta-iodobenzylguanidine myocardial scintigraphy. J Neurol Sci. 1998;155:60–7.
3. Orimo S, Ozawa E, Nakade S, Sugimoto T, Mizusawa H. ^{123}I-metaiodobenzylguanidine myocardial scintigraphy in Parkinson's disease. J Neurol Neurosurg Psychiatry. 1999;67:189–94.
4. Braune S, Reinhardt M, Schnitzer R, Riedel A, Lücking CH. Cardiac uptake of [^{123}I]MIBG separates Parkinson's disease from multiple system atrophy. Neurology. 1999;53:1020–5.
5. Taki J, Nakajima K, Hwang E-H, Matsunari I, Komai K, Yoshita M, et al. Peripheral sympathetic dysfunction in patients with Parkinson's disease without autonomic failure is heart selective and disease specific. Eur J Nucl Med. 2000;27:566–73.
6. Watanabe H, Ieda T, Katayama T, Takeda A, Aiba I, Doyu M, et al. Cardiac ^{123}I-meta-iodobenzylguanidine (MIBG) uptake in dementia with Lewy bodies: comparison with Alzheimer disease. J Neurol Neurosurg Psychiatry. 2001;70:781–3.
7. Yoshita M, Taki J, Yamada M. A clinical role for[^{123}I]MIBG myocardial scintigraphy in the distinction between dementia of the Alzheimer's-type and dementia with Lewy bodies. J Neurol Neurosurg Psychiatry. 2001;71:583–8.
8. Mitamoto T, Miyamoto M, Iwanami M, Hirata K. Cardiac ^{123}I-MIBG accumulation in Parkinson's disease differs in association with REM sleep behavior disorder. Parkinsonism Relat Disord. 2011;17:219–20.
9. Tijero B, Gómez-Esteban JC, Lezcano E, Fernández-González C, Somme J, Llorens V, et al. Cardiac sympathetic denervation in symptomatic and asymptomatic carriers of the E46K mutation in the α synuclein gene. Parkinsonism Relat Disord. 2013;19:95–100.
10. Nishioka K, Hayashi S, Farrer MJ, Singleton AB, Yoshino H, Imai H, et al. Clinical heterogeneity of alpha-synuclein gene duplication in Parkinson's disease. Ann Neurol. 2006;59:298–309.
11. Hasagawa K, Stoessl AJ, Yokoyama Y, Kowa H, Wszolex ZK, Yoshita S. Familial Parkinsonism: study of original Sagamihara PARK8 (I2020T) Kindred with variable clinicopathologic outcomes. Parkinsonism Relat Disord. 2009;15:300–6.
12. Nagayama H, Yamazaki M, Ueda M, Nishiyama Y, Hamamoto M, Katayama Y, et al. Low myocardial MIBG uptake in multiple system atrophy with incidental Lewy body pathology: an autopsy case report. Mov Disord. 2008;23:1055–7.
13. Mori K, Iwasaki Y, Ito M, Mimuro M, Yoshida M. Decreased myocardial uptake of meta-iodobenzylguanidine in an autopsy-confirmed case of corticobasal degeneration with Lewy bodies restricted to the sympathetic ganglia. Clin Neurol. 2012;52:405–10 (in Japanese).
14. Orimo S, Oka T, Miura H, Tsuchiya K, Mori F, Wakabayashi K, et al. Sympathetic cardiac denervation in Parkinson's disease and pure autonomic failure but not in multiple system atrophy. J Neurol Neurosurg Psychiatry. 2002;73:776–7.
15. Orimo S, Ozawa E, Nakade S, Hattori H, Tsuchiya K, Taki K, et al. [^{123}I] meta-iodobenzylguanidine myocardial scintigraphy differentiates corticobasal degeneration from Parkinson's disease. Intern Med. 2003;42:127–8.

16. Orimo S, Amino T, Ito Y, Takahashi A, Kojo T, Uchihara T, et al. Cardiac sympathetic denervation precedes neuronal loss in the sympathetic ganglia in Lewy body disease. Acta Neuropathol. 2005;109:583–8.
17. Orimo S, Amino T, Yokochi M, Kojo T, Uchihara T, Takahashi T, et al. Preserved cardiac sympathetic nerve accounts for normal cardiac uptake of MIBG in PARK2. Mov Disord. 2005;20:1350–3.
18. Orimo S, Uchihara T, Nakamura A, Mori F, Ikeuchi T, Onodera O, et al. Acta Neuropathol. 2008;116:575–7.
19. Orimo S, Takahashi A, Uchihara T, Mori F, Kakita A, Wakabayashi K, et al. Degeneration of cardiac sympathetic nerve begins in the early disease process of Parkinson's disease. Brain Pathol. 2007;17:24–30.

MIX
Papier aus verantwortungsvollen Quellen
Paper from responsible sources
FSC® C105338

If you have any concerns about our products,
you can contact us on
ProductSafety@springernature.com

In case Publisher is established outside the EU,
the EU authorized representative is:
Springer Nature Customer Service Center GmbH
Europaplatz 3, 69115 Heidelberg, Germany

Printed by Libri Plureos GmbH
in Hamburg, Germany